HEREDITY, EVOLUTION, and SOCIETY

Second Edition

HEREDITY
EVOLUTION
and SOCIETY

I. Michael Lerner / William J. Libby
UNIVERSITY OF CALIFORNIA, BERKELEY

W. H. FREEMAN AND COMPANY
San Francisco

The cover is an artist's conceptualization of 4.5 billion
years of the evolution of life on Earth. (Illustration by
Ray Salmon, after *Geologic Time,* U.S. Geological Survey
publication.)

Library of Congress Cataloging in Publication Data

Lerner, Isadore Michael, 1910–
    Heredity, evolution, and society.

    Bibliography: p.
    Includes index.
    1. Human genetics–Social aspects. 2. Human
evolution. I. Libby, William J., joint author. 1932–
II. Title.
QH438.L47   1976   573.2'1   75-33968
ISBN O-7167-0576-1

Printed in the United States of America

9 8 7 6 5 4 3 2 1

To the memory of
Th. Dobzhansky   1900–1975

CONTENTS

PREFACE

TO THE FIRST EDITION

This book has grown out of a course in genetics that I have been teaching for several years to students not majoring in biology. The course was designed to satisfy so-called breadth requirements like those in many universities and colleges which force students in the humanities and arts to suffer an exposure to natural sciences. All too often this exposure consists of fact-laden introductory courses in physics, chemistry, and biology more suited for students intending to pursue these subjects in greater detail later. In the past, biology majors and nonmajors alike have been, in effect, "marched through the phyla." With the development of molecular biology the subject matter has been changed but the method remains much the same: the students are now marched through the Krebs cycle.

In my view a different approach to the biological education of a nonscientist is required. It is not simply the presentation of facts that is objectionable; it is difficult (sometimes impossible) to explain principles without resort to facts because abstractions without illustrations have no substance. The difficulty lies more in the selection of facts to be presented. For the student who must live through the last third of the twentieth century, the most important facts are those that have social implications; those bits of information that demonstrate the involvement of every human being in the ethical, social, and political problems of this age of science, problems that are multiplying in geometrical progression in the wake of scientific and technological advances.

Essential details of biology that have daily import in a person's life (what do kidneys do?) should be imparted in high school or earlier. Compulsory laboratory exercises in which nonscientists perform with foreknowledge of outcome do nothing but give wrong ideas of the methods and purposes of scientific endeavor. There is no need for an educated layman to know or to remember the sequence of the stages of mitosis, the order of the geologic eras, or the wavelengths of different kinds of radiation (although such information does appear in this book). Similarly, although much technical language and, on occasion, scientific jargon, is unavoidable in presenting technical ideas, both facts and vocabulary must be servants of ideas and principles and not their masters.

There can be a great deal of disagreement over what emphasis a single-term course on the social implications of biology should take. Problems of ecology, including conservation and pollution, or problems centering on food supply could, no doubt, claim high priority. Allowing for a personal bias stemming from my education and research activity, it seems to me, however, that genetics and evolutionary thought may call for a higher one.

First of all, organic evolution through natural selection is the most important biological generalization of the century preceding ours. The shift that it generated from a typological mode of thought to a statistical one has overwhelming social significance, for example, in connection with stereotype images of racial differences.

Second, the discovery or deduction of the formal mechanism of hereditary transmission, Mendelism, formed a cornerstone of modern biology. Last, the deciphering of the language of biochemical communication within the cell and its relation to biological communication between generations was the most important experimental breakthrough of the biology of our day.

It would be protesting too much to insist at length that the interweaving of these three scientific advances and others accruing from them is having ever-growing effects on the individual, the family, and society. They are the advances that can supply the answers to the kind of questions that Francis Crick suggests intellectuals should be concerned with: "What are we?" "Why are we here?" "Why does the world work in this particular way?" They are the advances that pose novel questions calling for decisions from all levels of existence from that of an individual to that of a world-wide society. Genetics and evolution seem to me therefore to be the main biological areas of concern to the informed layman. Within these fields the choice of topics for a single course is so vast that it has to be made in a rather arbitrary fashion. In part, it is simply dictated by the writer's interests and knowledge. But the choices of particular topics may be defended on other grounds. For example, although I have referred to the molecular revolution in biology as most important, it appears to me that the details of molecular genetics, or of the growing field of biochemical developmental genetics, or, in general, the biology of lower organisms are less pertinent to the purposes of this book than, let us say, the peripheral subject of population explosion. I see no need to defend the particular subjects chosen for discussion, except, perhaps, for the biographical vignettes. Haldane, Galton, and Chetverikov have been singled out in this manner not because they are more important than Darwin and Mendel, but because of the socially significant overtones of their life or work.

The sequence of topics may need an explanation. It was chosen by trial and error to engage the attention and stimulate personal involvement of majors in psychology, economics, history, linguistics, and forty other departments. The progression of subjects is designed to feed tidbits of information of special concern to each of the diverse groups in turn, without losing the interests of any. The book then is structured with this intent in mind.

A word is also necessary about the handling of references. I believe that the names or identities of the architects of modern genetics and evolutionary theory should form part of the cultural equipment of a university graduate (a conviction that I find in my teaching experience shared only by history majors). But it is clearly impractical in a text of this kind to give more than a handful of names or to list all the sources drawn upon. In spite of the fact that many of the ideas presented, some of the expressions used, and most of the data included are derivative. I have therefore cited by name in the body of the text only a relatively small number of direct or indirect contributors to its contents. At the end of the book, however, is a list of scientists and others scholars whose work has been referred to without citation, and the Teacher's Manual accompanying this book contains a bibliography.

As a further prefatory word to both students and instructors, I want to comment on the problem of up-to-dateness. We are currently living in an era of tremendous scientific explosion. While the human population is doubling once every 35 years, the number of scientists and engineers in the United States is doubling once every ten years, although this does not give grounds for believing that scientists will soon outnumber people. The rates at which costs of research and development are doubling vary from once every two years in China and Japan to about once every five years in the United States, Great Britain, and Canada. Over 90 percent of all scientists who ever lived are alive today. The number of scientific journals has risen in the last two centuries from ten to 100,000. Now, while such statistics may not reflect completely the actual growth of knowledge, they do impose severe limitations on the possiblity of any one single person staying up-to-date in any but a very narrow area of information. In some fields it is even difficult for a teacher to keep up with reviews, if he is also to continue with his own research. Because of this and because it takes time to produce a book, some material to be found within these covers is not the latest and may have been superseded. But it is my hope that the students, after using this book, will be sufficiently acquainted with the principles and prospects discussed to appraise the validity and import of new discoveries that may come to their attention.

Just as the beginning of this century was the age of physics and the middle of it seems to be the era of biology, the concluding years of the century may be expected to be the age of the behavioral sciences. And as knowledge advances along the hierarchy of organization from subatomic particles, atoms, molecules, macromolecules, cells, organs, individuals, to socially organized groups, increasingly complex decisions have to be faced by members of society. In many areas, conflicts between individual and social values have already arisen. Many of us find such developments as conditioning of the mind or social control over the human gene pool or even predetermination of sex of unborn children repugnant. Yet these developments are technically possible and have to be considered. Our biology, our psychology, and our values have evolved over a long period of time to serve in stable or in very slowly changing physical and cultural environments, rather than the swiftly transforming ones of today. As Robert Oppenheimer, among others, has pointed out, in traditional society, culture, including ethics and religion, acted as a homeostatic stabilizing force. Now culture has become an instrument of rapid change. Reasoned decisions have become much more difficult in the absence of historical guidelines. Indeed, there are pessimists, such as Max Born, who think that science and technology have already destroyed the ethical basis of civilization. Scientific attitude, he maintains, creates doubt and skepticism towards both unscientific knowledge and the natural unsophisticated actions on which human society depends and without which keeping society together is impossible. This, I submit, is an unwarranted voice of desperation. There cannot be too much knowledge. But decision-making machinery in the atomic age should not ignore information about human beings and the world around them. Even if ethical principles are not deducible by the rational methods of science, it seems obvious that, wherever possible, consequences of alternatives must be considered before choices are made. It is impossible to foresee all the decisions related to genetics that the users of this book will be called upon to make in their lifetimes. But at least one of the purposes of education is to prepare students to make the more or less obvious ones. This is the aim of the course that I have been teaching, and this is the main goal of the present book.

Finally, a note on the boxes and the use of the boldface type: Descriptive material and most of the tabular material has been segregated into inserts called boxes, some of which include illustrations. Subject matter of peripheral relevance to the central topics has also been so treated. The important technical terms and names usually make their first appearance in **boldface type.** These are the terms and names that the student is expected to remember. In the index the numbers of pages on which they appear are also **boldface type,** for ready reference to their definitions. *Italics* are used for species names, for the introduction of technical terms of transient significance, and for emphasis.

Berkeley and Stanford
May 1968

I. MICHAEL LERNER

PREFACE

TO THE SECOND EDITION

Since 1968, there have been some changes in the course at Berkeley. Libby has replaced Lerner as course instructor, and as a result has assumed the main burden of revising the book for this second edition. The Berkeley "breadth requirements" have been relaxed, bringing fewer students to the course against their wills. Some of the changes in the second edition have been made in response to what the students who do come seem to want (although all well-motivated students surely do not want every fact and concept we have included in this book).

Since 1968, our society has continued to change, and there have been many changes in our knowledge and mastery of heredity and evolution. We have tried to incorporate consideration of some of these changes in the second edition. We have had some additional thoughts on the effectiveness of genetic engineering, and some of its problems. The world is running out of resources, and we have expanded the coverage of population and resources in a new chapter. We have become more concerned with questions of behavior, and have amplified the treatment of this complex topic in a new chapter. We have found that *time* and *diversity* are two concepts that are badly handled by students and most people in our society. We hope that an appreciation of evolution is useful in correcting this.

Based on the many communications we have received from readers of the first edition, the following three comments may allow the second edition to be used more effectively. First, we received many requests for a glossary. After much thought and consultation, we decided against it. We have attempted to define most terms that would appear in such a glossary in the context of a discussion or an example. They are identified in the text by **boldface type,** and the pages on which such definitions appear may be located by the **boldface numbers** in the index. Second, the material in the boxes is designed to serve two purposes. In some boxes, it is our attempt to accommodate the diversity of background of our intended audience. For instance, Box 10.A (The Normal Distribution) can be ignored by those familiar with concepts of probability or statistics, but may be profitably studied by those not at ease with these concepts as they are reading Chapter 10. An understanding of the material presented in some boxes (for instance, in Box 12.B—Schizophrenia) is not important in order to understand the continuing development of the main text, but this material is presented as information additional to that given in the main text, for those particularly interested in the topic. Third, the order of chapters is by no means appropriate for everyone. In particular, those not familiar with molecular biology may find it useful to read Chapter 6 first. An advantage to reading the chapters in numerical sequence is that concepts and terms are rarely used before they have been defined. But by use of the boldface page references in the index, the definitions of unfamiliar terms and concepts can be located in earlier chapters if chapters are read out of sequence, or if some sections are skipped.

We have not attempted to bring the book fully up to date, as we might in a textbook intended for genetics majors. We hope the users of this book, students, teachers, and others, will not allow their knowledge to be truncated at the point of latest information contained in these pages. If that happens, then we have failed, for the focus of the course and of the book is toward the problems that will be encountered in the future, not those that have been solved in the past.

Berkeley   *I. Michael Lerner*
August 1975   *William J. Libby*

ACKNOWLEDGMENTS

I am grateful to Ralph W. Tyler and to O. Meredith Wilson, who succeeded him as Director of the Center for Advanced Study in the Behavioral Sciences, Stanford, California, for the opportunity to write most of the book in the relaxed and intellectually stimulating atmosphere of the Center. Many members of the Center's staff contributed to making my stay here profitable and enjoyable. In particular, I want to thank Mrs. Irene Bickenbach who won against odds the battle with my handwriting.

I am grateful for many suggestions from colleagues and friends who have read and commented on all or parts of the manuscript. Much that may be good in the book is due to them, but they should not be held responsible for what is not: I am afraid I did not accept all of their suggestions. It is a particular pleasure to express my appreciation for the help received from Th. Dobzhansky, B. A. Hamburg, and D. O. Woodward, who read all of the first version of the manuscript for the first edition.

Acknowledgment of sources of illustrations are given in the captions. I am grateful to all who supplied me with pictures. I also want to thank the University of British Columbia for permission to include several paragraphs in Chapter 21 (of the first edition) that first appeared in a publication under its imprint.

Since the preparation of the first edition, Dee Baer, Garrett Hardin, Val Woodward, and Dan Zohary have taught the course at Berkeley as visiting professors. Each has brought something to it, and thus to this edition of the book, as have the teaching assistants drawn from the graduate students of the Berkeley Genetics Group.

We particularly thank the many students, teachers, colleagues, and others at Berkeley and elsewhere who have taken the time to comment on most of the topics included and many of those omitted in the first edition. They have found typos, errors of fact, errors of interpretation, and additional evidence in support of, or in opposition to, positions taken. Their notes ranged from a few lines to many pages of closely reasoned critique. We read them all, and have incorporated much of their information and many of their ideas in the second edition, though responsibility for what is in this book remains entirely ours.

We are also grateful to the colleagues and friends who have read parts of the second-edition manuscript. In particular, we express our appreciation to Yan B. Linhart, who read and helped revise all of the second-edition manuscript.

I.M.L. and W.J.L.

HEREDITY, EVOLUTION, and SOCIETY

1
INTRODUCTORY

In this chapter, a brief rationale for tying the concepts and developing knowledge of evolution and heredity to society's concerns, and an explanation of how these facts and concepts and the background fundamental to them are organized in the remainder of the book, are offered.

1.1 ORGANIC EVOLUTION

The concept of **organic evolution** is the most important biological generalization in history. It is highly relevant to our personal well-being, psychology, social organization, and future as a species—as well as to our world outlook and curiosity about ourselves and our immediate and cosmic environments.

A one-term course of lectures, discussions, and readings can cover only in broadest outline what such a potent generalization is based on, and what its implications may be, especially if many of the students are starting with little knowledge of biology. One approach is to organize such a course as one about evolutionary thinking rather than one on details of evolution.

This book, then, is designed for a course on evolution appreciation, following a format often used in courses on the fine arts. The details of the evolutionary process and the intricacies of hereditary transmission on which it is based will, at best, be only sketched in. What will be emphasized is what evolution is, what its mechanisms are likely to have been, what its future course may be. Above all, the significance of evolution to human thought, human experience, and human affairs will be presented.

It may be asked why a book entitled *Heredity, Evolution, and Society* begins with a reference to evolution rather than to heredity. Now, almost all biologists agree that organic evolution is a reality, and that humans (the currently dominant species on this planet), and all other existing kinds of life, were not always the way they are now, but descended with modification from pre-existing forms. The concept of evolution stresses the idea that Earth was not always as it is, but has a historical past. One of the features of the process of evolution is that it embraces a historical continuum, in which there are no sharp borders. Thus, it is possible to distinguish nonliving material from living organisms in a general way, although the exact point at which one turned into another is a matter of somewhat arbitrary opinion. Similarly, the precise point in history at which creatures that can be described as human beings first appeared on Earth is a matter of definition.

To many, evolution is self-evident. Yet it should not be forgotten that until recently there were states in the United States in which the law forbade the teaching of evolution. Indeed, it was only in 1970 that the Mississippi Court of Appeals overthrew the last of the state anti-evolution laws. Less than a year later, the California Board of Education suggested that Yahweh and Adam be given equal time with Darwin in the classrooms. In 1925, the attention of the world had been drawn to Tennessee, where John Scopes was fined for teaching Darwin's theory in defiance of that state's anti-evolution statute. In 1973, Tennessee stepped back toward 1925, passing a legally more subtle statute which requires that wherever Darwin's theory is taught, other theories including (but not limited to) the Genesis account must also be included in the curriculum. The vote was 28 to 1 in the state senate, and 54 to 15 in the house. National television was present for the senate vote, which was conducted without debate to protect the legislators from looking "like barefoot Tennesseans." A Nashville paper commented editorially: "If the senators are such a source of embarrassment to themselves, think of what they are to the rest of the state." But surely not all of the state. There are still many people in the Western world who share the belief of the Irish theologian Bishop James Ussher (1581–1656) and a later English divine, Bishop Lightfoot (1828–1889), that the world was created just as it is in six consecutive 24-hour days, starting precisely at 9 A.M. on the twenty-third of October of the year 4004 B.C. (arrived at by adding the ages of Biblical patriarchs).

The evolutionary outlook denies that the origins of today's world were as simple and straightforward as that. It is based on the evidence that Earth and all its kinds of inhabitants were not the result of **Special Creation,** but were produced by a complex, tortuous, and enormously long sequence of events.

Knowledge of the past, and of the processes that have led from the beginnings of life on Earth, has not yet been comprehended in full detail. Our understanding of even some of the major aspects of the evolutionary process is vague and speculative. Indeed, all we have as yet are a few islands in a vast sea of ignorance. But the evidence that life has evolved is overwhelming. In part this evidence is based on the geological record and on observations of present forms; in part it stems from actual experiments. And evolution is also, given the basic facts of genetics and an understanding of mathematical probability, a logical necessity, which is where genetics enters into the picture.

1.2 GENETICS

K. M. Ludmerer's 1972 book, *Genetics and American Society: A Historical Appraisal,* opens with the sentence: "Perhaps no science in modern times has had so great a social impact and has been so enmeshed in diverse social issues as genetics."

Genetics deals with the fundamentals of evolutionary change, including the reasons and the mechanisms behind it that cause humans to differ from elephants, and humans and elephants to differ among themselves. For organic evolution to have occurred, three attributes of matter are essential, and given the three it is an inevitable process: capacity for **excessive reproduction,** i.e., more offspring than are required for maintenance of population size; capacity for change, which leads to **variation** producing diversity between individuals; and capacity for continuity between generations, or **heredity.** In other words, what is needed for evolution to proceed is the conservative force of heredity and the radical force of variation, and **genetics** is precisely defined as the study of these two forces. It is also said to be the science of *biological communication* between generations, investigating the problem of how biological information is transmitted from parent to offspring. Thus a single cell, depending on the recipe (information) it contains, may give rise to a redwood tree or to a human being, each with a particular color of bark or hair and, subject to the environment in which each develops, with a particular spread of limb or a particular kind of intelligence and temperament. This information comes from the previous generation, which in turn obtained it from its ancestors, although parts of the recipe may have been changed in the course of transmission.

In organisms that reproduce sexually, the parents convey the necessary instructions through **germ cells,** or **gametes,** which unite to produce the **zygote,** or initial cell of the offspring. The details of the mechanism of information transmission have been worked out only recently, and represent a marvel of compactness and precision. The material in the single cell, the zygote, carrying all of the instructions for development into an embryo and then into a particular human being, occupies only 8 cubic microns and weights about 6/1,000,000,000,000 gram (a micron is approximately 1/25,000 of an inch and a gram is about 1/28 of an ounce). To make this clearer, the total hereditary material in the approximately four billion zygotes that developed to produce the present human population of Earth—coding for their hereditary composition, resemblances, and differences—weighed about 24 milligrams and would fit into a drop of water. We shall review in due course the general features of the transmission mechanism. Meanwhile we shall be directing ourselves to the somewhat broader aspects of evolution, in order to place the significance of genetic processes into its proper perspective.

Some of the facts about genetics and evolution seem, at present, valuable only for the satisfaction of intellectual curiosity. Others have intensely practical values for medicine, public health, agriculture, law, industry, and social relations. Recent scientific advances have enormously magnified our ability to direct the course of organic evolution of the flora and fauna around us, and, at least potentially, of ourselves. As Garrett Hardin has put it, "Believe what you will of evolution in the past: but you had better jolly well believe it will take place in the future if you hope to make political decisions that will give your descendents a reasonable chance to exist. The principles of evolution are inescapably relevant to the analysis of man's predicament."

1.3 ORGANIZATION OF THE BOOK

This book consists of six unequal parts that, on occasion, are not sharply separated but merge one into another. The first part, comprising the next four chapters, basically deals with the broad panorama of organic evolution. It considers the contribution of the most important figure in the history of evolution, **Charles Darwin** (1809–1882), whose *Origin of Species,* published in 1859, led to a major revolution in human thought. Also, some elementary principles of biology are surveyed, properties of living matter examined, and the process by which life originated on Earth and gradually transformed itself into the complex web of organisms existing today, is broadly examined. Darwin's place in history, the impact of Darwinism on human thought and action, and the evidence that he adduced for evolution, as well as some more recently discovered facts, are presented in general outline.

The actual procession of evolutionary events is dealt with very summarily. Description of the specific changes that have occurred in the many billions of years of the existence of our universe, the five or more billion years of Earth's existence, or the more than three billion years during which there has been life on Earth, are given only the briefest consideration. In general, throughout the book more emphasis is given to processes and factors operating on the microevolutionary level (that is, within a single species, as we shall define the term) than to those on the macroevolutionary one (that is, between such classes as mammals and birds or such higher categories as vertebrates and invertebrates). The last chapter in this part is concerned with human evolution and biological history, and with special human properties.

The second part covers the informational mechanisms of the cell and the methods of intergenerational communication. The questions discussed relate to the language in which this information is transmitted, what happens to the directions received, how they are copied in successive generations of cells and organisms, and how the language of transmittal is translated into production of substances determining resemblances and differences between individuals. This part merges, in Chapter 7, into the third part, which is devoted to the phenomena of Mendelism, the formal mechanics of heredity, named after the founder of the science of genetics, the Augustinian monk **Johann Gregor Mendel** (1822–1884).

In Chapters 8 through 12, the discussion centers on the laws of inheritance established by Mendel and later extended by many others. The hereditary determination of human sex is considered as a model Mendelian trait, in a somewhat extended discussion of various genetic aspects of sex in humans and other organisms. More complicated features of inheritance are taken up next, involving the action of and interaction between the units of inheritance first described by Mendel. A discussion of the inheritance of rather more complex traits than he studied and of the general problem of interaction between heredity and environment follows. Particular emphasis here is laid on the interaction of nature and nurture in the determination of the complicated characters which contribute to intelligence and behavior.

The fourth part is introduced by Chapter 13 on the genetics of populations, and continues with the consideration of evolutionary forces determining the structure of populations. The subject of selection, introduced in the section on Darwinism, is approached at a more formal level and its interaction with mutation, the important force producing variation, is considered. This part concludes with Chapter 17, which deals with augmented sources of genetic variation, such as radiation, caffeine, and pesticides.

The fifth part is devoted primarily to human inheritance. The bases of human diversity are examined with particular weight being given to the genetic problems of race, including some attention to the biology of the American black. A sampling of a variety of inherited human traits is given, containing examples from medical genetics, following which the inheritance of blood groups is discussed. The last chapter in this section deals with the genetic consequences of various human mating systems.

The final part is concerned with social, ethical, and political issues in genetics, including methods of possible and potential manipulation of human genetic resources, with the penultimate chapter describing how genetics and politics have become intertwined in contemporary society.

1.4 PURPOSE OF THE BOOK

Before turning to the specific material just outlined, an additional word on the general pedagogical attitude from which this book has been written may not be amiss. In science there are basically two kinds of instruction: (1) giving information, and (2) teaching how to obtain and evaluate information. As a rule the second kind is the more important, if it can be used. One reason for this is that the first kind can only provide knowledge of what is currently thought to be correct. The example of Simon Newcomb, an astronomer, considered during his time as one of the greatest living American scientists, is instructive. In October, 1903, he published an article proving, on the basis of all known physical facts, that heavier-than-air devices cannot fly. On 17 December 1903, the Wright brothers made their first flight.

In general, the history of science is a history of errors corrected. While we can assert that knowledge today is closer to some kind of reality than it was 100, twenty, or ten years ago, it shall be still closer ten, twenty, or 100 years from now, provided we choose to use our tools and our mental equipment wisely. In an introductory survey course it is not feasible to use exclusively the second, and better, kind of instruction. The sources of information are too vast and too widely scattered, and many of them are in highly technical language. Some remedy is provided by suggestions for additional reading. But essentially the best that can be done is to present what is known today, perhaps indicating along the way what appears to be fact and what is speculation. In addition, an attempt should be made to engender an outlook combining curiosity with skepticism. Those who are now students, and who remain alert and receptive, will be deluged for the rest of their lives with a cataract of information about further scientific advances. Hence, a background for judging new information must also be provided.

Ideally, then, absorption of the material covered in this book should be accompanied by development of the capacity to be informed, and to act intelligently on new information. In our age, we are facing a very dangerous part of our long journey. In the past we often traveled blindly, but the development of science and technology has imposed on us an increasing necessity for careful choice. As we strip the veils from what were previously considered the mysteries of God or Nature over which we had no control, decisions are bound to weigh heavier on us. Not too long ago in most societies, whether a given couple would have ten, three, or no children was not up to them but to divine will. It may be fully expected that in a few years a couple will not only be able to decide on the number, but also on the sex, and perhaps other characteristics, of each child. And this is an example of only one of the powers that we are to have. The graver ones, including decisions of whether life on Earth is to continue at all, need to be wielded on the basis of an approach involving rational thinking, sensing, feeling, and believing, and a most scrupulous evaluation of the consequences of our actions. It is in the spirit of developing such an outlook, as well as to provide reliable information, that the following chapters are written.

2

LIFE

Before considering some of the details of evolution and of genetics, it may be useful to review some biology as background. In this chapter, we discuss some of the important properties of life, and criteria for distinguishing living from nonliving. How life is organized is presented both in terms of biological organization, and in terms of the taxonomic hierarchies that humans have constructed. We give some particular attention to the things that distinguish the human species from others, and end with speculation about how life may have originated on Earth, and about the likelihood that it exists elsewhere.

2.1 PROPERTIES OF LIVING MATTER

Many criteria have been suggested to distinguish living matter from nonliving. Life is not, and probably cannot be, an entity separable from a physical system that is alive. Rather, it is a condition of a physical system. The problem of separating living from nonliving is different from the problem of separating living from dead. Defining human death is a serious practical and ethical question today. Medical successes with organ transplants have made it clear that organs and tissues can continue to live for many years after the organisms, or persons, of which they originally were a part have died.

The distinction between "living" and "nonliving" becomes difficult to make when physical systems that are near the lower bound of "living" are considered. It is likely that there is a point on the scale of organization where a clear-cut distinction cannot be made. This is essentially a matter of import only to linguistic purists, philosophers, and theologians. For our purposes we can list a set of properties that will define life.

One is *capacity for self-reproduction.* This refers to the production of reasonable replicates, of similar but not (generally speaking) identical offspring.

A second is the *capacity to respond to environmental stimuli.* It can vary from simple irritability, such as an amoeba's movement when touched by a foreign body, to a human's conscious change from light summer clothes to winter flannels (of course, machines can also react to external stimuli).

A third is the *capacity for metabolism,* for the binding or releasing of energy. If an organism obtains all of its energy and nutrients from inorganic sources, it is called an **autotroph,** or "self feeder." **Photosynthesis** is a process in which plants bind energy from light as they synthesize sugars from water and carbon dioxide:

$$H_2O + CO_2 + light \rightarrow O_2 + sugar \text{ containing potential energy.}$$

If an organism obtains some of its energy and nutrients from organic sources, as when animals eat plants, we call it a **heterotroph,** or "other feeder." Both autotrophs and heterotrophs respire to release stored energy in a form they can use. A typical **respiration** reaction is:

$$sugar + O_2 \rightarrow H_2O + CO_2 + kinetic \text{ energy.}$$

The photosynthesis and respiration reactions, as well as myriad other biological reactions, are mediated by **enzymes,** protein molecules which act as catalysts, that is, which promote a chemical reaction without being themselves consumed.

A fourth property is the *capacity to grow.* This is the capacity to differentiate, as well as to increase in size. Chemical and physical processes in nonliving things tend to decrease organization; to mix things up. Living organisms create organized structures and systems from less organized or unorganized materials. Furthermore, living organisms typically are (to a degree) self-maintaining, in that they can repair or replace parts of themselves that are damaged or worn out. To do this, they use energy to transform materials into appropriate organized structures, following the instructions provided by the informational apparatus of their cells.

Finally, there is the *capacity for perpetuated change,* or **mutation.** A mutation is a change in the informational apparatus of cells. When such a change takes place in a cell that will eventually give rise to gametes, the mutation is said to be **germinal** and is transmitted to the next generation. The "memory" of the change is *intergenerational*.

Is fire alive? It surely reproduces, as when a match lights many candles. A forest fire increases in intensity in response to the wind rising or the humidity dropping. It combines organic substances and O_2 to produce H_2O, CO_2, plus energy in the forms of heat and light. But it does not normally create more organized structures from less organized materials, and perhaps most important, it cannot mutate. A flame that is changed cannot impart that change to its offspring flames, because it has no informational apparatus or intergenerational memory that can be altered. Fire cannot evolve. But things that are "living" can, and do.

(Are computers alive? In 1975, by these same criteria, the answer is still no.)

A second kind of mutation, a **somatic mutation,** is a change in a cell that is not destined to give rise to gametes (say, a liver cell or a leaf cell). Cells carrying somatic mutations will pass on the changed instructions to cells descended from them, but these changes will disappear with the death of the individual. The memory of the change here is *intragenerational*. It should be obvious that in sexually reproducing organisms only germinal mutations are of significance to evolutionary processes, although somatic mutations may be important to an individual. For instance, if a change in instructions renders a liver cell unable to manufacture an essential enzyme, the cell may not survive, and in rare cases the individual may not survive as a result, but no evolutionary change will result.

One further property is sometimes erroneously ascribed to living matter: some kind of directedness or urge of organisms to convert part of their environment into their own likeness. The school of thought classed as "objective vitalism" assumed that life involves some sort of "élan vital" or "entelechy," forever impossible to analyze by reduction, that is, in terms of lower levels of organization. The failure to produce life from all its nonliving ingredients via spontaneous generation indicated that some "life force" was needed, beyond a particular combination of physical materials and energy states. The problem of repeated spontaneous generation is discussed more fully in Section 2.6. Suffice it to say here that there is today little evidence that supports vitalistic views (however, see Section 5.2). At best, we can describe this "urge" as an empirically discovered analogue to the first law of motion: living matter left alone and given the building materials will tend to increase itself indefinitely. As Bertrand Russell said: "every living thing is a sort of imperialist seeking to transform as much as possible of its environment into itself and its seed."

2.2 LEVELS OF ORGANIZATION

Organic evolution refers to changes in the living world. To begin to understand evolution, we must include a consideration of inorganic evolution, that is, the evolution of nonliving matter, or the changes in our universe before the appearance of life. We also have to consider social or cultural evolution, which treats

BOX 2.A HIERARCHY OF ORGANIZATION

This table outlines the levels of organization of matter and the scientific disciplines devoted to them. In reality these disciplines may overlap to a greater extent than shown. This is particularly true of the various branches of genetics.

Level of organization	Field of study	Branch of genetics
Subatomic particles	Physics	
Atoms	Chemistry	
Molecules		
	Biochemistry	
Macromolecules[1]		Molecular and biochemical

Cell organelles[2]
  Ribosomes
  Mitochondria
  Endoplasmic reticula
	Cell biology	

Nucleus containing chromosomes		Cytogenetics
Plastids and other organelles		
Cells	Cytology	
Tissues	Histology	Developmental
Organs	Anatomy	
Multicellular organisms	Physiology[8] Morphology[9] Embryology[10]	Formal[16] Special[17] Behavioral
Families		
		Population
Demes[3]	Ecology[11]	
		Ecological
Social colonies[4]		
	Systematics[12]	Quantitative
Species		Mathematical
	Paleontology[13]	
Communities[5]	Behavioral sciences[14]	
Higher taxa[6]		
Local biota[7]		
World biota	Biogeography	
Universe biota	Exobiology[15]	

[1]Giant molecules of the order of 100 Angstroms in diameter (see Box 2.B).
[2]See Box 2.C.
[3]Equivalent to populations.
[4]Such as a termite colony.
[5]Includes such ecosystems as, for example, one including a plant–caterpillar–bird–decay-bacteria–plant food cycle.
[6]Taxa are classifications that range in inclusiveness from subspecific races and varieties through species, genera, families, orders, classes, phyla, and kingdoms.
[7]The totality of various organisms living in the same region.

[8]Study of function.
[9]Study of form.
[10]Study of development.
[11]Study of interrelation of organisms with each other and with their environment.
[12]Sometimes called taxonomy.
[13]Study of fossil forms.
[14]For the study of humans these include anthropology, psychology, and sociology.
[15]As yet a speculative endeavor (see Section 2.7).
[16]Classified by the organism studied: human, mammalian, plant.
[17]Such as radiation genetics.

of changes in social organisms determined by information transmitted either between members of the same generation or between generations by means of instruction and learning. Some social information may be transmitted in biological ways, as for instance, among various social insects. Within the framework of this book we shall allude to inorganic evolution only very briefly and to social evolution primarily with reference to humans. Our major emphasis will be on organic evolution, since our concern here is primarily with life processes.

BOX 2.B SCALE OF SIZES

This illustration shows the relative sizes of objects from single atoms to the largest living organisms (see also Figure 2.2). Note that the scale used is an exponential one: each successive division represents a tenfold change. Thus, in the illustration shown, the relative change between one meter and ten meters is as great as that between ten meters and one hundred meters. As may be seen, one micron is a thousandth of a millimeter, and an angstrom is one ten-thousandth of a micron.

OBJECTS / MAGNITUDES / VIEWING INSTRUMENTS

Cells	Organisms	Macromolecules	Molecules	Meters	Milli-meters	Microns	Angstroms	VIEWING INSTRUMENTS
			H₂				1	2.5 MeV electron microscope
			Amino acids				10	Electron microscope
		Proteins				.01	100	Scanning electron microscope
	Viroid	Genes						
PPLO	Virus					.1	1,000	Light microscope
		Chromosomes						
Red blood cell					.001	1	10,000	
Bacterium					.01	10		Simple lens
Human ovum								Naked eye
Paramecium					.1	100		
Frog ovum	Fruit fly			.001	1	1,000		
	Mouse			.01	10			
Acetabularia				.1	100			
	Human			1	1,000			
Laticifer	Whale			10				
	Redwood			100				

THE ILLUSTRATION WAS ADAPTED FROM SIMPSON ET AL.

Before turning to evolution we should consider the hierarchical fashion in which the universe is organized. This is illustrated in Box 2.A, which shows the levels of organization of living matter from particles to the total cosmos of living things. Box 2.B may be helpful for visualizing the range of physical sizes at the various levels from molecules through organisms. Box 2.B would have a similar appearance if we substituted time for size. Biochemical reactions and changes in molecules occur within cells in fractions of seconds. The development from fertilized egg through embryo to a hatched, born, or germinated juvenile takes only minutes for some organisms such as marine worms, about a day for fruit flies, three weeks for chickens, and nine months for both humans and redwoods. The length of a generation varies from a few minutes in viruses to over a decade in humans. Life-span of some kinds of RNA molecules is only minutes long, fruit flies live a few weeks, mice a few years, humans several decades, and bristle-cone pines over four millennia. The complex interaction of populations resulting in the evolution of major new forms is a process that may take place quickly, but more typically requires thousands or even millions of years.

2.3 THE CELL AND ITS COMPOSITION

The **cell** may be viewed as the basic unit of life. While viruses (Figure 2.1) are smaller and simpler than even the smallest cells (Figure 2.2), they are not to our knowledge necessary components of more complex life forms. Box 2.C presents a considerably oversimplified diagram of a cell. Figure 2.3 is an electron micrograph of the interior of a cell, and Figure 2.4 is a photograph of the very large human egg cell, its first polar body, and the much smaller human sperm cells.

FIGURE 2.1
An electron micrograph of one of the larger and more complex viruses, T4 bacteriophage, enlarged about 125,000 diameters. T4 may seem less interesting to many people than, say, a polio virus, but it has been of great use to geneticists in their attempts to refine knowledge of the nature of the gene. Whether viruses are, indeed, living organisms is debatable. Although they reproduce and mutate, they do not grow or metabolize. Because they can direct synthesis of new viruses only by using the apparatus of their host cell, it seems unlikely that they originated before living cells. They are, perhaps, either degenerate descendents of bacteria, or they may have originated from cells as detached pieces of nucleoprotein. (Courtesy of Michael F. Moody.)

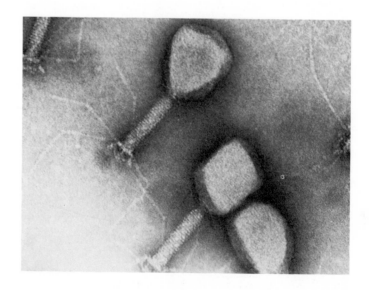

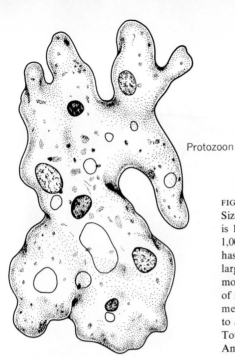

Protozoon   Mammalian   Bacterium   PPLO
                   tissue cell

FIGURE 2.2

Sizes of various cells. A protozoon, with a diameter of 0.01 centimeter, is 10 times bigger than a tissue cell, 100 times bigger than a bacterium, and 1,000 times bigger than the smallest pleuropneumonia-like organism, which has a diameter of 0.1 micron (= 0.00001 centimeter). Other exceptionally large cells include *Acetabularia*, a single-celled green alga that grows more-or-less vertically to a height of several centimeters, and the laticifers of several plant species, which are multinucleate cells that may grow 10 meters long with ramifications running through the branches of the plant to a total length many times that. (From H. J. Morowitz and M. E. Tourtelotte, "The Smallest Living Cells." Copyright © 1962 by Scientific American, Inc. All rights reserved.)

BOX 2.C THE CELL

This illustration provides a generalized schematic representation of a plant cell. There are differences between plant and animal cells (for example, the latter do not have thickened cell walls or plastids), which, however, are not relevant to our purposes here. Only some of the structures of a typical cell are shown and labeled.

Chromosomes are cell organelles containing protein and nucleic acids (see Chapter 7).

The *endoplasmic reticulum* is a network of membranes connected to other cell inclusions and is involved in lipid formation. The **ribosomes** are the sites of protein synthesis, and are frequently associated with the membranes of the endoplasmic reticulum.

Mitochondria are self-reproducing energy power-houses that also contain DNA.

The **nucleus** is the cell organelle that contains the chromosomes, carriers of genetic messages. The material outside the nucleus is referred to as the **cytoplasm.** The *nucleolus* is a major site of ribosomal RNA manufacture.

Plastids are self-reproducing bodies, some of which are concerned with photosynthesis. These also contain DNA.

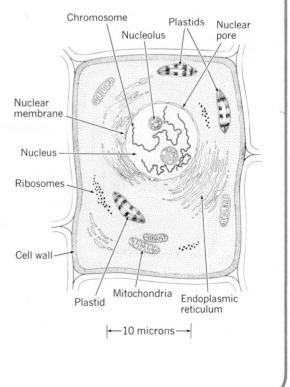

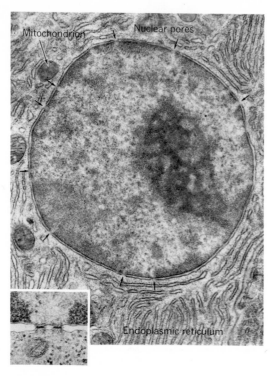

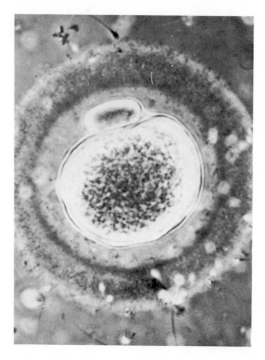

FIGURE 2.3

Electron micrograph of a nucleus surrounded by the cytoplasm. Arrows point to the pores in the nucleus (detail in lower left insert), which permit passage of material to the cytoplasm. Note the dots, which are the ribosomes lining the endoplasmic reticulum. (Courtesy Don W. Fawcett.)

FIGURE 2.4

A photomicrograph of a human egg. Note the tadpole-shaped spermatazoa surrounding the outer membrane, and the first polar body just extruded (see Box 7.A). (Reproduced, with permission, from L. B. Shettles, *Ovum Humanum.* Hafner Publishing Co., New York.)

Even as viewed with as crude a tool as a light microscope, cells appear as exceedingly complex aggregates of subunits. Chemically, their intricacy is awesome. Of the various kinds of materials in cells, the most abundant is water. Others include lipids (fats and related substances), carbohydrates, and inorganic materials, but the most important from our standpoint are **proteins** and **nucleic acids.**

There are two kinds of nucleic acids: **deoxyribonucleic acid,** usually referred to as **DNA,** and **ribonucleic acid,** designated **RNA.** Their significance lies in their functions of storage, transmission, transcription, and translation of genetic information. They are the carriers of hereditary instructions for the manufacture of the proteins and other macromolecules that give any particular organism its distinguishing characteristics. Indeed, the specificity of each and every living being is determined by the particular combinations of the various molecules that its cells can manufacture. We shall return to a more detailed consideration of nucleic acids and their functions in Chapter 6.

Proteins are large molecules consisting of folded chains of **amino acids,** which contain an amino group (NH₂) and a carboxyl group (COOH).

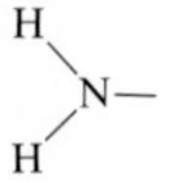

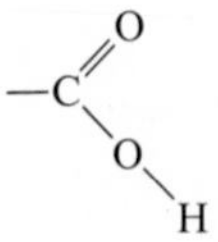

There are twenty common amino acids that primarily concern us. They are built on the model:

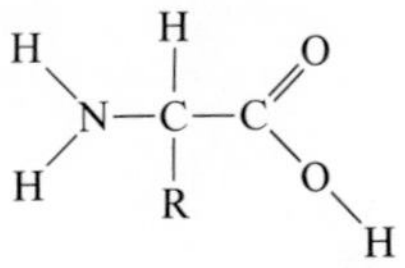

where R stands for an atom or group of atoms. (These molecular structures are shown in two dimensions here. Molecules as they exist in cells have three-dimensional structure, and the specific configuration of a molecule is often of crucial importance to its activity.) When R is only a hydrogen atom, we have the simplest amino acid, glycine:

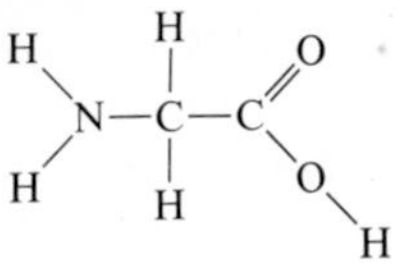

which contains only ten atoms. Others, for example tryptophan, have more complex structural formulas:

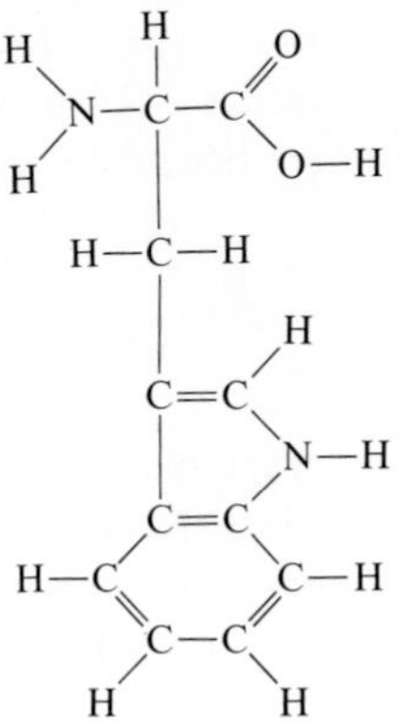

A series of amino acids linked through peptide bonds that connect the amino group of one to the carboxyl group of the next

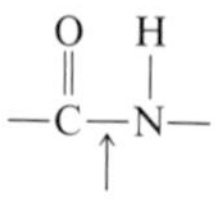

$$\begin{array}{cc} O & H \\ \parallel & \mid \\ -C\!-\!N\!- \\ \mid \end{array}$$

is known as a **polypeptide.** Amino acids become linked in chains by dehydration synthesis (the name does not imply that there has been a drying-out of amino acids, but rather derives from the fact that a molecule of water may form from the hydrogen and oxygen atoms that are removed). This is carried out as an enzyme-catalyzed reaction involving RNA:

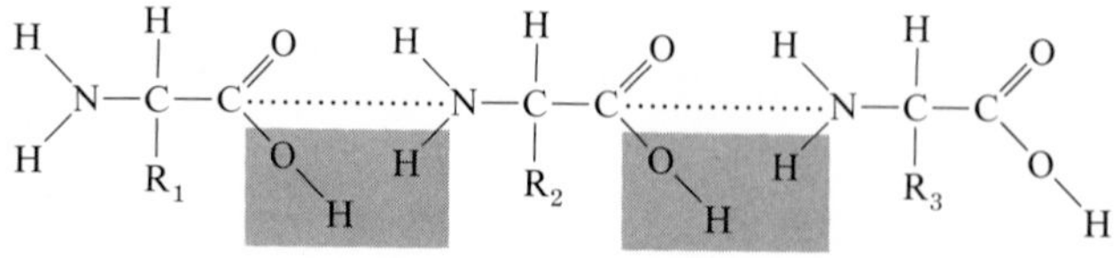

The so-called primary structure of a protein refers to the particular sequence of amino acids in the polypeptide chains, or chain, of which it is composed. Its secondary structure depends on how the chains are held together, and on the three-dimensional orientation of the amino acids in the chains. The tertiary structure is the folding and convoluting of the chains, and the quaternary structure depends on the number, kinds, and proportion of chains put together.

Proteins are present in cells both as separate molecular entities and combined with other large or small molecules. An example of the latter is the combination of large nucleic acid molecules and proteins, forming *nucleoproteins.* A small iron-containing heme group is included in each of the four chains making up a **hemoglobin** molecule, which is another example; hemoglobin is a blood component whose function is the transport of oxygen. The number of possible and actual different proteins and protein compounds is enormous, and their structural complexity is exceedingly great. Human hemoglobin (Figure 2.5) is a relatively simple molecule compared with most proteins manufactured by RNA in cells, but it would be impractical to give its structural formula here because it contains 9,512 atoms:

$$C_{3032}H_{4816}O_{872}N_{780}S_8Fe_4.$$

The very simplest independently living cell known is the so-called pleuro-pneumonia-like organism (PPLO), which has a diameter of 0.1 micron and a weight of 5/1,000,000 gram (one-billionth that of an amoeba—see Figure 2.2). It contains some 1,200 molecules; ordinary cells that are visible under the light microscope may have a quarter of a million protein molecules alone, each made

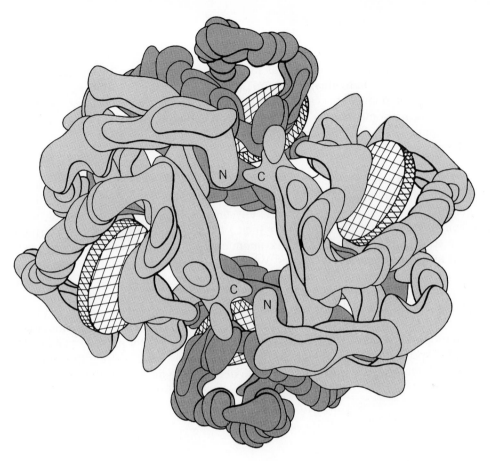

FIGURE 2.5
Hemoglobin molecule, as deduced from X-ray diffraction studies, shown from above. The
irregular blocks represent electron-density patterns at various levels in the hemoglobin molecule.
The molecule is built up from four subunits: two identical alpha chains (light blocks) and
two identical beta chains (dark blocks). The letter "N" identifies the amino ends of the two
alpha chains; the letter "C" identifies the carboxyl ends. Each chain enfolds a heme group
(disk), the iron-containing structure that binds oxygen to the molecule. (From M. F. Perutz
"The Hemoglobin Molecule." Copyright © 1964 by Scientific American, Inc. All rights reserved.)

up, on the average, of perhaps 20,000 atoms. It is easily seen how far from each
other in complexity are atom and cell—and the other items listed in Box 2.A.

More complex organisms have cells organized in tissues, such as muscle or
epidermal tissue. Cells in each type of tissue show characteristic differences of
shape and appearance, which are the result of differentiation in the course of
development. At even higher levels of complexity, the development of cells and
tissues is orchestrated by genetic regulating mechanisms to form organs, such as
a heart or leaf, and these must all work in concert to allow such complex organisms as corn, and corn-borers, and humans, to function.

2.4 THE SPECIES

Above the level of single organisms in the hierarchy of organization are aggregates or groups of individuals. A group defined only by the fact of its aggregation is a **deme,** or population. Of major biological significance is the *gamodeme,* or interbreeding population, as this is the unit of evolution in sexually reproducing organisms. Such a population that has evolved special adaptations to its particular environment is an *ecogamodeme,* or *ecotype.* In some cases, an ecogamodeme is properly called a race, a concept to be more fully considered in Chapter 18. All ecogamodemes that can successfully exchange genes without detriment to the offspring are grouped in an *ecospecies.* This means that members of the different ecotypes, or races, can successfully mate without special difficulties, and their offspring, and for that matter their offspring's offspring, are fully fertile. The ecospecies approximates the taxonomic classification **species,** which is used and misused in a variety of senses. Whatever its original meaning in Latin was, medieval philosophers (for example, St. Thomas Aquinas) considered it to refer to a real, discrete, and immutable unit. Even now neo-Thomist theologians (for instance, Pope Pius XII) talk of animal, vegetable, and mineral as being the three species. In answer to the question, "does species-unit deserve to be a fundamental philosophical concept?" the biosystematists Camp and Gilly replied: "The concept of species, or *kind,* as a unit, has become so firmly entrenched in the mind of man—so much a part of his awareness, so necessary to his basic philosophy—that

BOX 2.D THE ESTIMATED NUMBERS OF LIVING SPECIES

Mammals	3,200
Birds	8,600
Reptiles and amphibians	6,000
Fish	20,000
Total chordates[1]	39,500
Invertebrates[2]	1,050,000
Total animals	1,100,000
Plants[3]	325,000
Protists[4]	75,000
Total species[5]	1,500,000

[1]See Box 2.F.
[2]Including 850,000 species of insects, of which more than 250,000 are beetles. J. B. S. Haldane commented: "God must have truly loved beetles, he made so many of them."
[3]Probably an underestimate.
[4]See Box 2.E. Species classification of protists that reproduce vegetatively is not consistent with the "reproductive community" concept applied to species of sexually reproducing organisms. However, the more we learn about protists, the more we appreciate that genes are at least occasionally recombined by some form of "sexual" exchange between two cells of different lineage.
[5]Some biologists propose that there are as many as 10 million species in active existence today. It is also estimated that between 50 million and 4 billion species have existed on Earth at some time or another. It is clear from this that most species that have existed on Earth are today extinct.

it remains only for the systematist to interpret this unit." Thus, we are discussing "species" in moderate detail, not only because it is an important biological concept, but because our philosophical ideas about species sometimes interfere with our understanding of the dynamic nature of evolution. A species complex, or *cenospecies,* is composed of those ecospecies that can exchange **genes** to a limited degree. (We will return to the gene in Chapter 6. Temporarily, we will define it as the unit of inheritance.) Matings between such ecospecies are not as successful as matings within ecospecies, and the offspring of such between-ecospecies matings are frequently impaired in reproductive capacity. Members of different cenospecies may mate and successfully produce offspring, but gene exchange is prevented because these offspring, or the offspring's offspring, are completely sterile. This is the most tenuous level of reproductive compatibility, and puts such groups in the same *comparium,* roughly equivalent to the taxonomic **genus.** Higher taxonomic orders (see Boxes 2.E, 2.F) are based on morphological and biochemical patterns of similarity and difference, not reproductive compatibility. Thus, members of different genera should not be able to mate and produce offspring if the taxonomists have done their work correctly.

The species, among these classification levels, is the main focus of our conceptualization of biological diversity. This is at least partly because the species level is exceedingly important in the continuum of evolution. In the eighteenth century, the Swedish botanist Carolus Linnaeus (1707–1778) devised the currently used system of classification of living things. The term species was applied by him to groups of plants or animals that were defined as being sufficiently similar to each other to have descended from common ancestors of special creation, as not interbreeding with other such (distinct) groups, and as being constant in all but insignificant characteristics. The Linnaean system calls for identification of each such group by the name of the genus followed by a "specific epithet" for the species. Thus, for example, the human species is called *Homo sapiens,* the fruit fly used extensively in genetic experiments, *Drosophila melanogaster,* and cultivated rice, *Oryza sativa.* The name of a genus is often abbreviated to a single letter (for example *Gallus gallus,* the name of the domestic chicken, may be written *G. gallus*).

Implicit in the eighteenth and early nineteenth century usage of the term species is the Platonic notion that there is for every species a prototype and that the variations observed among individual humans, fruit flies, or rice plants are due to imperfections and deviations from the ideal. This **typological** approach is completely inconsistent with evolutionary thinking. It may provoke repercussions in social attitudes because it gives rise to stereotypes, often in caricature form, for subspecific groups, such as the jew in Nazi mythology or the black among bigots in white America.

Students of genetics and evolution generally reject this approach, replacing it by a more dynamic, populational view of living systems. There is no prototype for an evolutionary species. The average of a species does not sufficiently describe it for most characteristics, but equally or more important is the variation in the characteristic within the species. For instance, we could say that human adults average about $5\frac{3}{4}$ feet tall. But it is useful to add that it is possible to encounter normal adult humans $4\frac{1}{2}$ feet tall, and others more than 7 feet tall. Species do change because of the hereditary transmission of genes for various properties that deviate from the mean. A single species may split into two or more under the influence of evolutionary forces (see Section 4.3); the formation of a new species by splitting or transformation is called **speciation.** In some groups, the processes of speciation are now in progress. When this is happening, it is difficult to decide whether one species exists, or two. If two, individuals with intermediate characteristics may fit about as well in both, but not wholly satisfactorily in either. This worries the typological classifier, who wants a species to be a conceptually discrete box, into which any individual either clearly fits or doesn't fit. Such biological messiness does not worry the evolutionist, however, who brings a **populational** or **statistical** approach to problems of classifying dynamic, not static, aggregations of organisms.

There are, according to **George Gaylord Simpson,** three provisions in the evolutionary meaning of species: (1) community of inheritance among the members, (2) capacity of the genes to spread throughout the group, and (3) inhibition of the gene spread to other groups. Perhaps a fuller comprehension of the concept of species is provided by the definition proposed by **Theodosius Dobzhansky,** and particularly appropriate to biparental organisms. Dobzhansky views a species as *the most inclusive population in time and space representing a discrete reproductive community.*

The reference to time simply means that the unit maintains its form from generation to generation even if its components change. This is analogous to the life of a given organism, in the course of which the cells of its body will be continuously replaced, though its identity remains the same. Or to take another example, a city such as Paris is classified as a particular identifiable community and political unit, although it may grow and contract, and its inhabitants, its social institutions, and even its landscape, change continuously over the course of history.

The notion of reproductive community refers to the fact that members of a given species share common descent and interbreed. "Discrete" relates to the idea that different species do not share the same gene pool (see Section 4.3), nor are they able to exchange genes. Thus, horses and asses are separate species. Although they can mate with each other, the offspring of such a mating, called mules or hinnies, are sterile and therefore do not pass their genes into either of the parental populations (see Box 14.C). Finally, the term "most inclusive" denotes that a species contains all potential interbreeding populations, which in fact may not really interbreed. Thus, St. Bernards are not likely to cross with Chihuahuas, but they belong to the same species, *Canis familiaris,* because the two breeds may exchange genes through intermediate breeds.

The number of living species described to date is shown in Box 2.D. There are many more species that have become extinct in the course of the history of life on Earth than there are living ones. Indeed, some scholars have estimated that more than 99 percent of all species that ever evolved are no longer in existence.

2.5 PROPERTIES OF THE HUMAN SPECIES

We are not concerned here with the detailed characteristics of various kinds of living beings. Box 2.E will suffice to indicate the major differences between the three different kingdoms. Having an anthropocentric outlook, we may, however, want to have a look at our own position in the animal kingdom. Box 2.F gives a simplified taxonomic classification.

BOX 2.E CHARACTERISTICS OF LIVING KINGDOMS

The kingdom is the most inclusive taxon used in biological classification (see Box 2.F). There is no general agreement among biologists about the correct number of kingdoms; it will serve our purposes to divide all living things into three—protists, plants, and animals.

Most **protists** are unicellular organisms that are simple in structure (or, perhaps we should say that they are acellular, in the sense that they are not further divided). Some types of protists live in conjoined colonies; some types are autotrophs and others are heterotrophs. In older systems of classification, which still have adherents, the various types are viewed either as plants (for example, algae) or animals (for example, amoebae), but resemblances among the various protists are greater than resemblances between them and members of the other kingdoms. Some biologists consider the protists to constitute three separate kingdoms: the eukaryotic (containing membrane-bound nuclei) amoebae, ciliates, and flagellates; the eukaryotic fungi; and a group without such nuclei (prokaryotes) including blue-green algae, bacteria, and viruses. The kingdom of protists (if we accept that there is only one) comprises twelve phyla (the phylum being the next most inclusive major taxonomic level—see Box 2.F).

Plants are, with a few exceptions, autotrophic multicellular organisms that capture energy by photosynthesis. They are usually immobile. Their cells have walls and their reproductive cycle is complicated, producing separate independent or semi-independent male and female structures or organisms. There are two phyla of plants.

Animals are multicellular heterotrophs. None of them have the apparatus for photosynthesis. They are usually mobile and, except for the sponges, have nerve cells coordinating different parts of the body. Their cells do not have walls. Animals have two very important attributes that plants exhibit only in much less emphatic forms: (1) An animal has a fixed developmental pattern; each individual's life history goes through a set of stages characteristic for its species. (To appreciate the difference between animals and plants being suggested here, we may consider that a plant growing under marginal conditions may produce one flower very early, while the same type of plant growing under ideal conditions may grow luxuriantly and then produce 100 flowers. By contrast, a female puppy cannot be expected to reproduce when very young, nor will she develop 100 wombs if well fed and cared for.) (2) Animals possess in high degree and in a myriad of forms, the property of **homeostatis,** which is the capacity for self-regulation, adjustment, and balance of function. (For example, sweating in response to high temperature is a homeostatic mechanism of warm-blooded animals.) The animal kingdom is subdivided into nine phyla.

BOX 2.F CLASSIFICATION OF *Homo sapiens*

If visitors from outer space were instructed to bring back specimens of the species *Homo sapiens,* they might find the following key useful for identifying the creatures they sought. The classification given is very incomplete; for instance, the infra-order is not given, and only a few of the characteristics appropriate to the given taxa are listed.

This classification is based primarily on anatomical features. With the development of biochemical genetics, it will become possible to develop a similar taxonomic table based on differences in protein structure. Similarly, it should eventually be possible to establish as precise a scheme on the basis of behavioral traits.

Taxon	Identifying characteristic
Kingdom: Animalia	Non-photosynthetic. Ingest food.
Phylum: Chordata	Mobile. Have a rod of elastic cells as part of the internal skeleton.
Subphylum: Vertebrata	Adults have a vertebral column.
Superclass: Tetrapoda	Four-limbed.
Class: Mammalia	The young are suckled.
Subclass: Theria	Young are liveborn.
Infraclass: Eutheria	Do not have pouches.
Order: Primates	Have flat nails.
Suborder: Anthropoidea	Tailless.
Family: Hominidae	
Genus: Homo	Erect. Have large brains.
Species: sapiens	

It is arguable whether the human species has achieved the "uppermost rung on the evolutionary ladder." Perhaps all species or organisms surviving at the same point in time are equally advanced, each with its own successful evolutionary strategies to meet the problems of the environments it occupies. Humans are surely not most advanced under all criteria. For example, many mammals exceed humans in the perception of odors and sounds, many birds have superior vision, some animals receive signals to which humans are impercipient, and most plants can manufacture sugars from sunlight, air, and water within their bodies. But humans have many unique and near-unique biological attributes that, among other things, allow them to acquire abilities through cultural and technological evolution that other organisms have acquired through biological evolution. It is beyond the scope of this book to give here a comprehensive list of unique and near-unique characteristics of *Homo sapiens.* However, a selected partial listing is still worthwhile.

Competition in occupying similar ecological **niches** (that is, exploiting similar environments) is strongest between similar species. Hence modern *H. sapiens* may have displaced competing forms with similar properties and requirements, so that some human attributes are now unique. Yet many other human attributes are also attributes of existing relatives. For instance, humans used to be defined as tool-making apes, until it was found that other apes are capable, to some slight extent, of this feat. Similarly, the statement that the distinct human characteristic is educability by other humans is no longer valid. And while it has been said that "the animal knows, of course; but certainly it does not know that it knows," there is no firm assurance of such certitude, since awareness is a subjective phenomenon. Hence, the following discussion must be viewed in relative, not absolute, terms.

In their morphology, anatomy, and physiology, humans have many striking properties: They are the only running mammalian bipeds (the kangaroo, for example, is a leaping mammalian biped); their legs are longer than their arms; the size of the human brain is both absolutely and relatively greater than that of any other primate, and its structure is more complex; the sexual receptivity of human females is not restricted to a particular time of the month or year; and humans are generally not as hairy as their living close relatives. There are even more interesting psychological and behavioral differences, many of which Darwin recognized. We list some under five headings:

Control of the physical environment. The invention of clothes, the ability to refine metals, and the use of a variety of artifacts provide humans with a non-physiologically controlled ability to maintain a steady state. Humans can therefore exploit environments for which their biological makeup is not suitable. Indeed, this is probably the reason that humans are the dominant species on Earth and can utilize nearly all available niches. They can also manufacture objects not found in the natural world and maintain them against dissolution. Thus humans make vertical structures from reduced metals, while most nonliving natural objects are sloping, and metals in nature are mostly in the oxidized state.

Educability. In the human species there is an unusually long period in which the child is dependent on the care and protection of adults. This not only makes low fecundity and high survival rate a reasonable strategy for species survival, but it also has the important effect of providing a long period during which the child can learn from adults. Whereas chimpanzees and even mice can learn by observing others of their species perform invented novel acts, the human species may be unique in having an ability to learn how to learn. In addition, humans strike a balance between acceptance of authority (don't put your fingers in the light socket!), which increases individual survival, and rejection of authority, which is responsible for much of the continual expansion of knowledge.

Capacity for communication across space and time. Many animals may communicate across space, signalling for example by scent, sound, or behavior. But ability to communicate nonbiological information not only to contemporaries, but to remote descendents (though, alas, not to remote ancestors) is found only in humans. This **time-binding** ability is exceedingly important and complex. It makes cultural evolution possible. Acquired cultural attributes, such as knowledge and custom, are transmissible to succeeding generations, whereas acquired biological ones, such as practiced skills, are not. This general ability is connected, as cause or effect, with a great many human features: the development of symbolic spoken and written language; the use of words for things, and the use of words for ideas; the ability to abstract; the possibilities of lasting influence of single individuals on the species; the so-called property of *displacement,* that is, speaking of things in the past or future, or of things imagined.

Consciousness of self. We are also very likely the only living beings who know we have a history as a species, and recognize its historical dimension. We are the only creatures known to practice burial or ritual disposal of our dead, and we do so and have done so in all our races and cultures. We are probably the only creatures that know that we as individuals are mortal, which may be why we have developed a sense of the sacred (as well as a sense of the scared).

Capacity for conscious direction of evolution. We already use this ability in many ways. For instance, in agriculture we produce new kinds of crops and modify many economically useful animals; in medicine, new forms of antibiotic-producing protists; and for our pleasure, new kinds of ornamental plants, new breeds of dogs, and fast-running race horses. Our capacity for directing our own evolution will be discussed in the last part of this book.

2.6 THE ORIGIN OF LIFE

Having briefly considered the present status of human development in the course of evolution, let us now turn to the beginnings of the process, the origin of life on Earth. There have been numerous speculations on this subject. They may be summarized under six headings.

Divine creation. This hypothesis involves a supernatural agency, and hence is outside the realm of science. It is neither provable nor disprovable by experiment, but there is nothing to prevent anyone who so chooses from believing that the other theories are included in this one.

Life on Earth always existed. The geological and chemical record provides grounds for rejecting this hypothesis.

Cosmic dust. This is the theory that primitive life forms arrived on Earth from outer space. Material resembling nucleoproteins has been found in meteorites, but there remains the question whether unprotected living matter could have survived long periods in space or passage through the radiation belts surrounding Earth. The theory is difficult to disprove; its acceptance would still leave the question of how life originated elsewhere.

Unique spontaneous generation. This is the most likely hypothesis and there is much evidence for it. Nearly every step in the process can be accounted for. Under this theory, there was a single successful origin of life on Earth, and all present living forms trace back in unbroken lineage to this unique beginning. The tenets of some eastern religions, which view trees, and ferns, and fish, and snakes, and humans, and all living things as kin, thus may have considerable biological basis.

Continuous spontaneous generation. This hypothesis was widely held in the nineteenth century. It was disproved by the French chemist **Louis Pasteur** in 1859–61, by showing that spontaneous generation did not occur under conditions thought favorable to it, and that the life which appeared in rich broths arose only from pre-existing life.

Repeated spontaneous generation. On a cosmic scale, this is likely. But it seems unlikely that different evolutionary lines on Earth were founded by separate events of spontaneous generation, each starting with chemicals derived only from inorganic chemicals. The universality of the genetic code, the commonality of many biologically active chemicals, and population theories of competition and niche occupation, all argue against this hypothesis. However, the death of organisms and new spontaneous generation from their remains is a plausible if perhaps trivial hypothesis, which the counter arguments do not refute. Such a new organism might be spontaneously generated from the DNA, RNA, and protein debris of the dead organism, and would thus have these chemicals in common with established life forms. It is most likely that such a new line would initially be a parasite on the species of organism that gave rise to it.

Thus, it is generally accepted that all living matter today comes from pre-existing life. Following Pasteur's work, this was accepted as having always been true until, in the 1920's, the British polymath **J. B. S. Haldane** (Box 2.G) logically pointed out that unless either the first or second hypothesis (divine creation, or life always existed) is correct, spontaneous generation must have occurred at least once in the past. A few years earlier, the Russian biochemist **A. I. Oparin** had formulated an explicit theory of how spontaneous generation might have happened and why it is no longer happening.

Much experimental work has been done since. At present it is generally considered that given (1) a certain range of temperatures, (2) the presence of a number of primitive compounds, (3) protection from existing heterotrophs, against which newly originated life is defenseless, and (4) a reducing atmosphere (one lacking oxygen—since life originated, oxygen has accumulated in the atmosphere from organic processes such as photosynthesis), then a successful origin of life is highly likely.

According to one hypothesis, the process is presumed to have taken place in the following sequence. After Earth was formed, either by evolution from a nebula, or by capture by the Sun of a gaseous dust cloud, or in some other way, simple compounds of hydrogen, oxygen, carbon, and nitrogen began to be formed. These included

H—O—H　　O=C=O　　H—N—H　　H—C—H (with H above and below C)
water　　*carbon dioxide*　　*ammonia*　　*methane*

In a reducing atmosphere, and with energy available from ultraviolet radiation, these could start forming slightly more complex compounds, such as

H—C(=O)—O—H　and　H—C(H)(H)—C(=O)—O—H

formic acid　　*acetic acid*

These substances have, in fact, been synthesized in laboratories using as energy sources intense ultraviolet radiation similar to that which reached Earth before an atmosphere (which shields the surface) developed, electric discharges similar to lightning, searing heat comparable to that of volcanic eruptions, or shock waves similar to thunderclaps or meteor concussion.

BOX 2.G J. B. S. HALDANE (1892–1964)

Haldane was one of the most stimulating and erudite men of modern science. His many books and articles deal with cosmology, genetics, biochemistry, evolution, animal behavior, politics, religion, the social order, and many other topics. He was one of the last of men of universal culture—a polymath—equally at home with Dante's *Divina Commedia* and the Koran in the original language on the one hand, and mathematics, astronomy, biology and behavioral sciences on the other (music was the one blank area among his interests, since he was tone-deaf). His father, J. S. Haldane, was a foremost physiologist of his day, and his uncle, Lord Haldane, a philosopher who served as Minister of War in a Liberal government and later as a Labor party Lord Chancellor. Haldane was educated at Eton and Oxford, receiving degrees in classics and mathematics simultaneously.

After serving in World War I, he turned to evolution, genetics, and biochemistry. For over forty years he continued at Cambridge and London to produce an astonishing output of books and theoretical papers in a variety of disciplines. He was much preoccupied with problems of social justice, was an excellent popularizer and journalist, and served for some ten years as chairman of the editorial board of the British communist *Daily Worker*. Between the time of the Spanish Civil War and his retirement from the chair of biometry at University College in London, he was a brilliant apologist for Marxist thought. This philosophy was gradually displaced in his writings by Hinduism, some aspects of which he accepted and others of which he criticized severely. In 1957 he migrated to India, where, while continuing his teaching and research in genetics, he adopted the Indian

form of dress and vegetarian diet, and became an Indian citizen. The photograph reproduced here by courtesy of Dr. Krishna Dronamraju, one of his students, was taken in this last period of his life.

There were elements of eccentricity, hot temper, and bohemianism in his personality, and he could willingly be unbelievably rude. But essentially he was an abstemious, courteous, kind man in whom, at times, perhaps, the ratio of wisdom to intelligence fell below the point considered desirable by many who admired him, and who was always in conflict with his environment. As a scientist he was essentially a theoretician, most of his actual experimental work being done with himself as a subject (he suffered severe spinal injuries while studying escape systems in submerged submarines). The all too few references to his work scattered throughout this book do but very little justice to the fertility of his ideas in the human quest to understand nature.

After the first complex compounds formed on Earth further syntheses under similar conditions could have produced increasingly complex precursors of organic compounds such as the amino acids, and it has also been suggested that polypeptides could have arisen directly from condensation products of such gases as ammonia and hydrogen cyanide. Indeed, in 1953, the American chemist Stanley Miller subjected a mixture of methane, ammonia, water, and hydrogen to an electric discharge (simulating lightning) and obtained glycine (see Section 2.3) and alanine (another amino acid), as well as other materials.

Little by little, a sort of soup of these various materials probably came into being. From what is known of chemistry, it is not difficult to see how nucleotides (see Chapter 6), which are the building blocks of nucleic acids, could arise. Today amino acids, polypeptides, and carbohydrates, plus energy-transporting substances essential for the operation of a cell, energy-binding components of chlorophyll, and nucleic acids have all been synthesized in the laboratory under conditions simulating those thought to have prevailed on the early Earth.

Once nucleic acids had been created, the instructional apparatus for organisms was available. The incompletely understood step is how an organic shell was formed, creating a cell-like organism. However, it is known that complex molecules can develop surface layers and boundaries like soap bubbles by concentration into droplets in water solution. These droplets can increase in size, split, and perhaps eventually acquire a complex self-duplicating apparatus. Thus the origin of organic molecules, their aggregation into droplets, and their evolution of unit-replicating mechanisms appear able to account for the original spontaneous generation. Whether this happened only once or several times we shall probably never know. Perhaps a number of such beginnings occurred (Figure 2.6). But

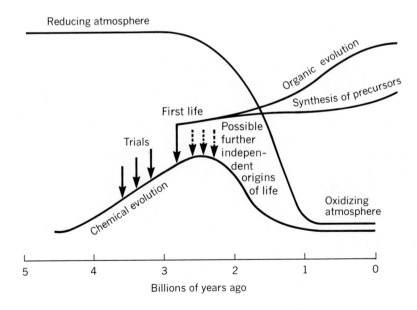

FIGURE 2.6
Probable origin of life and
course of evolutionary events.
(Redrawn with modification
from J. Keosian.)

BOX 2.H THE ORIGIN OF SECONDARY HETEROTROPHS

The following is a speculative, but possible, sequence of events. An early organism, X, was heterotrophic. To survive it needed (among other things) c, a substance it could not manufacture itself. The substances a and b were common in the environment, and as an occasional event they combined spontaneously to produce c.

X lives on c

A mutation in X created an organism Y that had the genetic capacity to produce an enzyme that catalyzes the reaction $a + b + \text{light energy} \rightarrow c$. Y could then live on the abundant a and b, and thus had a large advantage over X, which still depended on the scarce c in the environment. It is likely that Y completely replaced X, an example of natural selection (see Chapter 3).

Later, an advanced form of Y (with a number of favorable mutations incorporated since the primitive X first mutated to form the autotrophic Y) sustained a mutation in a gene controlling the enzyme producing c, so that this new organism that we call Z could no longer produce its own c. Z was then capable of obtaining sufficient c from existing Y's, either by eating them or their c-containing waste products. Z was thus a secondary heterotroph, living on c from Y.

A dynamic equilibrium then became established, with the numbers of Z depending on the availability of c from Y, and the numbers of Y controlled by the predation of Z.

BOX 2.I TIME SCALE OF EVOLUTION

In 1921, it was established that more distant galaxies are receding from us faster than nearer ones, and that the universe is expanding. Based on this expansion from a common point, or big bang, the age of our universe since that beginning is calculated at present (i.e., August 1972) as 17.7 billion years, and of the galaxies as 12–15 billion years. Earth is about 5 billion years old, and life on Earth started between 3 and 4 billion years ago. The following table gives a summary of Earth evolutionary events on the scale of a single calendar year.

Years ago	Event	Relative date
5 billion	Origin of Earth	January 1
3–4 billion	Origin of life	March 14–May 26
600 million	Marine invertebrates appear	November 18
350 million	Land plants appear	December 5
205 million	First dinosaurs appear	December 16
65 million	Last dinosaurs die out	December 27, 6 A.M.
1–2 million	Humans appear	December 31, 9:30–10:15 P.M.
5000	Early civilization	December 31, 11:59 and 28 seconds P.M.
30	First atomic bomb explodes	December 31, 11:59 and 59.81 seconds P.M.

once living matter became established on Earth, the atmosphere began to contain oxygen, and the reducing conditions necessary for this kind of spontaneous generation no longer existed. Furthermore, as heterotrophs appeared and increased in number, imperfect new organisms would have been less likely to establish viable lines: the newly arising forms would have been quickly displaced or consumed by advanced forms of life that had already evolved through millenia of trial-and-error selection.

Indeed, the first life on Earth was probably heterotrophic, feeding on precursors of organic matter. From it, photosnythetic autotrophs could have arisen, in turn giving rise to secondary heterotrophs. One suggestion that has been made about this process is illustrated in Box 2.H.

It is not possible to date precisely the events described, but it is likely that they occurred somewhat more than 3 billion years ago, since fossil bacteria of that age are known to exist (see Chapter 4 for dating methods). The length of time it took for human beings to evolve from primordial living matter, and the relative span of our civilization, is illustrated in Box 2.I.

It is not very likely that we shall ever know for certain how life did originate here. All that experiments will permit us is to infer how it might have originated. It is also possible that information on the existence of life elsewhere than on Earth will throw some light on the question. This is the subject matter of the still highly speculative new science of **exobiology.**

2.7 EXOBIOLOGY

Much effort and time is being spent in preparation for seeking and investigating life outside our own planet. There is considerable debate whether this work can yield useful results. As will be shown presently, there is a very high probability that life exists elsewhere in our universe, but there is no certainty that meaningful communication with any sentient extraterrestrial beings can be established soon— or ever—by any means within our command. It is not very probable that there are humans or humanoids elsewhere, except in the view of extreme determinists. They claim that, given conditions similar to those that existed on Earth, similar forms of life, including humans, would inevitably have evolved. How easily and to what degree each person accepts these possibilities involves his or her concept of infinite, as applied not only to our immediate universe, but to what occupies the space outside of it.

Life, as we know it, required the development of self-replicating information mechanisms such as is provided by nucleoproteins. As we have seen, this development can proceed under certain conditions. On Earth, these conditions included a reducing atmosphere, free water, and a restricted range of temperature, perhaps averaging from $0°C$ to $80°C$, and in general not dropping below $-100°C$. (If temperature is too low, reaction rates are too slow; if temperature is too high, precursors of organic molecules break down.)

Looking first at our own solar system, we can eliminate all but two of the heavenly bodies as possible places, in addition to Earth, where life as we know it might have developed. The Moon, Mercury, and the asteroids have no atmosphere and no surface liquid. Jupiter, the Jovian moons, Saturn, Uranus, and Neptune have reducing atmospheres, but they are too cold, judging from their atmospheric temperatures (their surface temperatures are not known). Whether the enormous pressures on Jupiter and Saturn might create conditions for a form of self-replicating life beyond the present comprehension of our science remains intriguing speculation. Venus has a surface temperature of more than 400°C in some places, but a cooler upper atmosphere, and polar caps that cannot be excluded as possibilities, although they seem highly unlikely ones. Mars is left as a serious candidate.

The atmosphere of Mars has no detectable oxygen, much ultraviolet radiation that could serve as energy, and a temperature range from about 30°C in the summer daytime to $-40°C$ in the winter. Whether enough water vapor is present for life is still debated. Seasonal changes in color that could be interpreted as polar-cap melting have been observed. All evidence, however, suggests that if life is present on Mars, it is represented only be microscopic forms. The evidence from the Mariner spacecraft missions is not clear-cut and leaves open the question of whether there is life on Mars. The United States Viking project will attempt to land instruments that should be capable of providing an unequivocal answer.

Several experiments are contemplated. One involves incubating Martian soil samples in solution and monitoring changes in turbidity that would indicate growth of some living organism. Another entails use of refined chemical methods to determine whether nutrient solutions added to the soil produce the type of changes that accompany growth of microorganisms. Still another experiment is based on checking whether radioactively labeled carbon dioxide is incorporated into organic compounds, or conversely, whether radioactive carbon-14 is released in the form of CO_2. An important requirement of these experiments is making sure that the probes do not carry with them terrestrial microorganisms that could establish themselves on Mars and hopelessly becloud the inquiry regarding existence of native Martian life. Similarly, if any material were returned from Mars, precautions against contamination here would have to be taken. It is not out of the question that Martian life is completely different from ours (for instance, silicon-based instead of carbon-based) and, if brought to Earth, might be very difficult to control.

The probability of life outside the Solar System has also been seriously examined. Only a little is known about exact conditions prevailing beyond our system, but some deductions may be made. One essential for life to develop on a planet is that the planet have an orbit that is not excessively eccentric, since an eccentric orbit would produce a range of temperatures greater than that under which life can exist. The likelihood of finding planets with appropriate orbits in any particular solar system depends on the origin of that system: if by collision, the right kind of orbit is unlikely; if by gaseous condensation, highly probable.

From what we know of nearby systems, they are not promising habitats for living matter. Within five parsecs (approximately one hundred trillion miles, or about 16 light years) of our solar system, there are 41 stars. Our nearest neighbor, Alpha Centauri (4.25 light years distant), does not have suitable planets. Indeed, among our close neighbors, only Epsilon Eridani (10.8 light years away) and Tau Ceti (11.8 light years distant) have.

The farthest galaxy yet discovered, 3C123, found in 1975 and 8 billion light years away (light travels about 6 trillion miles per year), may be five or even ten times the size of our own galaxy, the Milky Way. Although, in the Milky Way alone, there are at least 100 billion planetary systems like ours, and claims have even been made that within the 10,000 cubic parsecs proximal to Earth some 200 billion habitable planets exist. This leads many astronomers and biologists to the view that it is almost certain that life has developed on many of these. Yet, the distances are so great that we may never be able to detect such life or communicate with any sentient beings that may have evolved on such planets. A short-term attempt has already been made to search with medium-short radio waves for signals from outer space. Project "Cyclops" of the United States Space Agency has recently investigated further methods to the same end, and conferences to promote international cooperation in such searches have been held. It is not improbable that living beings on some planets have a technology that would permit them to contact us, but the whole matter is beset by innumerable uncertainties. We do not know, for instance, how long civilizations last. It may be possible that the luminous outbursts called supernovae, which could be exploding stars, record the destructions of civilizations, or that we ourselves will become a self-destroying civilization before we learn if life is to be found elsewhere.

3

DARWINISM

Darwinism is not synonymous with the concepts of evolution, as they are understood today. Yet the theses that Darwin proposed had such seminal and far-reaching effects that an analysis of Darwinism, as a pivotal focus for understanding the concepts of evolution, is a good way to start.

3.1 PRE-DARWINIAN VIEWS

The outstanding landmark in the history of evolutionary thought was the publication in 1859 of Charles Darwin's *On the Origin of Species by Means of Natural Selection and the Preservation of Favoured Races in the Struggle for Life.* As is often the case with important generalizations, his central thesis, natural selection, was not entirely new (see Box 3.A). But his work appeared when the intellectual time was ripe, and as he marshalled overwhelming amounts of convincing evidence, the idea caught fire. It spread and made numerous converts. Little by little, Darwinism became an important and central part of human thought. There have been setbacks in this intellectual development, and opposition was organized not only in Victorian England, but in twentieth century America. Some of the more recent conflicts have been the Scopes Trial in Tennessee in the twenties, court tests of laws against teaching evolution in Arkansas in the sixties, and attempts to require that equal space or time be given alternative (*read* religious) theories of creation in Texas, Colorado, and California classrooms and textbooks in the sixties and seventies. However, today it is unusual to find informed persons who do not accept that, with or without Divine Design, evolution is the way the human species arrived on Earth.

Four periods may be distinguished in the history of Darwinism:

1. The pre-Darwinian period was essentially speculative, with primary consideration being given to the question of whether evolution occurred at all.

2. Darwin and his immediate followers lived in the observational period, that is, when observations were gathered as evidence, and his proposed theories of evolution were defended and attacked on the basis of such observations.

3. The early part of the present century was devoted to continued accumulation of observations, but with greater emphasis on experiments, and with a more rigorous examination of objections to the notion of natural selection.

4. During the remainder of this century, the synthesis between his ideas and the newly established laws of hereditary transmission was attained. The rediscovery of Mendel's laws erased many of the difficulties with Darwin's scheme of evolution. The recent and profound investigations in molecular biology and biochemistry have not only removed much equivocation about evolution having happened, but have made this process a compelling idea. They also gave us a highly probable account of how life could have originated in the first place. Of course, there is a great deal yet to be learned about details, some of which may forever remain speculative.

The first of these periods goes back to the beginnings of the human search for understanding of man and the world, which to Western culture means back to the classical Greek philosophers. The Greeks proceeded in their rationale from *a priori* philosophical grounds, from observation by the unaided eye, and from logical arguments and so-called common sense. Hence many things that we know to be false were reasonable to some of them: e.g., continuous spontaneous generation and the **inheritance of acquired characters.** In general, there is a tendency to read more science into the views of the Greeks than is really there and to credit them with pre-vision. For instance, Democritus, who lived in the fifth and fourth centuries B.C., is sometimes called the father of atomic theory, although there is hardly any resemblance between the atom he talked about and that of current nuclear physics.

Among Greek philosophical writings, from the sixth century B.C. on, speculations are found about life originating from the sea. They also include mythical ideas about the adaptive changes involved in the transition from the aquatic to the terrestrial form of existence. Fossils were recognized by at least one Greek philosopher as being animal remains. Views on the nature of change, the basic tenet of evolution, ranged from the idea that all change is merely an illusion of the senses to the idea that everything is always in flux. A vague notion of organic

BOX 3.A SOME PREDECESSORS OF DARWIN

In addition to Linnaeus (Section 2.3), who was the father of modern classification of living creatures but a believer in special creation, the following may be mentioned among pre-Darwinian students of natural science.

FRANCIS BACON (British, 1561–1626) was a firm Aristotelian and thought that species were immutable. He is generally credited with the revival of scientific inquiry, but many philosophers and biologists now believe that deduction rather than the inductive Baconian method is the appropriate way toward scientific discovery.

WILLIAM HARVEY (British, 1578–1657) is often credited with the first biological experimentation and the discovery of the circulation of blood, although he had some forerunners in these investigations.

GEORGES L. L. BUFFON (French, 1707–1778) was, perhaps, the first true evolutionist, recognizing the importance and ubiquity of variation, and suggesting, with reservation, an evolutionary process based on the inheritance of acquired characters.

ERASMUS DARWIN (British, 1731–1802) was the grandfather of Charles Darwin and a speculative evolutionist. His main theses were that Earth's history was longer than specified by Bishop Ussher's chronology (Section 1.1) and that all life came from a common source. By the end of the eighteenth century the general idea of evolution was becoming relatively popular.

At least three different men wrote of natural selection without, however, arriving at a firm statement of its role in evolution. Two of them were British. They were Edward Blyth, who thought that natural selection, by discriminating against variation, would lead to immutability of species; and Patrick Matthew, whose clear discussion of the phenomenon was hidden in a treatise dealing with naval timber and architecture. The third was Charles Wells, a South Carolina physician, who wrote of the idea as a commonplace fact in a paper on a white female, part of whose skin resembled that of a Negro.

JEAN BAPTISTE DE LAMARCK (French, 1774–1829) advanced the first comprehensive theory of evolution. One of the mechanisms he invoked for the process was the inheritance of acquired characters, which is now referred to as **Lamarckism,** although he was not the first believer in it. The usual example of Lamarckism is the evolution of giraffes, whose ancestors were assumed to have acquired their long necks by stretching to reach upper leaves of trees, and to have transmitted the acquired length to their progeny. We now know that it is more likely that giraffes who happened to have longer necks could obtain more food and thereby had an advantage over others, enabling them to leave more offspring. If their long necks were even in part due to a difference in their hereditary endowment, their offspring, more numerous than that of the others, would receive genetic instructions for the formation of longer necks through the gametes. Thus, the average neck length of the next generation would be increased. This cumulative process is **natural selection.**

THOMAS R. MALTHUS (British, 1766–1834) suggested in his "Essay on the Principle of Population" that humankind multiplies geometrically while the means of subsistence do not. Although Darwin credited his reading of this essay with generating in his mind the notion that an average individual produces more offspring than can survive, thus permitting selection to occur, there is evidence that Darwin discovered the principle of natural selection before reading Malthus. He had also read Blyth, Matthew, and Wells.

GEORGES CUVIER (French, 1769–1832) was a defender of special creation and Lamarck's opponent in the evolutionary debate of the day. He recognized that fossils were the remains of extinct forms of life, following the discovery in 1791 by an English surveyor, William Smith, that different layers of rock contain different kinds of fossils. Cuvier's explanation was the theory of **catastrophism**—that is, that in the past, life on Earth had been decimated or destroyed several times, as in the Biblical account of the flood, and new kinds of organisms somehow appeared after such catastrophies.

CHARLES LYELL (British, 1797–1875) provided an alternative to catastrophism with the theory of **uniformitarianism,** which held that historical changes on Earth were not due to a series of catastrophes but to the same gradual changes as may be observed today.

evolution has been identified as existing as early as the fifth century B.C. It held that the various living beings arose by a fortuitous combination of parts, and that those which were badly put together did not survive. It thus contained the germ of the principle of natural selection.

The most important scientific influence was that of Aristotle (384–322 B.C.), the creator of natural history and of the logical method exclusively used in science until recent years. He was a teleologist, that is, a believer in intelligent design and in the idea that processes in nature are directed toward certain ends. Teleology was rejected by modern scientists until the recent discoveries that, contrary to the situation in physical science, many biological processes are based on feedback and are, indeed, end-determined; for example, the production of enzymes by cells, which will be described in Chapter 6. Aristotle's views and his classification of plants and animals were basically nonevolutionary, and, in many ways, his authority inhibited development of evolutionary ideas for a millenium and a half or more. (It is a problem by no means restricted to the ancient Greeks, that advanced thinkers tend to accrue authority, and that very authority then inhibits further advance.)

There is little to be said about the post-Aristotelians, the medieval scholastic philosophers, because they contributed mostly to metaphysics and to moral, rather than natural, philosophy. Some of them insisted on the notion that conditions of life were immutable. The view was taken that the world was always as it is now and that both biologic and social relations and structures were static. Interpretation by previous authorities, which Dobzhansky calls the "comfortable certainties of the traditional medieval world," was replaced only over a long period of time by scientific questioning.

A thorough history of pre-Darwinian thinking is, of course, impossible here. Box 3.A provides some highlights. R. C. Lewontin viewed Darwinism as "the percussion cap for a charge already set." The details of how Darwin (1809–1882) and, independently of him, **Alfred Russel Wallace** (1823–1913) crystallized the idea of natural selection are perhaps deserving of a chapter or more. As a young naturalist, Darwin (Figure 3.1) made a trip around the world on H. M. S. *Beagle,* in the course of which he accumulated a vast store of observations. Combined with his reading and deductive cogitation, they led him to the formulation of his theory. Wallace, who studied the natural history of Malaya, apparently had an intuitive flash about the role of natural selection in the evolutionary process, and in 1858, wrote to Darwin about the idea. This led to the reading of papers written by the two men, although they were not themselves present, at a historical session of the Linnean Society of London, followed by the publication in the next year of *Origin of Species.* The 1,250 copies of the first printing of the book were sold immediately. It is a curious fact that the annual report of the Linnean Society records that "nothing of significance happened in 1858."

A large number of followers were quickly attracted to Darwin and the idea of evolution via natural selection. They popularized his ideas and added to his observations. There were many critics, and various objections were raised, some reasonable and some based on lack of comprehension. We shall omit much of

FIGURE 3.1

Charles Darwin is frequently shown as an old, even sickly, man, as the picture on the right portrays. (Portrait by the Hon. John Collier, from the National Portrait Gallery, London.) The picture at the left (from an 1840 portrait by George Richmond, Time/Life Picture Agency) is Darwin at age 31. Four years earlier the epic expedition of the cartographic ship *Beagle*, with young Charles aboard as naturalist, had been completed. (It is true that Darwin was sick often enough during the voyage, but he was also tough enough to transform himself from an unsure amateur biologist to a competent professional in the course of the voyage, which lasted from 1831 to 1836.) The center picture (from *Atlas of Evolution* by Sir Gavin de Beer, Elsevier Nederland B. V.) is Darwin at about the time his work on *Origin of Species* was nearing completion.

the history of post-Darwinian developments. But before discussing the essence of Darwinism, we shall consider the nature of the revolution produced by Darwin and its relation to other scientific revolutions.

3.2 SCIENTIFIC REVOLUTIONS

Historians of science differ about how science develops. Traditionally, it has been assumed that scientific progress is always gradual and slow, and that in a historical continuum there are no critical points or revolutions. Recently, the view has been advanced that there are, indeed, revolutionary discrete moments in the history of ideas, such as are exemplified in the works of Aristotle, Descartes, Newton, Maxwell, and Einstein, and in concepts such as relativity or that the Earth is round, or that it (or the Solar System) is not the center of the Universe. Certain people, via their ideas, have had a profound and far-reaching impact on our views of ourselves and of our position in the Universe.

It is possible to find examples both of advances that took a long time to come to fruition, and of sudden advances of which it may be said that they shook the world. We may look at some of these sudden advances, choosing them from different disciplines in order to place in perspective the magnitude of the Darwinian one, which, it may be argued, eclipsed all others in significance.

The Keplerian revolution was in the realm of astronomy. A number of contributors to it are listed in Box 3.B. It was the beginning of the discovery that humans are not the center of the Universe, and it substituted the **heliocentric** for the **geocentric** view of our Solar System. It is now generally accepted that Earth, far from being the center of the Universe, is one of innumerable planets and no more distinguished than billions of others.

Though this revolution affected both science and general intellectual activities, it was probably less significant to human society then Darwinism. It is a defensible view that the Darwinian revolution had even more important social effects than the revolution dealing directly with society that was initiated by Karl Marx (1818–1883) roughly at the same time. Marx, and his associates in the fields of political and economic thought, introduced momentous notions about the nature of motivation in human history and about the historical nature of social change. They believed they had discovered fundamental laws of the development of

---

BOX 3.B THE COSMOLOGICAL REVOLUTION

Although we may speak of the "Keplerian revolution" or the "Copernican revolution," at least six men made important contributions to the change in thinking about Earth's place in the Universe.

NICHOLAS COPERNICUS (1473–1543), a Pole, first suggested that Earth is merely a planet revolving around the Sun, rather than the center of the Sun's orbit.

TYCHO BRAHE (1546–1601), a Dane, observed what were presumably exploding stars, or novae, a phenomenon which suggested to him that the Universe is not a static but a changing place.

GIORDANO BRUNO (1548–1600), an Italian, proposed that the Universe always existed and was not created as described in Genesis. The Inquisition could not agree less and burned him at the stake.

GALILEO GALILEI (1564–1642), another Italian, brought experiment to natural philosophy, helped establish the view of Copernicus, and changed the basis of our questions about nature from "why" to "how."

JOHANNES KEPLER (1571–1630), a German, proved that Copernicus was right, and worked out the laws of planetary motion.

ISAAC NEWTON (1642–1727), an Englishman, generalized the laws of motion that prevailed in cosmology for over two hundred years. Curiously enough, he was a believer in special creation and in Ussher's calculations.

The list illustrates, among other things, the fundamentally international character of scientific progress. In addition, many of the philosophers and scientists of this period moved from one country to another, further breaking the bounds of nationalism.

societies. The political effects of Marxism have indeed been exceedingly far-reaching. Yet it is arguable whether these principles have been of great importance at a practical level, or that they are fundamental scientific or intellectual principles in the way evolution by natural selection is. (Most communist philosophers would probably disagree with this evaluation.)

The outlook of Darwinism is that humans were not created masters of all they saw; that they were not from the time of their arrival the center of activities on this planet; that they have become the dominant species by a blind self-correcting process and not by design; that their antecedents were just as humble as those of the rest of living beings. In brief, it led to the rejection of **anthropocentrism.**

Today it may be fashionable to say that Darwinism and the religious outlook, in particular the Judeo-Christian and Moslem traditions, are not in conflict with each other. This is probably correct, but only because in the face of the overwhelming impact and evidence of Darwinism most religions retreated from literal scriptural interpretations of the world's origin and of humanity's historical position. But it would be correct to say that in Darwin's time organic evolution through natural selection and the established church were in opposition.

If we think of new ideas as if they were new mutations, we can make an analogy with natural selection. Over time, if the new are better, they sometimes (not always) replace the older ones. One may identify three groups of concerned people during such a period of possible intellectual evolution:

I	II	III
Adherents of the old dogma, who will not or cannot change	Adherents of the old dogma, who reject it and accept the new	Recruits who are taught the new dogma by members of Group II

Old dogma is rarely overthrown. Rather it is buried—literally outlived—as members of Group I die, retire, or otherwise lose power, and Groups II and III become the effective majority. Group II is critical, for on the number and persuasiveness of its members depends whether a change will in fact be made, and how long it will take. For this reason, it is important that, as a student matures, education increasingly stresses how to acquire and evaluate information and ideas, rather than providing a set of then-known facts and concepts.

In general, there is no compelling reason for religion (in the sense of ultimate concern for values and ethical systems, rather than revelation of the historical past) and science to be at odds. One view is that religion, dealing with absolutes, cannot deny verifiable facts, and must adjust itself to scientific progress. And the process of reappraisal has to go on continuously. If contraception has proved a bit of a stumbling block, one may imagine the theological problems that would be provoked by discovery of life on other planets, extension of organ replacement to the brain, or creation of human beings in a test tube (see Chapters 22 and 24).

As an aside, there is a sort of lesson in the United States today, where roughly ten times as many people make a living from astrology (a creed accepted on faith and spiritual sensing) as from astronomy (a science based on logic and physical sensing). There may similarly be many times as many people making a living from special creation, and its sometimes associated hellfire and damnation, as from evolution and its cool view of the meaning of life.

The Darwinian theory also produced Social Darwinism. This was an outlook wherein "struggle for existence" and the incorrect term "survival of the fittest," were used to justify social inequalities and unethical commercial practices, aggression, colonialism, imperialism, racism, and military expansion. It is unfortunate that evolutionary thought has not always worked on the side of the angels.

Since Darwin's time, some other revolutions have occurred. Whether Sigmund Freud (1856–1939) was right or wrong, his view of humans as not entirely rational animals has had repercussions on our behavior, evaluations of personal relationships, and social structure. It is perhaps too early to assess the ultimate consequences of the psychological revolution, which is no doubt incomplete. The same is true of the revolution in physics. Thus, the demonstration by Albert Einstein (1879–1955) that observation is a function of the observer's position and of matter-energy equivalence, the formulation of the complementarity principle (that there are mutually exclusive ways of seeing phenomena), and the other developments of the new physics, may possibly take us to the stars. This revolution may also enable us to blow ourselves up, making evolutionary thought somewhat irrelevant.

One of the most rapid revolutions in our view of ourselves may now be in progress, perhaps catalyzed by the December 1968 flight of Apollo 9. In their Christmas Eve telecast from the first manned orbital flight to Earth's Moon, Borman, Lovell, and Anders shared a view of Earth from space with a significant proportion of Earth's human population. Shortly thereafter, the concept "Spaceship Earth" exploded from the worried conversations and writings of a few ecologically oriented professionals, and became general public property. We suspect that a conception of evolution and selection has made us better able to deal with the problems of continued existence on this beautiful, but small and lonely planet. The effects of Darwinism, in all our understanding (incomplete as it is) of ourselves and our world, are with us every day.

3.3 THE ESSENTIALS OF DARWIN'S REASONING

Darwin's theory sets forth the idea that all modern living forms descended with modification from pre-existing forms, which, in turn, had ancestral forms. The modifications were undirected. That is to say, contrary to a teleological view, they arose without reference to the needs of the plants and animals bearing them. Because those organisms that were better adapted to their environments left more offspring than those that were not, modified forms might replace, over a period of generations, the nonmodified ones. This process of natural selection led to the origin of more and more complex forms capable of exploiting diverse ecological niches. If it is possible to speak of progress in evolution (there are differences of opinion on this point), the progress consisted of increasing individual adaptability, so that increased homeostatic (Box 2.E) powers developed.

Darwin's definition of natural selection began with the premise that there are restraints on the number of offspring that can survive in a given time and place. These restraints include predation, limitations of food supply, such physical factors as climate, and disease. Today we know a variety of other population-regulating mechanisms, which themselves evolved by means of natural selection.

On the basis of this premise Darwin said: "As many more individuals of each species are born than can possibly survive, and as, consequently, there is a frequently recurring struggle for existence, it follows that any being, if it vary however slightly in any manner profitable to itself, under the complex and somewhat varying conditions of life, will have a better chance of surviving, and thus be naturally selected. From the strong principle of inheritance, any selected variety will tend to propagate its new and modified form."

Elsewhere he continued: "This preservation of favorable individual differences and variations, and the destruction of those which are injurious, I have called Natural Selection, or the Survival of the Fittest." Actually, the term "selection" was, as Darwin made explicit, a metaphorical term, and did not imply a conscious choice either by the organism or by an external agency, except for **artificial selection,** in which humans exert control over the development of plants or animals by choosing certain individuals for breeding (Chapter 14).

Careless application of the word "survival" may also be seriously misleading. Darwin's central idea was differential reproduction. Clearly, to reproduce at all, an organism must survive for some minimum period of time, and organisms that survive longer generally have more opportunities to reproduce. But survival *per se*, without successful reproduction, is not sufficient for evolution, or even for the continuation of the species. Early in human history, differential survival before and during the reproductive years was an important component of the natural selection affecting human evolution. In much of the modern world, most children survive and achieve normal reproductive age, and thus at present differential survival is only a small component of human natural selection. Rather, natural selection in humans is now mostly associated with differential reproduction among the adults. In the Marshall Islands, for example, only 25 percent of the women produce approximately 75 percent of the babies.

Finally, the superlative term "fittest" is an incorrect overstatement. Confusion between "fittest" (or better, the comparative term "fitter") and *fitness* has caused difficulty in interpreting Darwin's ideas. Because of this term, anthropomorphic value judgments have been attached to the process of selection. Yet Darwin did not have this in mind at all. Neither strength of character, nor moral goodness, nor any attributes of physique or size, nor high intelligence, nor even long life *per se* cause an individual to produce more than the average number of offspring, or to be fit in Darwin's sense. Indeed, often organisms that are totally undistinguished by any physical or mental standards, organisms that exhibit average dimensions for various properties, these are the ones most successful in propagating themselves. In a group adapted to its environment, more often than not, it is the mediocre that survive and reproduce and not the exceptional. The Darwinian fitness of George Washington, of Beethoven, of Lenin, of Albert Schweitzer, and of Leonardo da Vinci, none of whom had children of record, is zero, whereas the Darwinian fitness of a half-witted, crippled, social misfit, who has fourteen grown children, is very high.

Darwin's description of natural selection may now be paraphrased:

1. in nature, individuals differ in various characteristics;
2. their differences are, in part, determined by hereditarily transmissible factors;
3. whenever these differences are associated with differences in fitness, that is, the relative success in leaving offspring surviving to reproductive age, the characteristics of the more fit individuals will occur in succeeding generations to an increased extent.

Thus, changes in characteristics of successive generations are determined at least in part by the inequalities in reproductive rates of individuals differing in hereditary endowment for those characteristics. Such endowment is referred to as the **genotype:** the genetic instructions that an individual may transmit to its descendents. The carriers of the various genotypes may be selected or rejected on the basis of their **phenotypes,** that is to say, the expressed characteristics that make them more, or less, able to leave surviving progeny in any particular environment. *Natural selection, then, is essentially the differential reproduction of the different genotypes.*

Natural selection by itself cannot fully account for evolution. As Darwin realized, some apparatus of hereditary transmission and some variation are necessary. The first property is now understood very clearly, as a result of the work of Mendel and his successors, and of workers in molecular genetics. The second property in large part derives from the phenomenon of mutation, which will be discussed in Chapters 16 and 17, and to a number of other processes, including occasional hybridization between species. Some of the other sources of evolutionary effects are **isolation,** which prevents interbreeding between species, and population size, to which we shall return in Section 4.3.

3.4 DARWIN'S EVIDENCE

In the hundred-odd years since the publication of Darwin's book, a tremendous amount of evidence on evolution has accumulated, including experimental observations made at a level of refinement not even dreamt of by him. Let us first consider a few examples of the data Darwin gathered. In the next section, a few examples from the post-Darwinian age will be given.

The first type of evidence Darwin used was that from the geographical distribution of organisms, which he observed on his *Beagle* voyage. Consider the flora and fauna of oceanic islands that have arisen from the sea a comparatively short time ago. Whether you prefer the hypothesis of special creation or of evolution, it is reasonable to assume that the organisms inhabiting such islands have descended from individuals that arrived there from the mainland. From the first hypothesis, the expectation is that they would have remained the same as their ancestors. From the second, because of their isolation and the mechanisms of mutation and selection, it is predicted that many lineages of such organisms would have changed into forms unique to that locality. Indeed, this is what is observed. For instance, rats on many islands are distinct from related mainland rats. On the Azores, a group of volcanic islands some 1,450 kilometers off Portugal, flying animals, such as birds, a bat, and butterflies, are essentially unmodified European species. But of the 74 species of beetles, 14 are not known elsewhere; of the 69 species of land snails, 32 are peculiar to these islands; and of the 480 species of plants, there are 40 species related to but different from any on the mainland.

Of the various localities visited by Darwin, the most important to him was the Galapagos Archipelago (Figure 3.2). This is a group of twenty small islands about 950 kilometers off the coast of Ecuador that arose by volcanic eruption some million years ago. Darwin noticed that the plants and animals native to these islands were generally similar to, yet, in peculiar ways, strikingly different from, species on the South American mainland. Many of them also differed from island to island. For instance, people familiar with the islands could identify which island a giant tortoise came from by the distinctive shape of its shell. It has since been observed that the land and marine iguanas of the Galapagos are clearly related, although they occupy very different niches, and the marine iguanas from different islands have different colored skins. Most important to Darwin were the 13 species of finches, which were similar to those he had recently seen in Peru. Although apparently of common origin, they had developed remarkable differences suiting them for the variety of niches at their disposal on the islands. In

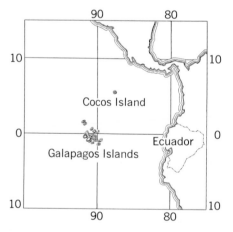

FIGURE 3.2

The Galapagos are some 600 miles west of Ecuador. One of the species inhabiting these islands, the woodpecker-finch, has evolved the beak (but not the long tongue) of a woodpecker. Rather, it has evolved the behavior of carrying a twig or cactus spine and using it to dislodge insects from bark crevices. (From D. Lack, "Darwin's Finches." Copyright © 1953 by Scientific American, Inc. All rights reserved.)

particular, they differed in size and shape of beak, some being adapted to cracking hard seeds, others to eating softer vegetable parts, still others to catching insects, and perhaps most remarkable, one to holding a cactus spine for poking insects out of holes and crevices (Figure 3.2).

The second kind of evidence Darwin relied upon came from comparative anatomy. From the hypothesis of special creation there is no reason to expect anatomical resemblances between various forms of life. If we, however, postulate an evolutionary process, transitional types are to be expected. Figure 3.3 presents an example of anatomical similarities and changes in structure in the course of evolutionary diversification.

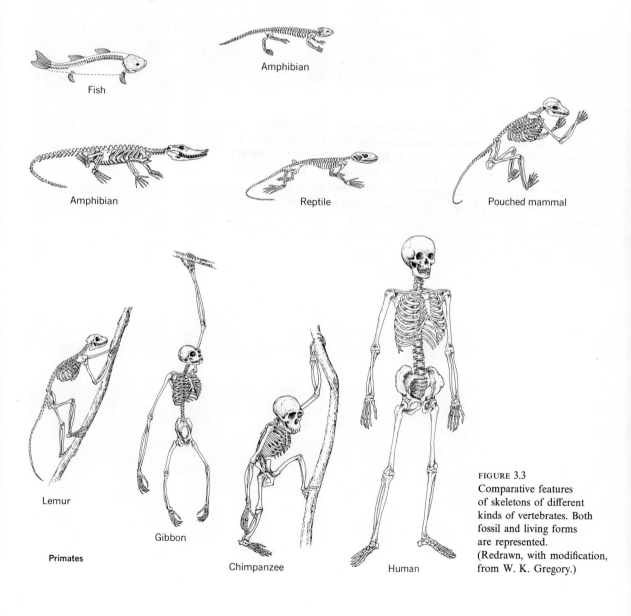

FIGURE 3.3
Comparative features of skeletons of different kinds of vertebrates. Both fossil and living forms are represented. (Redrawn, with modification, from W. K. Gregory.)

Another type of evidence for evolution was found in comparative embryology. Similarities between young forms related to each other are greater than those between adults. All animals develop from single cells that resemble each other very closely. As each individual grows, it gradually assumes the form characteristic of its own species, and thus diverges from individuals of other species. Figure 3.4 pictures the stages of development of human embryos, which show a remarkable likeness in the early stages to embryos of other vertebrates.

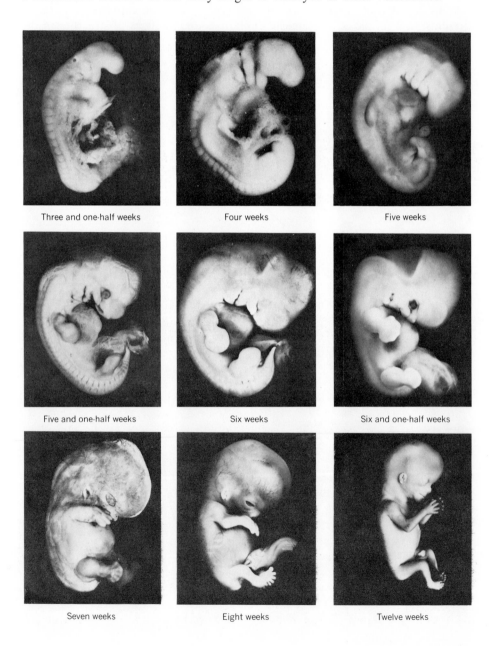

Three and one-half weeks   Four weeks   Five weeks

Five and one-half weeks   Six weeks   Six and one-half weeks

Seven weeks   Eight weeks   Twelve weeks

FIGURE 3.4
Photographs of
human embryos.
(By G. W. Bar-
telmez, courtesy
of A. S. Romer.)

The most massive testimony to evolutionary theory was found in the fossil record studied by paleontologists, which supplies an increasingly clear picture of evolution. Until the birth of genetics, paleontology was the source of the most convincing evidence for the principle of evolutionary change. Fossils in rock layers of different ages show a distinct progression of change from simple to more complex forms of life. An incomplete but consistent history of life on Earth has become available through the use of a variety of geologic and chemical methods for dating the remains of extinct forms (the latter developed since Darwin's time). We shall discuss the fossil record in the next chapter.

Finally, Darwin drew heavily on analogy between natural and artificial selection. He discussed domesticated plants and animals at length. He emphasized that domestication provided a form of experimental evidence in support of the importance of selection in the creation of new biological forms. Still, he was not fully aware of the dynamics of these processes. The better understanding of them that we have today lends more weight and precision to the case for evolution. We shall discuss examples of artificial selection in Chapter 14. Suffice it to say here that the experimental evidence now available supports and strengthens Darwin's general views on the matter.

3.5 POST-DARWINIAN EVIDENCE

Since Darwin's time much new information has accumulated to bear witness to his notions of evolution, to amplify them, and to make his generalizations more precise. Some of it falls into the categories described in the previous section; other evidence is based on our understanding of cytology, biochemistry, molecular biology, and comparative behavior, fields of study that were barely formulated—or didn't even exist—in Darwin's time.

One of the most spectacular and convincing examples is from the study of the **industrial melanism** of certain insects. About 70 species of moths have light and dark (melanic) variants. Until the middle of the nineteenth century, the former were the common kind, with only an occasional melanic, or dark, moth being found. Since then, however, it has been noted that in areas close to factories, the dark forms have become common, in some cases outnumbering the light moths. In the countryside, far from sources of industrial smoke particles, the light moths are still the common form, and dark moths remain rare. More recently, in industrial areas where pollution is being reduced, the melanic forms are reported to be decreasing in frequency.

A group of British biologists has carefully studied this phenomenon in nature, as well as in the laboratory. They found that the rate of survival of the two forms depended on how visible they are to the birds that eat them. The moths rest on tree trunks during the daylight hours. In unpolluted areas, light moths on tree trunks are practically invisible, while the dark forms are conspicuous. As a consequence, predation by birds tends to eliminate the dark forms. In industrially

FIGURE 3.5

Left: Peppered moths resting on a lichen growing on a tree trunk in an unpolluted area. The almost invisible light form is below and to the right of the melanic form. *Right:* The visible and invisible forms are reversed in a polluted area. (From the experiments of H. B. Kettlewell, University of Oxford.)

polluted areas, where tree trunks are covered with soot, the reverse is true. Figure 3.5 illustrates the camouflage in the two environments. The phenomena were not only observed and recorded on photographs and films, but quantitative experiments based on releasing and recapturing the two types of moth in the two environments were conducted. The hypothesis that natural selection was responsible for the increased frequency of the melanic forms in sooty environments, and their continued rarity in unpolluted regions, was fully substantiated.

Darwin and his followers relied heavily on reproductive compatability in analyzing evolutionary relationships near and below the species level (Section 2.4). It is possible this can now be further extended by new techniques that measure relative success of interspecific transplantation of ovaries and of hybridization of cells of different taxa in tissue culture. One of the most spectacular is the mouse-human hybrid cell cultures, which have been repeatedly created and studied for over a decade.

Another type of evidence unavailable to Darwin was found in the study of chromosomes. Chromosomes are the bearers of hereditary information (Box 2.C). Their number and structure are characteristic of each species. Just as the anatomical features of related organisms bear a resemblance to each other, so it might be expected that chromosome numbers and details of their internal organization would be similar in closely related species. Indeed, this is found to be the case in many groups of living beings studied.

Even when chromosome numbers in related species are not identical, the numbers are usually similar, or in multiples of some base number (presumably of a common ancestor). The chromosome numbers of various kinds of wheat are in multiples of seven (14, 28, 42). Those of chrysanthemums are in multiples of nine (18, 36, 54, 72, 90). Humans have 46 chromosomes, and among our closest living relatives, the chimpanzee and the gorilla have 48 and the gibbon has 44. A number of related species of *Drosophila* (the fruit fly) have from 6 to 12 chromosomes. Their characteristic shapes and sizes are such that it is possible to reconstruct from their number and appearance their **phylogeny** (evolutionary history), and thus the evolutionary relationships of the *Drosophila* species included in the study.

In the salivary glands of the larvae of *Drosophila* and some other insects there are giant chromosomes whose fine structure can be readily studied. It has been found that each chromosome is characterized by an orderly succession of cross bands and knobs (Figure 3.6). In some species the order of bands in certain sections of chromosomes is seen to be reversed from what it is in other species.

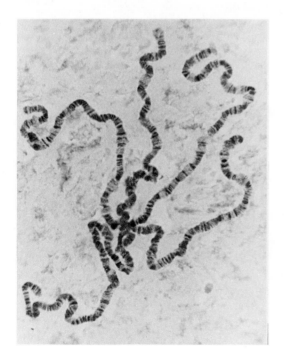

FIGURE 3.6
Salivary-gland chromosomes
of *Drosophila melanogaster.*
(Courtesy of Berwind P.
Kaufman.)

BOX 3.C IMMUNOGENETICS OF DOVE ANTIGENS

To determine the community of antigens between two species of doves, blood cells from the Pearlneck dove, containing their particular antigens, are injected into rabbits. These antigens then induce the production of specific antibodies in the rabbits. Blood cells from the Ring dove are added to serum from the immunized rabbits. Antigens in the Ring dove blood common to the two species combine with (agglutinate) the Pearlneck-induced rabbit antibodies and can be precipitated. The antibodies for antigens that the Pearlneck dove has and the Ring dove lacks, however, remain in suspension. This can be verified by adding again Pearlneck blood cells to the rabbit serum from which the first precipitate has been removed and obtaining a second precipitate after centrifugation. The relative amounts of precipitate observed in these two steps give an estimate of the community of antigens in these two species.

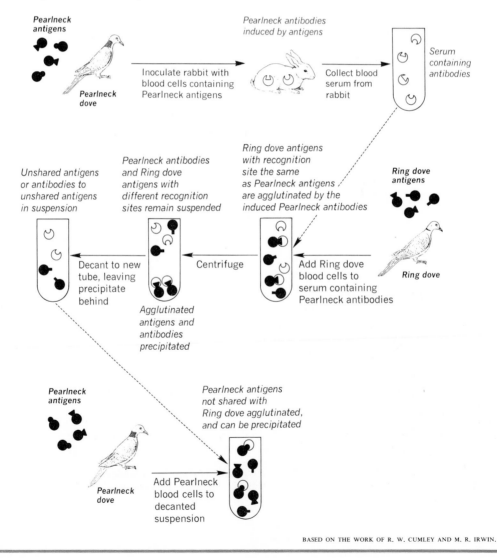

Pearlneck antigens

Pearlneck dove

Inoculate rabbit with blood cells containing Pearlneck antigens

Pearlneck antibodies induced by antigens

Collect blood serum from rabbit

Serum containing antibodies

Unshared antigens or antibodies to unshared antigens in suspension

Pearlneck antibodies and Ring dove antigens with different recognition sites remain suspended

Ring dove antigens with recognition site the same as Pearlneck antigens are agglutinated by the induced Pearlneck antibodies

Ring dove antigens

Ring dove

Decant to new tube, leaving precipitate behind

Centrifuge

Add Ring dove blood cells to serum containing Pearlneck antibodies

Agglutinated antigens and antibodies precipitated

Pearlneck antigens

Pearlneck dove

Add Pearlneck blood cells to decanted suspension

Pearlneck antigens not shared with Ring dove agglutinated, and can be precipitated

BASED ON THE WORK OF R. W. CUMLEY AND M. R. IRWIN.

Studying which *inversions* of chromosome segments are present in which species adds to the developing picture of species phylogeny. This type of information is particularly useful, since the necessary sequence of overlapping inversions can generally be unambiguously deduced.

Even more spectacular support for evolution has come from more recent studies in immunology, biochemistry, and molecular biology. **Immunology** is the study of the response of tissues and organisms to large foreign molecules. It may be recalled that each organism is characterized as possessing a specific combination of polypeptides determined by the genetic message received from its parents. Some of these polypeptides may be found in all members of the species (or higher taxa). Others are peculiar to the particular individual.

Normally, the body of an organism recognizes its own molecules and distinguishes them from those foreign to it. When large foreign molecules, known as **antigens,** are introduced into the body, it reacts in defense against these invaders by producing **antibodies** that inactivate the invaders. This is the mechanism by which an organism combats infectious disease; it produces antibodies that react with the antigens of viruses, bacteria, and other protists. An important point is that the antibodies are highly specific and each generally reacts with only one form of antigen.

Blood-group substances (discussed in greater detail in Chapter 20) can act as antigens. Thus if blood cells of type A are introduced into the bloodstream of an individual whose blood type is B, the anti-A antibodies will destroy the A blood cells by clumping (**agglutination**). On occasion, because of failure of the recognition system, a body may start manufacturing antibodies against its own proteins, giving rise to severe *autoimmune* diseases such as lupus. Treatment of this disease by *immunosuppressive* drugs, like cortisone, or by X-rays, lowers the capacity of the patient's body to produce any kind of antibodies, thus rendering it less able to defend itself against infection.

The more closely two individuals or two taxa are related, the more proteins they have in common. By a variety of techniques, one of which is illustrated in Box 3.C, it is possible to establish the relative amounts of antigenic substances shared by related species, and thus to estimate their phylogenetic relationships. Further consideration of phenomena of this type within the human species will be given in Chapter 18. Here it only need be said that more support for the theory of evolution is provided by this type of investigation.

It is also possible to study the variety of proteins found in the blood sera of various animals by biochemical techniques. Many proteins are characterized by particular net electrical charges that make them travel at different rates in an electrical field. When, for instance, the serum of a primate is given such **electrophoretic** treatment, it separates into some 19–25 different components (Box 3.D). Here, too, relationships among species can be elucidated by resemblances and differences in the patterns of movement of their proteins. Other techniques for separating proteins have also been used to obtain clues for the reconstruction of evolutionary history.

Methods of studying variation in proteins are continually being refined. It is now possible to analyze the polypeptide chains of proteins and to establish exactly what their amino acid sequences are. For instance, cytochrome c, an important enzyme involved in transporting electrons in the cell, differs from species to species in the particular amino acids at many of the 100-plus positions in its polypeptide chain. Furthermore, in the horse this enzyme consists of a chain of 104 amino acids, in yeast it has 108, and in wheat 112. Through use of the sequencing techniques, the order of the cytochrome c amino acids has been established for more than 40 different species. (See Section 6.5, Box 6.G.)

A mutation changes the genetic message. In some cases, it may order a cell to place, in a specific position in a given polypeptide sequence, an amino acid different from that found in the analogous position among the ancestral forms.

BOX 3.D ELECTROPHORESIS PATTERNS OF HOMINOIDS

Each picture was made in an experiment in which a drop of serum from the species was placed at one spot, and allowed to migrate in an electric field organized to separate the components both vertically and horizontally. Each blob in the illustrations is made up of one or more kinds of polypeptides (mostly proteins); if a blob does represent more than one polypeptide, then these have similar or identical shape and charge properties, which determine their migration in the electric field. The blobs labeled T are *transferrins* (proteins that transport iron), those labeled H are *haptoglobins* (proteins that bind hemoglobin of old and broken-down red blood cells), and those labeled G are *gamma globulins* (proteins of the antibodies). Note the strong resemblances between the serum components of humans and of the closely related primates. Note also the occasional differences (for example, in gamma globulins). While a single analysis of this type is not sufficient to establish relationships among species, the variety of new biochemical techniques combined with anatomical information about living and extinct forms should eventually enable us to make increasingly accurate reconstructions of the history of life on Earth.

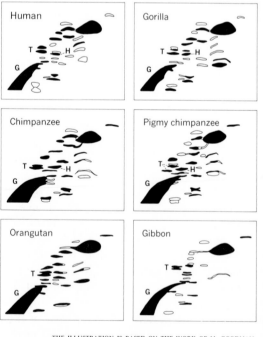

THE ILLUSTRATION IS BASED ON THE WORK OF M. GOODMAN.

These mutation events occur independently in different populations, or in different species. By comparing the sequence of amino acids along the polypeptide chain of one species with that in the analogous chain of another, it is possible to estimate how many mutational steps were taken from the point of divergence from a common ancestor. Even the number of years since the division into two species can be estimated, if we make certain assumptions about the rate of amino acid substitution. For cytochrome c, the number of differences in amino acids between several species is shown in Figure 3.7.

The comparisons of different organisms based on cytochrome c and other protein sequences generally make good intuitive sense. Worrisome among these comparisons was the finding that rattlesnake cytochrome c is more like human (and chimpanzee) cytochrome c than like any other known cytochrome (Box 6.G). (But in fairness to both rattlesnakes and humans, our cytochrome c does not differ as much from the cytochrome c of any of the other tested mammals as human and rattlesnake cytochrome c differ from each other. We should perhaps test more snakes.)

Another example of evolutionary relationships based on biochemical relationships between different species is illustrated in Box 3.E. As may be seen, there is a general correspondence between the number of differences in amino acids of the proteins examined and the estimates of chronological separation of the various species, although it is not perfect. If the two sets of figures are to be taken as true, then some variation in the rate of amino acid substitution must be conceded. There are generally certain positions in the chain of a polypeptide

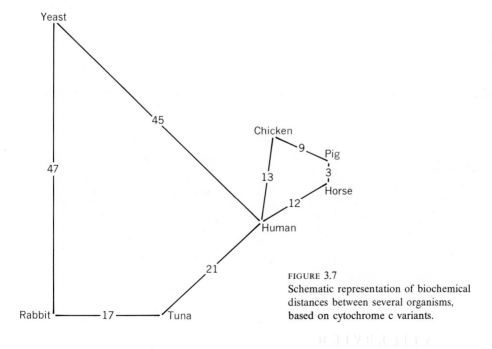

FIGURE 3.7
Schematic representation of biochemical distances between several organisms, based on cytochrome c variants.

BOX 3.E BIOCHEMICAL EVOLUTION

An example of determination of the evolutionary relationships of a number of mammalian species is given in a study by R. F. Doolittle and B. Blöm-back. They investigated the amino acid sequences in portions of the molecules of fibrinogen, a protein that functions in blood clotting. The two portions, fibrinopeptide A and fibrinopeptide B, contain chains of 17–19 and 13–21 amino acids in the species studied. In the diagram the numbers in lightface type give the percentage of correspondence between the different species with respect to the kind of amino acid found in each position of the chains (data for the two fibrinopeptides are combined here). The boldface numbers give the time, estimated independently of this study, since the divergence of the species in millions of years. A tabulation of the estimated number of years since any pair had a common ancestor against the percentage of amino acids common to the same position in their fibrinopeptide molecular chains indicates a close but not a perfect correlation:

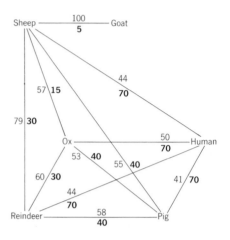

Time of divergence in millions of years	Percent correspondence in amino acids
5	100
15	57
30	60, 79
40	53, 55, 58
70	41, 44, 44, 50

The recent work of A. C. Wilson and V. M. Sarich has carried this line of investigation further. For example, there has been controversy over whether humans are more closely related to the African apes or to the old-world monkeys. The evidence from the fossil record and from comparative anatomy is equivocal. Comparisons of the amino acid differences between appropriate components of blood, such as serum albumins, transferrins, and hemoglobins, and data from DNA hybridization experiments, all agree in showing much greater differences between old-world monkeys and humans than between African apes and humans. These comparisons suggest that ancestors of humans and of the African apes diverged about 5 million years ago, and that the ancestors of this group diverged from those of old-world monkeys about 30 million years ago. Some 1969 data on alpha and beta hemoglobins are presented below. (The horse lineage diverged from that of the primates about 75 million years ago.)

Species compared	Number of amino acid differences
Human and chimpanzee	0
Human and gorilla	2
Monkey and human	12
Monkey and gorilla	14
Horse and human	43
Horse and gorilla	45
Horse and monkey	43

These estimates are far from precise, but the biochemical evidence does strongly suggest a much more recent divergence of human and African ape lines than of human and old-world monkey lines, even when such difficulties as unequal or changing generation times, different mutation rates, and different percentages of neutral amino-acid substitutions are considered in the calculations. Generation time may not in fact be a serious problem, although why it is not a problem demands more investigation. As an example, although there have been many more generations in the lineage of tree shrews than in the lineage of humans since the lines diverged, human and tree-shrew albumins, when compared with the albumins of four nonprimate carnivores, had immunological distances of 162 and 156. In other words, tree shrews and humans have diverged from the carnivores by similar amounts, and the rate of change correlates better with absolute time than with number of generations.

Wilson and his colleagues have continued to analyze the problem that genetic divergence between species (and between higher taxa) based on organismal criteria (anatomy, physiology, behavior) sometimes does not agree well with genetic divergence based on molecular criteria (electrophoretic—Box 3.D; immunological—Box 3.C; sequencing—Box 6.G; DNA hybridization—Box 3.F). For instance, M.-C. King and Wilson exhaustively reviewed the considerable organismal and molecular data available (as of 1974) on similarities and differences between humans and chimpanzees. They found humans and chimpanzees to be much more similar on molecular criteria than on organismal criteria. They further found evidence that molecular change has occurred in the two species at approximately equal rates since they diverged, but that far more organismal change has occurred in the human line than in the chimpanzee line. The length of each line in the illustration below represents the amount of change that has occurred since the human and chimpanzee lines diverged.

Wilson's group has suggested that in this and in similar cases, organismal evolution may be more influenced by changes in regulatory genes than by changes in structural genes. **Regulatory genes** are those genes which influence when the **structural genes** (the genes that code for the structure of proteins and other polypeptides) become active, and influence the level of their activity (See Boxes 6.C and 6.D). The evolutionary changes in gene regulation may occur either by changes in the positions of the regulatory genes, or by their mutation. It is mostly structural gene changes that are measured by electrophoretic, immunological, (and to a lesser extent) sequencing, and DNA-hybridization techniques. It is possible that small differences in the time of activation or in the level of activity of identical structural genes could considerably influence such things as embryo development, and thus create large organismic differences.

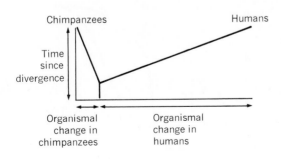

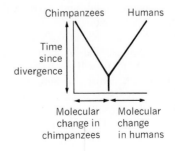

BOX 3.F DNA HYBRIDIZATION

The important things to know in order to understand this technique are: (1) that the genetic material, DNA, is a double-stranded molecule; (2) that the two DNA strands fit together and are complementary at many thousands of sites along the paired strands, the complementary sites being analogous to alternative types of plugs and sockets; and (3) that genetic differences between organisms are caused by differences in the composition of their DNA. This will be developed more fully in Chapter 6.

When a solution of extracted DNA is properly heated, the double-stranded molecules separate into single strands. Upon cooling, these complementary single strands encounter each other in the solution and reassociate to once again form double-stranded molecules (as if each plug had slipped back into its original socket). The rate of reassociation depends on how often complementary strands encounter each other, and thus on the frequency of complementary strands in a solution.

A heated mixture of DNA from two closely related species (much of whose DNA is similar or identical) will, upon cooling, associate into double-stranded molecules almost as quickly as dissociated DNA from a single species. Some of the molecules formed will be "hybrids," having one strand from each species. If DNA's from species that are not closely related are treated in this way, those DNA strands that differ sufficiently between the species can reassociate only with complementary strands of their own species. Thus, it will take longer for them to encounter a correct complementary strand, and reassociation will be slower for such mixtures.

The analysis can be carried further. If the taxa are sufficiently similar to permit formation of hybrid DNA, there will be imperfections in the pairing of such two-strand hybrids due to mutations that have independently occurred in one taxon but not the other (as if, in our analogy, a few plugs or a few sockets in the sequence had been replaced by ones with which the original complements could not now pair). It takes less energy to separate the two strands of a DNA

molecule that includes such pairing imperfections than to separate a comparable sequence of the same length that is exactly paired at every site along its length. This results in a lowering of the dissociation (melting) temperature upon subsequent reheating of about 2°C for every one percent imperfect pairing, which allows further quantification of the degree of DNA similarity in different taxa.

In the example below, DNA from *Drosophila melanogaster* was melted and then allowed to reassociate. Upon slow reheating, it was found that half of this DNA had remelted at 83°C (solid points). A similar experiment with DNA from a related species, *D. hydei,* gave an identical curve. When equal quantities of DNA from these two species were melted in mixture, allowed to reassociate, and then slowly reheated, the curve with the open dots was obtained, with half of the DNA melting by the time 63°C was reached. The difference of 20°C in melting temperatures indicates that there was mismatching at about 10 percent of the sites in the hybrid DNA molecules. When such percentages of DNA mismatches are compared with the number of generations (estimated from other sources) since two taxa separated, the two measures of evolutionary relationship agree well. It is worth noting that changes in the DNA molecule appear to occur more frequently than do amino acid substitutions in proteins.

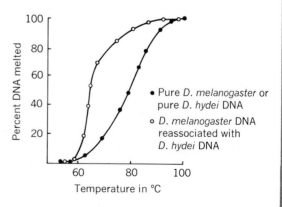

● Pure *D. melanogaster* or pure *D. hydei* DNA

○ *D. melanogaster* DNA reassociated with *D. hydei* DNA

BASED ON THE WORK OF R. LOGAN.

at which replacement of one amino acid by another impairs or destroys the function of the molecule; a mutational event at such a position is not, of course, raw material upon which evolution may operate. For those positions at which replacement may be tolerated, an average rate of 3–4 substitutions per amino acid position per billion years has been suggested in the evolution of cytochrome c, alpha hemoglobin, insulin C, and fibrinopeptide A; a somewhat higher rate has been suggested for beta hemoglobin.

Still another molecular method for studying similarities relies directly on similarities in the DNA of different taxa (Box 3.F). Additionally, techniques are being developed that allow direct comparisons of DNA and RNA sequences, which should provide even more precise measures of genetic similarity and difference.

The data from such molecular methods are still sparse compared to data from older types of taxonomic studies. It is encouraging that the relationships being developed from similarities and differences of related molecules have generally been in good agreement with classical biological opinion. Following such verification with well-known groups of organisms, it should be possible to use these molecular methods to determine relationships among groups with incomplete or missing fossil records, or whose modern relationships have not been well demonstrated in classical biosystematic studies.

3.6 SOME OBJECTIONS TO DARWINISM

Many objections to Darwinism have been raised in the course of more than a century of its history. Most of them carry little conviction in the face of the growing body of data and the increasing comprehension of evolutionary mechanisms. Some objections arose from a lack of understanding of the principles involved; others from the contradiction of traditional religious conviction. For instance, the English naturalist, Philip Gosse (father of the writer, Edmund Gosse), admitted all of the evidence adduced by Darwin. Then he proceeded to say that the Creator had made Earth six thousand years ago, giving it the appearance of having evolved in order to test man's faith in his Creator. One is tempted to ask why an omnipotent Creator should play such games.

Because evolution is the only consistent explanation of our accumulated observations on current and extinct life on Earth, it may be said that evolution has been demonstrated beyond all reasonable doubt—although dissipation of doubt based on faith or acceptance of dogma has by no means been accomplished.

We shall not take time to examine the varieties of objection to evolution except for two; one, somewhat trivial, the other (to which Darwin had no answer, but which was met after Mendel's laws were established), somewhat more serious.

The first one suggests that life is such a complex phenomenon that it is improbable that it arose on Earth. This argument is not a very convincing one, because evolution is indeed a mechanism for generating the highly improbable. It is becoming increasingly clear that life is the potential property of matter, so that given (1) enough time, (2) the presence of the conditions specified in Section 2.6, and (3) the sort of feedback mechanisms described in Section 2.6, life was likely, even probable, on a place like Earth. But even if this were not so, and there was much randomness in the process of evolution from inorganic matter to humans, the improbability argument still does not hold.

Consider an event which is so improbable that it can happen only once in a thousand times. The probability of it not happening on a single trial is then 999/1000. But if there are 1,000 trials, the laws of chance tell us that the probability of not happening drops to $(999/1000)^{1000}$ or 0.37. The probability of the event in 10,000 trials is 19,999/20,000: that is to say, it is practically inevitable.

Despite assertions to the contrary, it seems highly unlikely that a regiment of monkeys, provided with typewriters and unlimited time, would produce (along with much gibberish) all of the books in the British Museum. But we must distinguish probability before the fact and probability after the fact. For instance, the probability of a person being dealt a bridge hand consisting of 13 spades is roughly one in 54 quadrillion, a highly improbable occurrence that few bridge players ever witness. This probability, however, is no less nor more than any other hand specified *a priori*. Yet no one raises an eyebrow over hands containing a miscellaneous assortment of cards of all suits that have no greater probability of being dealt. It is not proper to compare, let us say, *Homo sapiens* to the one-suit hand (or any that is designated *a priori*), as is often done by disbelievers in human origin by evolution. Humans, and all other species, are more appropriately represented by the ordinary hands whose specifications are not known until they are dealt.

A better analogy might be to draw poker, where misfit cards are discarded, and new cards added with the possibility of improving an already good hand. An evolutionary precept that reduces the power of the improbability argument is that selection, far from being blind—operating entirely by chance, and begetting successful kinds of organisms as a result of improbable accidents—may be correctly described as a *creative* process. The meaning of this term, in reference to natural selection, is well illustrated by quoting Michelangelo on the process of creation. The opening lines of one of his best known sonnets say, in a somewhat free translation:

> *The best of artists has that thought alone*
> *Which is contained within the marble shell;*
> *The sculptor's hand can only break the spell*
> *To free the figures slumbering in the stone.*

In the same way, natural selection does not originate its own building blocks. From such blocks selection does create complexes; it solves in a diversity of ways the great variety of problems that individuals and populations face. Step by step, it builds entities of great complexity, ingenuity, and perhaps, beauty. It needs appropriate raw materials; it may not be able to make a silk purse out of a sow's ear; yet, interacting with other evolutionary mechanisms, it has created the hu-man species out of stuff that in its primordial stage may not have looked very promising.

The second objection weighed heavily on Darwin. He, like most of his con-temporaries, believed that the mechanism of inheritance was of a **blending** type. That is to say, the hereditary information from the two parents blended in the offspring, just as a glass of water mixed with a glass of red wine becomes a pink liquid. The difficulty with such a scheme of inheritance lies in the fact that half of the variation in a population would be lost in every generation by the blending process. Very soon, therefore, complete uniformity would be produced and nat-ural selection would run out of material on which it could operate.

Observation showed that the rate of occurrence of mutations is not sufficiently high to maintain the variation seen in nature. Darwin's need to overcome this objection led him to adopt the hypothesis that different organs of the body send particles through the blood as messengers to the **gonads,** the organs of reproduc-tion. As the body organs change under environmental influences, so would the messenger particles, thereby producing new variation. Essentially, this is a form of Lamarckism.

Darwin's remarkable cousin, **Sir Francis Galton** (1822–1911), some of whose contributions are described in Box 10.B, tested this hypothesis by injecting blood of rabbits of one color into females of another. According to Darwin's theory the offspring of the recipients, having a mixture of messenger particles, should have shown coat colors intermediate between those of the parent and the blood donor. The coat colors of the offspring were consistent with those of the parents, and showed no effect of the blood donors. Inheritance of acquired characters, however, was not fully dismissed, and was powerfully revived many years later (Chapter 23).

The problem of how variation is maintained became tractable when the Men-delian laws, showing that inheritance is **particulate,** were established. The units of hereditary information received by the zygote from one gamete do not blend with the units received from the other but retain their identity. They are reas-sorted in the following generation and are not contaminated by having been present in a zygote containing an alternative unit. The details of this process are considered in Chapter 7.

Under this scheme of inheritance, instead of one-half, approximately $1/N'$ (where N' is twice the number of effectively interbreeding individuals in a popu-lation) of initial variation is lost, and this figure is compatible with observed facts. The removal of this objection to Darwin's theory was of great importance in the development of current concepts of organic evolution.

Finally, there has recently been much interest in so-called "non-Darwinian" evolution. This is not really an objection to Darwinism, but a suggestion that variability may be maintained, and important populational (and higher level) differences may be established, by means other than natural selection. This will be pursued further in Section 6.5.

4

THE EVOLUTIONARY RECORD

At this point, we are ready to examine in broad terms the evolutionary process as revealed by the paleontological record, and to introduce the forces that shape evolutionary change.

4.1 THE FOSSIL STORY

Fossils are remains or impressions of living forms found in layers of sedimentary rock. They provide a clear, but usually incomplete, record of the past. The more recent the period under investigation, the more precise is the information available. This is not only because of the proximity of recent fossils to the surface, but also because the parts of organisms that may best be preserved are such hard ones as skeletons, which are of relatively late origin in the history of life on Earth. Nonetheless, bacteria, pollen, seeds, and imprints of soft parts are also found.

Paleontologists have learned to make complete reconstructions of many forms, though some biases and a certain amount of guesswork enter into them. Behavior, as well as the anatomical features of extinct forms of life, may be deduced from the fossil record; for example, certain animals have left tracks that indicate burrowing. Other details about the life of a particular organism, such as the kind of food an animal ate, may be inferred from the structure of its teeth and from coprolites (fossilized excrement). It has been proposed that it is even possible to reconstruct the body temperatures of extinct organisms by studying the chemical structure of preserved fossil proteins, which vary depending on the temperature at which they were synthesized.

Mistakes, disputations, and even hoaxes have accompanied the efforts to decipher the paleontological record. For example, an amateur English paleontologist, for motives unknown, reported an anomalous find, the remains of "Piltdown man." Many years later it was discovered that a human skull had been combined with an ape jaw to constitute this "fossil." To their credit, many professional paleontologists had refused to accept the find as authentic even before the hoax had been uncovered.

The oldest fossils of protists are more than three billion years old. Fossils found in Canada that are older than two billion years have been interpreted as being the remains of multicellular animals, but this has not been conclusively demonstrated. Only a limited fossil record before the Cambrian Period (some 500–600 million years ago) is available (Figure 4.1).

In early attempts to reconstruct the past, paleontologists dated fossils by assigning an age to them corresponding to the age of the rocks in which they were found. On the assumption of uniformitarianism, with allowances for earth movements and other geologic phenomena, the age of the rocks was calculated from known or inferred rates of sedimentation. In recent years, chemical methods have become available for obtaining the dates not only of the rocks but of the fossils themselves.

Carbon-14 (C^{14}), a radioactive isotope of carbon, originates in the upper atmosphere from the action of cosmic rays on nitrogen, which frees neutrons, transmuting the nitrogen into radioactive carbon. The C^{14} is unstable, and its disintegration, or radioactive decay, proceeds at a known rate. An equilibrium, based on the formation of new C^{14} and its radioactive decay, is established in the atmosphere. The ratio of C^{14} to ordinary carbon (C^{12}) in atmospheric carbon dioxide has apparently been constant for some time (about one atom of C^{14} to a trillion atoms of C^{12}). A living plant fixes (by photosynthesis) the two kinds of carbon atoms in this ratio in organic molecules, some of which are incorporated into the plant's structures. After the plant dies, no more carbon of any kind is incorporated into these structures. A C^{14} atom may become part of a plant fossil, or it may be converted to animal or fungus parts as it moves in a herbivore–carnivore–decay-organism food chain. After an organism dies, the radioactive decay of C^{14} in its structures continues while C^{12} remains stable, and thus the ratio of C^{14} to C^{12} drops at a calculable rate. By determining this ratio in a fossil, it is possible to establish approximately when the animal or plant lived. The rate of C^{14} disintegration is expressed by saying that C^{14} has a half-life of 5,760 years, that is, one-half of the C^{14} atoms decay in that period of time and thus a fossil

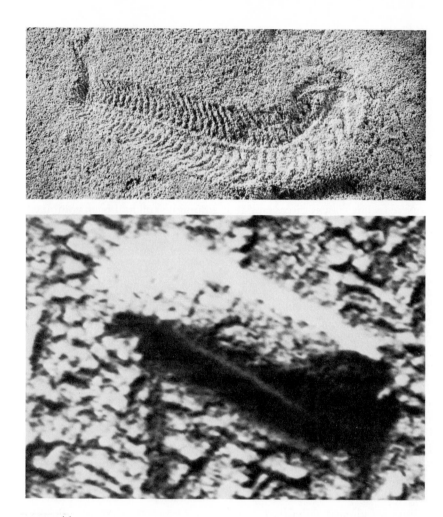

FIGURE 4.1

Top: Fossil worm from Australia, about 600 million years old, shown approximately twice the actual size. (Courtesy of M. F. Glaessner.) *Bottom:* Fossil of oldest known bacterium, one of two primitive forms of life preserved in a Precambrian rock formation in South Africa, appears as a raised capsular shape in this electron micrograph. What is seen is a carbon replica of the polished surface of a rock sample, shadowed with heavy metal. The organisms are some 3.2 billion years old and have been given the name *Eobacterium isolatum.* It is shown approximately 200,000 times the actual size. (Courtesy of E. S. Barghoorn.)

5,760 years old has only half the C^{14} radioactivity of a living organism. Carbon dating is generally acknowledged to be accurate for material up to 30 thousand years old, and some paleontologists confidently accept figures up to 70 thousand. A somewhat different method, based on shifts in isomeric forms of amino acids, appears to work well for periods from 40,000 to more than 100,000 years.

Carbon-14 is properly called the second hand of the cosmic clock, and it helps us measure those very interesting cosmic seconds when the human species was inventing civilization. There are other elements that might be thought of as hour hands or minute hands: they permit long-term dating by means similar to the C^{14} technique. For instance, the relative amounts of carbon-12 (the preferred isotope in photosynthesis) and carbon-13 in Transvaal sedimentary rock suggests that photosynthesis became common on Earth about 3.3 billion years ago. Other long-term dating is provided by radioactive potassium (K^{40}), which has a half-life of 1.3 billion years, and radioactive uranium (U^{238}), which has a half-life of over four billion years.

There are many general principles that have become established on the basis of the paleontological record. Eight particularly important to our concerns are:

1. Most fossil animals and plants belong to the same major taxa as do organisms living today, but, in general, they differ from the living species in many features.

2. Fossils found in a given layer of sedimentary rock generally differ in significant respects from those in other layers.

3. Phyla such as the chordates are not represented in the older layers by specialzed forms like those alive today, but by more generalized ones.

4. The more recent the layer of rock, the more resemblance there is between the fossils found in it and living forms.

5. The number of extinct types is enormously greater than the number of living ones (Section 2.3).

6. It is common to find in the rocks that mark the end of one geologic period many new forms that become dominant in the next. Since most geologic periods appear to have had environments very different from those of adjacent periods, their ends coincided with drastic environmental changes and the appearance of new niches. These produced new selection pressures and new adaptations by the surviving species to the changed external world.

7. The evolution of many present-day species can be reconstructed very fully. For instance, the evolutionary history of the modern horse is exceedingly well documented for the last 60 million years (Figure 4.2).

8. According to the fossil record, rates of evolutionary change seem to vary. There are bursts of species formation and long periods of relative species stability. However (see Section 3.5 and Box 3.E), the data on amino acid substitution in molecules such as cytochrome c are interpreted on the basis of a constant rate of change. This is not necessarily a contradiction, because many of the chemical traits used to study molecular evolution may be selectively neutral, whereas the occasional rapid evolution of new forms apparent in the fossil record is probably the result of selection for characteristics adaptive in a changed environment.

On the one hand, mammals, which appeared only 120–150 million years ago, have undergone their rapid expansion within the last 70 million years. On the other, there are many forms, the so-called living fossils, that appear not to have changed over very long periods of time. In New Zealand, a living lizard-like creature, *Sphenodon*, looks exactly like a fossil in rocks 175 million years old.

62

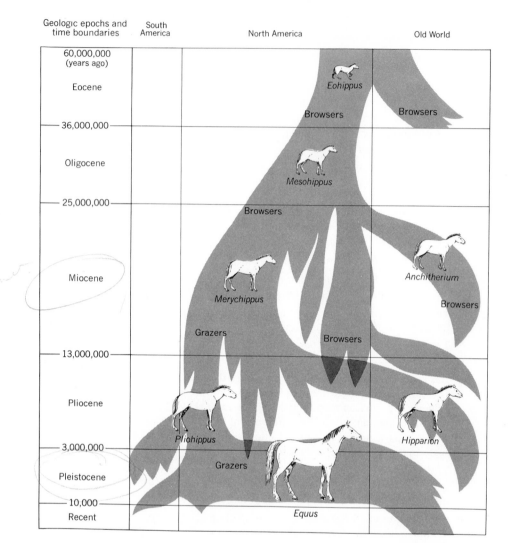

FIGURE 4.2

Simplified family history of the modern horse. In the Eocene, a very small horse, *Eohippus*, that had four toes on its front feet and three on its hind feet, occupied parts of present-day North America and Europe. From fossils of its teeth and other evidence we infer that it browsed succulent leaves and perhaps also ate fruits and seeds. The old-world populations had become extinct by the beginning of the Oligocene. The horse of the mid-Oligocene, *Mesohippus*, was larger, had a much bigger brain, and had only three long toes and a fourth, vestigial, toe bone. During the Miocene, the Old World was twice recolonized by browsing horses, each line of which became extinct. In North America, one Miocene line, *Merychippus*, had become adapted to grazing the increasingly abundant grass, and its three toes were greatly shortened. In the Pliocene, the central toe of *Pliohippus* had become a small hoof, and the other two toes had become vestigial. Grazers and browsers cohabited parts of North America for a time, but the browsers became extinct. Grazers recolonized Asia by crossing the Bering land bridge from Alaska, and spread through Eurasia and throughout

Geologic periods and time boundaries	Group of animals

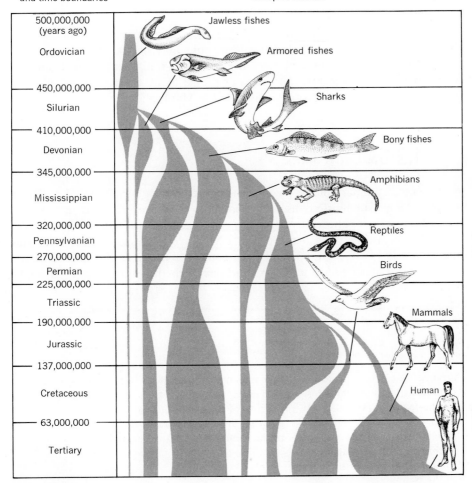

500,000,000 (years ago)	Jawless fishes
Ordovician	Armored fishes
450,000,000	Sharks
Silurian	
410,000,000	Bony fishes
Devonian	
345,000,000	Amphibians
Mississippian	
320,000,000	Reptiles
Pennsylvanian	
270,000,000	Birds
Permian	
225,000,000	
Triassic	
190,000,000	Mammals
Jurassic	
137,000,000	
Cretaceous	Human
63,000,000	
Tertiary	

FIGURE 4.3

Relative importance of various vertebrates since the Cambrian period. The vertical dimension is time in years past. The horizontal dimensions of the shaded areas indicate the relative importance of each group at the various times. (Redrawn from Garrett Hardin, *Biology: Its Human Implications.* W. H. Freeman and Company, Copyright © 1949, and J. Millot, "The Coelacanth." Copyright © 1955 by Scientific American, Inc. All rights reserved.)

much of Africa. Others colonized South America. During the Pleistocene, horses became extinct everywhere except in North America, where the size of the body and the size of the hoof greatly increased; with these changes, the modern horse, *Equus,* had arrived. Another colonization of South America failed, but a recolonization of the Old World by *Equus* was successful, and a good thing, for horses later became extinct in the New World. Horses still survived in both North and South America when the forbears of the American indians reached these lands, but none were left at the time of Columbus. The final recolonization was from Europe to the New World, this time by ship.

The opossum has a fossil counterpart 75 million years old. Live specimens of the ocean fish called coelacanths, which had been known as fossils, and were thought to have been extinct for 70 million years, were caught in 1938 and several times since then. There is also an animal, *Lingula*, which looks somewhat like a clam (but is not closely related to clams) whose genus may be 400 million years old (some paleontologists, however, place the fossil into a different genus from that of the living form). Recently, a freshwater protist fossil, apparently morphologically identical with living forms, was dated at 900 million years.

4.2 GEOLOGIC ERAS

A condensed summary of the paleontological record is given in Box 4.A. There are many finer subdivisions used by geologists than are indicated in the box, but these need not concern us here. The oldest good fossil material is from the Cambrian Period of the Paleozoic Era, when invertebrates, marine flora, and many protists became abundant. In the course of the next 300 million years, crustaceans, spiders, insects, and vertebrates made their appearances (Figure 4.3).

BOX 4.A THE GEOLOGIC TIME SCALE

Geologists usually arrange time scales to correspond to the layers of rocks; that is, the most recent intervals are at the top of a tabulation. It is, however, more customary for human beings to read from top to bottom. Hence in the various tables and figures of this book, excepting those directly reproduced from other sources, the most recent intervals are at the bottom.

The customary time classification corresponding to rock layers is into eras, which are subdivided into periods, and, further, into epochs. The units are named either descriptively (for example, Pleistocene Epoch from *pleisto* meaning most and *kainos* meaning new), or by the location at which the typical rock layers were originally found (for example, Cambrian Period, from Cambria, the ancient name for Wales). For our purposes the details of the kinds of life found in different layers are not important, and only a few names of the different subdivisions are given in the text.

Estimates of the age of the different subdivisions are undergoing repeated revision as more information is being accumulated. Hence there are discrepancies among the various ages published; but differences of several millions or tens of millions of years are not significant for us in the context of the five or more billion years of Earth's history.

Geological era	Millions of years ago	Expansion of major new forms
Origin of Earth's crust	4500	
Archeozoic or Azoic (archaios = old; a = without; zoon = animal)	4000	Protists
Proterozoic (proteros = early)	1000	Aquatic forms
Paleozoic (palaios = old)	350	Fish, land plants
Mesozoic (mesos = middle)	200	Reptiles
Cenozoic (kainos = new)	75 0.01	Mammals Humans

The plants invaded land, the first trees established themselves, sharks, amphibia, and, eventually, the first reptiles appeared. The end of the era saw a great increase in the number and diversity of reptiles, and the forests were largely ferns and horsetails.

There have been three important glaciations since the Cambrian: one possibly in the early stages of the Paleozoic Era, one in the Permian Period at the end of that era some 250 million years ago, and one only some thousands of years ago in the Recent Epoch of the Cenozoic Era. Their causes are not known, but it has been computed that a general reduction of only 3.0–4.5°C in Earth's atmospheric temperature would be enough to produce them. The associated environmental changes must have had a profound evolutionary effect, spreading plants and animals widely in retreat from the encroachment of the ice caps and affecting the kind of natural selection pressures to which they were exposed.

The Mesozoic Era was the age of reptiles. The first dinosaurs appeared at its beginning. Mammals and birds originated during the middle of it, and by the time the Cenozoic Era started, the dinosaurs were gone. After this, flowering plants, broad-leaved trees, and mammals were the ecologically dominant organisms over much of Earth's land surface.

Mezoic,
↓
Cenozoic.

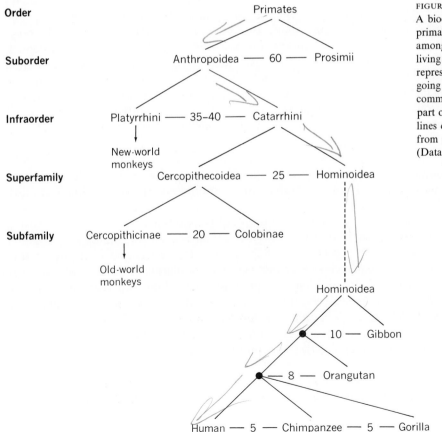

FIGURE 4.4
A biochemical pedigree of primates based on resemblances among serum albumins of living forms. The numbers represent millions of years, going back to the most recent common ancestor. The lower part of the figure differs from lines of descent constructed from morphological evidence. (Data from V. M. Sarich.)

The Cenozoic Era has been divided into two periods, the Tertiary, which lasted about 70 million years, and the Quaternary, about 3 million years long. During the Tertiary, the differentiation between monkeys, apes, and humans occurred (Figure 4.4). This, like much of the evolutionary process, involved adaptive diversification of form and function. Such diversification of mammals is schematically represented in Figure 4.5.

The immediate ancestors of modern humans appeared in the Pleistocene Epoch of the Quaternary. Use of tools goes back, probably, 1.75 million years. *Homo sapiens* appeared on Earth some 20–50 thousand or more years ago and arrived in America shortly thereafter.

4.3 EVOLUTION, AND THE FORCES THAT CHANGE GENE FREQUENCY

The debate over the reality or probability of evolution has generally been conducted in the absence of an agreed-upon definition of evolution. Among population biologists, the generally accepted definition of evolution is: **a change in gene frequency.** (A more precise definition is: "a change in *allele* frequency." **Alleles** are alternative states of a gene, originally caused by mutations that alter the genetic message, and thus usually alter the kind, amount, or timing of the polypeptide produced. See Chapter 6 and Section 7.2.) If this definition is accepted, the controversery, where it still exists, then shifts to: what levels of evolution constitute significant evolution, and whether higher levels of evolution can be explained by the same forces and principles as lower levels. In the following discussion, we will consider mostly the lower levels of evolution, sometimes called **microevolution,** where "a change in allele frequency" clearly suffices as an adequate definition.

We have noted what is necessary for organic evolution: the creation of genetic variation; mechanisms to transmit to descendents the revised instructions that produce genetically variant individuals; and differential reproduction, or fitness, of such variants.

One of the fundamental contributions of Darwin's revolution was the focusing of attention on variation within populations as a means of gaining insight about variation between populations. There are five forces recognized at present that change allele frequencies within and between populations, four of which are fairly well understood. In addition, two processes recombine and redistribute genes, influencing the genetic constitutions of populations without directly changing allele frequencies. It will be noted that these forces and processes are dynamic, and frequently work in opposite directions. This creatively serves to maintain variability within and between populations, thus making continued evolution possible.

Force 1. Mutation. The only source of genetic variability, given a single successful generation of life on Earth (Section 2.6), is mutation. Mutation contributes to variability in the **gene pool** of a population (the hereditary instructions passed

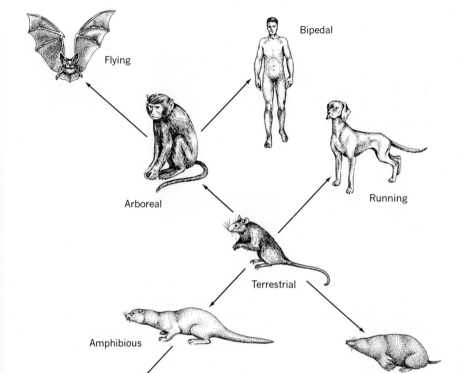

FIGURE 4.5
Adaptive diversification in
modes of locomotion and life
habits (precise relationships
between the various forms are
not intended).

on to succeeding generations) if it occurs in the germ line, i.e., in cells which are
or become gametes. There are at present two models for explaining how mutations contribute to genetic variability; they give different predictions about the
effect of mutation on variability between populations (Box 4.B). But note that
in each model, allele frequencies within populations are changed by mutation,
and variability within populations is increased.

Force 2. Migration.   It may be seen in Box 4.C that the effects of immigration
on gene frequencies, and on variability within and between populations, are
identical to the effects of recurrent mutation. There are important differences
between these two forces. Immigration only operates upon the genetic variability
that already exists, but mutation can create new genetic variability. Mutation
occurs at a low but relatively constant frequency; immigration may be rare but
can massively infuse a population with new variability when it does occur. Furthermore, mutations of different genes are generally independent, both as events
and with respect to the new alleles produced; immigration may simultaneously

introduce new alternative alleles of many genes that had been co-adapted to work well together in the donor population by generations of evolution.

Force 3. Selection. One of Darwin's great contributions was an understanding of the importance of selection in evolution, sometimes inaccurately characterized as "survival of the fittest" (Section 3.3). One problem raised by critics of Darwinism was that, if generation after generation the more fit survived and successfully reproduced while the less fit failed to do so, the result would be a reduction and finally elimination of genetic variability. Modern theory indicates that this would indeed occur with many generations of selection in a stable environment,

---

BOX **4.B** NONRECURRING AND RECURRING MUTATION

Molecular biology suggests that there are an enormous number of ways that a gene can mutate to something else (chapter 6); so many that identical mutations may be very rare events. Consider two populations that have different and (for simplicity) complementary allele frequencies for the a_1 and a_2 alleles of gene a. Suppose the alleles are equally likely to mutate to some alternative state (this is not, in fact, a correct assumption, but makes no serious difference in this example). It may be seen that under **nonrecurring mutation,** population I comes to possess five new alleles of gene a, which are different from any of the alleles in population II, and population II also acquires five new and unique alleles. Populations I and II have clearly become more different from each other. Each has become more genetically variable, and each has changed allele frequency. For example, population I now has less than 80 percent a_1, less than 20 percent a_2, and a small percentage of a_3, a_4, a_6, a_8, and a_{10}.

Although the structures of two mutated nucleic acids may differ, they may code for functionally similar or identical products (Section 2.3). For instance, two alternative structures of an enzyme may both permit the enzyme to function normally, while two other alternative structures may both cause the enzyme to be inactive. The nonrecurring mutation model for populations I and II becomes equivalent to the **recurring mutation** model if we assume that all alleles with even subscripts code for inactive structures of the enzyme and all alleles with odd subscripts code for active structures. In this case, populations I and II have clearly become more similar to each other (the frequency of a_1 is reduced in population I and increased in population II and the opposite happens for a_2). Allele frequencies have changed in both populations. It is not obvious, but genetic variability has also increased in each population. (It can be shown that genetic variability is maximum when the frequencies of the alleles are equal, and thus variability increases as the frequencies change, becoming more nearly equal.)

	Population I	*Population II*
Allele percentage before mutation	80% a_1 20% a_2	20% a_1 80% a_2
Nonrecurring mutation	4 a_1 mutate to a_4, a_6, a_8, and a_{10} 1 a_2 mutates to a_3	1 a_1 mutates to a_{12} 4 a_2 mutate to a_5, a_7, a_9, and a_{11}
Recurring mutation	4 a_1 mutate to 4 a_2 1 a_2 mutates to 1 a_1	1 a_1 mutates to 1 a_2 4 a_2 mutate to 4 a_1

BOX 4.C MIGRATION

Assume that populations I and II have been reproductively isolated from each other and that they differ both in the frequency and identity of alleles of the *a* gene. Case one is the one-way migration of a representative sample of population I to population II. The emigration from population I has no effect on allele frequencies or variability within population I, because the migrants remove alleles in the same proportions that characterize the population I gene pool. However, the **immigration** into population II has three important effects: (a) the population II allele frequencies are changed; (b) genetic variability within population II is increased, most obviously from the introduction of allele a_3, but also because the frequencies of all four alleles tend to change to more nearly equal values; and (c) the frequency of each allele in population II is changed in the direction of its frequency in population I, thus

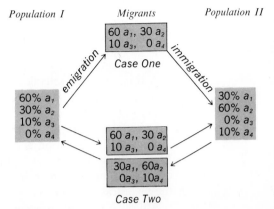

decreasing the genetic variation between the two populations. Case two, which is the two-way exchange of representative samples, creates effects (a) and (b) in both populations, and intensifies effect (c).

at least with respect to the genetic variability that responds to selection (see Sections 10.2, 10.3, and Chapter 14). The effects of such selection in stable environments would be (a) to change allele frequencies until the genetic variability responding to selection was eliminated; (b) thereby to reduce genetic variability within populations; and (c) because different characteristics would be selected for in different environments, to increase genetic variation between populations.

The evolutionary force of selection can be of several different types. If a population is highly adapted, most of its recurrent mutations have already been tried out and incorporated into the gene pool if favorable, or discarded if disadvantageous, in the given environment. Similarly, with such continuously variable characteristics as stature (as contrasted to characteristics forming discrete classes), it is likely that variation between individuals in adapted populations falls within a range optimal for the success of the population. Any recurrent mutations or genetic recombinations that produce phenotypes outside this range will be discriminated against by selection. The form of selection that operates against mutations or genetic recombination producing relatively unfit phenotypes is known as **stabilizing selection.** It works towards the maintenance and persistence of the population as is.

However, if an environment so changes that the average characteristics of the population are no longer within the optimum range, or if mutations previously untested under the particular conditions arise, selection becomes **directional.** It causes the composition of the gene pool or the ranges of phenotypic expressions (e.g., stature, disease resistance) to shift in some direction. Selection in a changed environment may involve what the American evolutionist **Ernst Mayr** has called a **switch** type of evolutionary change. This also enables the species to spread and occupy new niches.

A third form of selection is **diversifying,** which simultaneously favors two or more phenotypes and is possible in an environment that has multiple niches. This form of selection can lead to formation of subspecies characterized by different phenotypic ranges. If gene interchange between them ceases, that is to say, if reproductive isolation is established, they may then become separate species. This splitting of a species is one method of speciation. The other is replacement or transformation, which happened, for example, when *Homo sapiens* replaced the ancestral *Homo erectus.*

Although we have presented the likely effects of selection in a stable environment, we note that stable environments are rare. This is not only because the physical components of an organism's environment are subject to fluctuation and change, but also because the biotic components are evolving. L Van Valen has recently proposed his **Red Queen hypothesis.** The name derives from Lewis Carroll's *Through the Looking Glass:* "Now, here, you see, it takes all the running you can do, to keep in the same place." In essence, the hypothesis says that if a population does not continue to adapt at the same average rate as its biological competitors, it will decline in the number of niches it can successfully occupy, and if it lags long enough, it will become extinct. F. J. Ryan and colleagues impressively demonstrated this effect with the colon bacterium *Escherichia coli.* These bacteria normally live in the thermally stable, nutrient-rich, dark environment of the human colon. Ryan's experiments provided the bacteria with physical and chemical conditions that were perhaps even more stable: they were contained in laboratory apparatus designed to maintain a near-constant nutrient supply, population density, and temperature. Yet, in over 7,000 bacterial generations, Ryan found no exception to the rule that later samples of the population were more fit in competition with *any* sample of the population taken earlier. These populations were clearly running hard (continuously evolving in response to competition among themselves) to maintain their constant place in Ryan's stable environment.

Force 4. Drift. **Genetic drift** is a sampling phenomenon. As such, it can be mathematically described with precision, but in any given generation neither its magnitude nor its direction can be predicted with accuracy. One type of genetic drift, which is explained in terms of the **founder principle,** is illustrated in Box 4.D. Two initially identical groups of organisms may each found a different, separate population. Small departures from the average genetic constitution, which occur by chance among a relatively few founders of a population, are magnified into enduring differences between the larger populations they establish. A number of examples of founder effects in humans are presented in Section 19.3.

BOX 4.D DRIFT

In order to begin to understand the possibly large evolutionary significance of a trivial chance event in a small population, consider two widely separated abandoned fields, each with a group of four plants of genotypes *AA, Aa, aA,* and *aa* on its edge. No other plants of this species are in the neighborhood of either field. Genotypes *Aa, aA,* and *AA* produce blue flowers; *aa* produces white. The percentages of the two alleles, *A* and *a,* in each founding population are 50 percent *A,* 50 percent *a* (four alleles of each). Before the pollen sheds, one blue-flowered plant in each population (note the crossed-out plants in the diagram) is killed by independent and random causes. In this case, by chance, one *AA* and one *Aa* plant are killed. Pollination occurs at random within each population, and 900 successful plants are established from these parents in each of these two fields.

Perhaps the most obvious effect is that white-flowered plants are nearly twice as abundant in field 2 as in field 1. As long as each population maintains a large size, and neither selection nor migration changes the allele frequencies, field 2 will continue to have about twice the proportion of white flowers as field 1. If we had looked at another characteristic with variation controlled by different alleles of another gene, we might have observed this second characteristic in field 1 and

in field 2 to be about the same, or to be changed to different degrees in the same direction in the two fields, or to be changed in opposite directions. Thus, in this situation, it can be reliably predicted that sampling events, or drift, will cause two such populations to become different in many of their characteristics that are genetically variable. It cannot be reliably predicted which characteristics will change, nor the direction or exact amount of the change in any one population. Furthermore, once different, it is very unlikely the two populations will ever be identical again, unless they merge and freely interbreed. Thus, while the details of the differences can't be predicted, the *condition* of *being* different is not only predictable, but it is permanent.

This effect is related to population size, and to a lesser degree, to initial allele frequency. The smaller the population, the greater the effect of drift on changes in allele frequency. For most higher (i.e., diploid) organisms, the effect per generation is predicted by the expression $(pq)/(2N)$, where p and q are frequencies of the alleles, and N is the number of breeding individuals. As a rule of thumb, when N is 20 or less, drift effects become biologically significant, and more important as N is smaller. In our example, we achieved spectacular one-generation effects with a random N of 3 from a total population of 4.

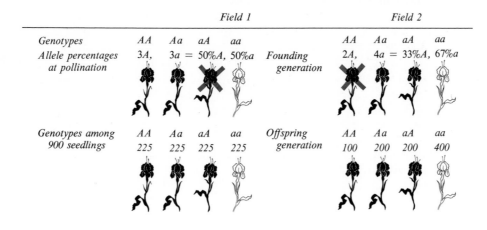

	Field 1					Field 2			
Genotypes	*AA*	*Aa*	*aA*	*aa*		*AA*	*Aa*	*aA*	*aa*
Allele percentages at pollination	3*A*,	3*a* = 50%*A*, 50%*a*			Founding generation	2*A*,	4*a* = 33%*A*, 67%*a*		
Genotypes among 900 seedlings	*AA* 225	*Aa* 225	*aA* 225	*aa* 225	Offspring generation	*AA* 100	*Aa* 200	*aA* 200	*aa* 400

The principal effects of drift are: (a) allele frequencies are changed, sometimes very quickly and significantly; (b) genetic variability within populations is generally decreased, as allele frequencies tend to move away from central values (if the frequency of any one allele becomes 1.0, the alternative alleles have been lost and all genetic variation for that gene disappears in the population unless restored by mutation to, or immigration of, another allele); and (c) since the direction and amount of allele-frequency change is unpredictable, at least some characteristics will change in opposite directions or in unequal amounts in different populations, and thus the variation between populations will be increased by drift.

Force 5. Segregation distortion.   Sometimes called "meiotic drive" (Box 7.A), the term is applied when one or more alleles are consistently more frequent in the gamete pool forming the subsequent generation than they were in the parental gene pool. For instance, in some lines of "tailless" mice, among the effective sperm of males with genotype t_+t_i, only 12% are t_+ and 88% are t_i instead of the expected 50% of each. The causes of such distortion are not well understood, nor are its evolutionary effects. It is clear that such segregation distortion can (a) quickly change allele frequencies not only of the genes directly subject to such distortion, but also of alleles of other genes that by chance are located on a part of a chromosome near a gene undergoing such distortion. The effects of segregation distortion on within- and between-population genetic variability are not clear. It is possible (b) within-population variability will be increased by rare and somewhat deleterious alleles being raised to higher frequency by a "hitchhiking" association with a segregation-distortion allele. It is similarly possible (c) that different alleles in different populations will benefit from this hitchhiking phenomenon, thus increasing variability between populations. Of the five forces that change allele frequency, segregation distortion is probably the least important in causing significant evolutionary change, although it cannot be wholly dismissed.

The following summary table is presented with the disclaimer that these are general, or most-probable, effects on variability. Examples of changes in opposite directions can be cited for almost every case.

	Effect on variability		*Importance of effect*	
Force	*Within populations*	*Between populations*	*Short-term*	*Long-term*
1. Mutation				
Recurrent	Increases	Decreases	Negligible	Source of all genetic variability
Nonrecurrent		Increases		
2. Immigration	Increases	Decreases	Can be large	Probably large
3. Selection	Decreases	Increases	Usually moderate	Very large
4. Drift	Decreases	Increases	Can be large	Probably moderate
5. Segregation distortion	Increases (?)	Increases (?)	Probably small	Probably negligible

This table provides a framework for understanding how dynamic interactions between these forces that change allele frequency may operate to keep allele frequencies at intermediate values within populations, or how allele frequencies may change relatively quickly in response to events or situations causing or requiring evolutionary change. For instance, the between-populations column illustrates possible interactions, particularly between immigration and selection, that serve to keep similarities among populations of a species, while providing for differentiation among such populations, perhaps leading to the formation of new species.

Process 1. Recombination. In the apparatus of particulate inheritance, units of genetic information can be reshuffled (Chapter 7), giving rise to an enormous diversity of possible genotypes (Box 4.E). Just which genotypes are actually produced, by chance, out of the generally astronomical number possible due to recombination is one of the elements of force 4, genetic drift. That recombination

---

BOX $4.E$ RECOMBINATION

As a somewhat oversimple example of the power of recombination to generate genetic variability, consider a self-pollination mating system, which limits us to a maximum of two alleles per gene. A plant with genotype *Aa* may produce offspring with three genotypes: *AA, Aa,* and *aa.* If that plant has a genotype with two genes capable of segregating and recombining, *AaBb,* then it may produce offspring with nine genotypes: *AABB, AABb, AAbb, AaBB, AaBb, Aabb, aaBB, aaBb* and *aabb.* With three segregating genes (*AaBbCc*) in the parent, 27 offspring genotypes are possible:

AABBCC, AABBCc, AABBcc, AABbCC, AABbCc, AABbcc, AAbbCC, AAbbCc, AAbbcc,
AaBBCC, AaBBCc, AaBBcc, AaBbCC, AaBbCc, AaBbcc, AabbCC, AabbCc, Aabbcc,
aaBBCC, aaBBCc, aaBBcc, aaBbCC, aaBbCc, aaBbcc, aabbCC, aabbCc, aabbcc.

For higher numbers of segregating genes, this type of calculation becomes cumbersome. The number of offspring genotypes in this system is predicted by 3^n, where *n* is the number of genes with two alleles in the parent. A parent with 10 segregating genes can produce offspring with 59,049 different genotypes. Only a few plants, such as the giant sequoia, are capable of generating that many offspring. With 15 segregating genes, the number of possible offspring genotypes is 14,348,907, and perhaps only eucalyptus trees can generate this number of offspring in a lifetime of seed production. Yet many more than 15 genes with two alleles are carried in the average individual, be it human or tree. The potential to create genetic diversity not only exceeds biological possibility, but the numbers of genotypes possible exceed rational comprehension. In mating systems other than self-fertilization, even greater diversity can be generated. Where more than two alleles exist per gene, the number of possible genotypes is $[r(r + 1)/2]^n$, where *r* is the number of alleles per gene. To begin to get some idea of the numerical possibilities, consider that ten genes that each have four segregating alleles can recombine to form ten billion genotypes.

created this great genetic diversity each generation was not understood by Darwin or by his early critics, and helps to explain how force 3, natural selection, can continue to operate over long periods of time—i.e., because genetic diversity is repeatedly renewed.

Process 2. Selection of mates. Patterns of selection of mates by type are known as **mating systems.** Neither recombination nor a change in mating systems change allele frequency *per se,* but both influence the way in which the genes are arrayed within populations. At this point, we will consider three types of mating systems: (a) assortative mating; (b) disassortative mating; and (c) random mating. These will be discussed more fully with reference to humans in Chapter 21.

Assortative mating means that similar individuals tend to mate, or are mated, with each other. The general effect of this tendency is to increase the *genotypic* diversity (a function of the array of genotypes) within populations. In plant and animal breeding, this increases the effectiveness of selection, which may reduce the *genetic* diversity (a function of the number and frequency of alleles per gene) within the population. In both artificial and natural populations, this may or may not be a good idea.

The term disassortative mating describes the opposite tendency. Unlike individuals mate with each other, a process that reduces the genotypic diversity but generally maximizes the genetic diversity in the population. In the short run, selection is not as effective in such a mating system. Enforced mating between different mating types (e.g., in humans, between males and females) is a strategy adopted by most animals, and some plants and protists, that serves to introduce at least an element of disassortative mating.

Random mating strictly defined means that each individual in a population has an equal chance of mating with any other individual, including itself. Operationally, it means that neither assortative nor disassortative mating is very important. Its effects on genotypic and genetic variability within populations are intermediate between those of assortative and disassortative mating.

As an example, consider the three surviving parental plants in field 1 of Box 4.D. Under random mating, they produced 225 *AA,* 450 *Aa,* and 225 *aa,* or 3 blue-flowered plants for each 1 white-flowered plant (flowers of genotypes *AA* and *Aa* are both blue, and phenotypically indistinguishable). Self-fertilization is the most powerful form of assortative mating. If the three plants of field 1 were self-fertilized, and if each then produced 300 offspring, on the basis of Mendelian expectations (to be covered in Chapter 7), we would expect that parent *AA* would produce 300 *AA,* parent *Aa* would produce 75*AA,* 150 *Aa,* and 75*aa,* and parent *aa* would produce 300 *aa.* Allele frequencies are not different from those of the progeny in the random-mating example, but the expected genotypic array in the offspring population is 375 *AA,* 150 *Aa,* and 375 *aa,* or only 1.4 blue-flowered plants for each 1 white-flowered plant. Under disassortative mating, the blue-flowered plants would be mated with the white-flowered plant: *AA* × *aa* would produce all *Aa; Aa* × *aa* would produce 50 percent *Aa,* 50 percent *aa.* Note that the genotype *AA* would not exist in this offspring population. (Allele frequency would change, not as a direct consequence of the mating system, but because more alleles of the rare type would be forced into the offspring population. This is in fact a case of density-dependent selection, which may be an important force in maintaining genetic variability.)

We may at this point briefly summarize what has been presented so far as a description of the genetic basis of the evolutionary process. Nuclei of cells contain instructions for the synthesis of the different proteins that characterize species and individuals, coded in the form of nucleic acids. These instructions also duplicate themselves through a series of cell divisions. Errors, or mutations, in the coded genetic message, may occur. If they arise in cell lines that do not produce gametes, these mutations may affect the phenotype of the individual, but will have no evolutionary consequences, because they will not be transmitted to the next generation. Such mutations are called somatic.

If, however, the mutations are included in gametes, they may be passed on to the next generation. Such germinal mutations are the raw material of evolution. They furnish the variability on which selection can operate. Favorable mutations are those that increase the fitness of their carriers in the environment in which they live. As natural selection takes its course, they will tend to become the predominant genetic messages, even replacing the genes they mutated from.

Other evolutionary forces that may change the composition of the gene pool include introduction of alleles from other populations (immigration). A further effect is associated with the size of the reproducing population. The composition of the gene pool of a generation produced by only a small number of parents will depend to an important degree on chance.

Changes in environment can be responsible for changing the magnitude or the direction of selection. Indeed, if the environment were completely stable, evolution would cease after the various species had reached their optimal adaptation levels.

The results of the various evolutionary pressures operating on a population can lead either to the maintenance of the population as is (by stabilizing selection); to its gradual transformation into a genetically different population, or even a different species (by directional selection); or to its splitting into two or more populations, or species (by a variety of mechanisms including diversifying selection). Reproductive isolation, which prevents gene interchange between the diversifying groups, is generally a requirement for the process of speciation to go to completion.

The evolutionary forces discussed are the ones that operated in the past. They are still operating today, in many instances responding to human intervention, which, of course, is part of the environment. They are expected to continue in the future.

Views have been expressed that the levels of adaptation reached by most extant species today have approached their optima. Further major evolutionary changes seem unlikely to those who hold this opinion. But the fact is that the environment is constantly changing, and is likely to continue to change—not only in conformity with the Red Queen, but also because humans are repeatedly introducing new factors of evolutionary significance. Industrialization leading to melanism, discussed in Section 3.5, is but a single example. Pollution, urbanization, increased radiation, pest control, public-health and medical practices, and increased human mobility, which carries in its wake increased mobility of plants and animals, are a few others.

If humans were wiped off Earth completely, *Homo sapiens* might not evolve again from the species that survived. Major specialized adaptations are generally irreversible. The other species may no longer have the genetic variability that would permit them to be ancestors of such an evolutionary line. But a new dominant form would very likely arise.

In the absence of such extinction, evolution in humans most certainly is going on at the microevolutionary level, as will be shown in Section 7.4. It is even possible that evolutionary changes will lead, let us say, to *Homo superior* or *Homo continuus* (see Section 8.6).

Evolution as a discipline has a historical outlook. It *explains* the past, but is not necessarily capable of *predicting* the future. Just because we understand how evolutionary changes could have taken place, we cannot necessarily forecast the future. We have no way of foreseeing what the dominant species on Earth will be 150 million years hence, should Earth last that long, although we can describe the dinosaurs that dominated life 150 million years ago. But we must realize that because of cultural advances, and because of the accelerated rate of human cultural evolution, a new factor has entered the picture: our species has the potential capabilities of putting evolution under intelligent management. This has already been done with crops and domestic animals, and the possibility that this power can be extended to control the evolution of the human line may be at hand. We shall return to this idea in later chapters.

5

HUMAN EVOLUTION

To conclude the part of this book concerned with the evolutionary process, we present some of what is known about the evolution of our own species, and some speculation based on that knowledge.

5.1 HUMAN ANCESTRY

Our knowledge of human evolution is based on the fossil record, on the present-day physical and chemical structure of primates as studied in the laboratory, on the behavior of primates in natural environments, and on general theories of evolution as studied in the laboratory, in the field, and by means of mathematical and biochemical models.

In spite of recent findings, the time and place of origin of the order Primates remain uncertain. Present evidence indicates that the primate taxon belongs among the oldest documented divisions of placental mammals. The evolution of primates, which has been in progress for over 60 million years, took place mostly in the tree canopy of forests. Some of the features common to most primates are interpreted as being consequences of this arboreal existence:

1. Free mobility of limbs and digits, with limb movement not restricted to one plane.
2. Dexterity, involving a nervous system (providing precise and rapid control of muscles) and hands capable of grasping (having nails rather than claws, and extensive surfaces of skin on the palms to provide friction).
3. A tail as an organ of balance or "fifth limb" for grasping.
4. Upright body posture and ability to rotate the head through a large angle.
5. Eyes that look forward, with an overlapping field of vision, and a brain that can interpret such stereoscopic images in three dimensions.
6. Eyes and brain to receive and interpret low levels of illumination and different wavelengths of light (i.e., color).
7. A bony protection for the eyes.
8. Few progeny (generally only one infant is born at a time), which are given a great deal of adult attention throughout a long period of development.
9. An intermediate body size.

It is readily apparent that these and other primate characteristics would have been highly selected for among mammals that chose to leap from limb to limb, or otherwise pursue an active arboreal life.

Human ancestry clearly lies among the old-world primates of Miocene and Pliocene times (Figure 5.1), a period extending from about 25 million to three or four million years ago. In 1871, a dozen years after publication of *Origin of Species*, Darwin published *The Descent of Man*. Only two incomplete parts of "human" fossils, a skull cap and a lower jaw, were known to him. His ideas, based largely on a comparison of living primates, have held up remarkably well in the light of present knowledge. Since then, numerous finds of hominoid and anthropoid fossils in Africa, Asia, and Europe have permitted the reconstructions of the sequence in Figure 5.1 and in the illustrations to follow.

Paleontologists and anthropologists frequently dispute the proper classification of fossil remains. The discovery of a new humanlike fossil, however incomplete, is a rare and precious event. It is understandable and even necessary to name each such find. But there is a tendency for such names to take on separate biological significance, and to emphasize differences instead of similarities. This tendency seems to intensify the more the fossils resemble modern human parts, resulting in a multiplication of fossil "species" and even "genera" closely related to modern *Homo sapiens*, with the accompanying necessary conclusion that all but us have gone extinct. Yet enormous variability is characteristic of the human species today and it seems likely that such variability was also present among our ancestors. S. L. Washburn and J. Lancaster have put it thus: "The more a primate is bipedal, tool-using, and hunting, the less likely the form is to speciate and the more likely it is to occupy wide areas with only racial differences."

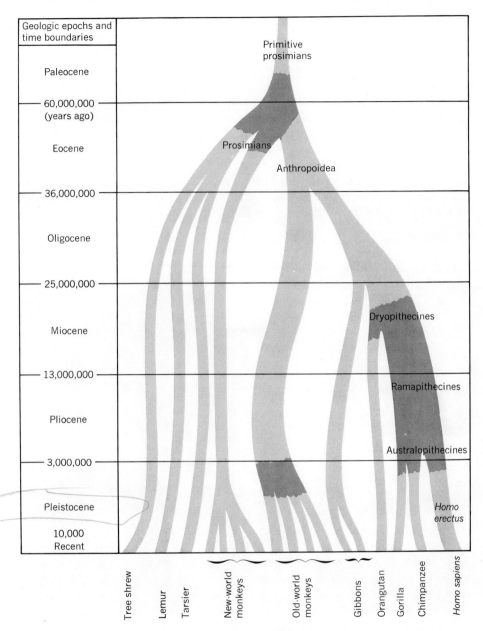

FIGURE 5.1

Tentative phylogenetic relationships of the primates, with detail presented in the line leading to *Homo sapiens*. Note that the time scale in years on the vertical axis is not linear. The areas in this phylogenetic tree with the darker shading indicate periods during which important and interesting groups must have diverged, but about which unambiguous and detailed information is still lacking.

Thus, Figure 5.1 was drawn with a conscious omission of detail, failing to locate *Australopithecus africanus* or *A. robustus*, or Peking, or Steinheim, or Neanderthal, or Cro-Magnon, or Solo, or Rhodesian man, or other named fossils, either as direct ancestors of *Homo sapiens*, or as representatives of distinct side branches that have failed.

Some primates that existed in what is now East Africa about twenty million years ago (which may have been about the time of separation between the gibbons on the one hand, and the ancestors of humans and apes on the other) are known as proconsul, or *Dryopithecus africanus* (Figure 5.2). Many related fossils have been found, including specimens of other species of *Dryopithecus* from as recent as eight million years ago. They appear clearly apelike, and whether the early dryopithecines (proconsul and other species of *Dryopithecus*) contributed to human ancestry is not known at present.

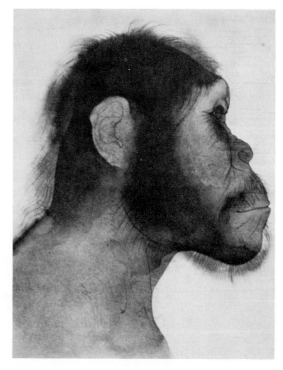

FIGURE 5.2
Reconstruction of proconsul, made from a 1948 find in the Olduvai Gorge in Tanzania. (Reproduced by permission of the Trustees of the British Museum, Natural History.)

FIGURE 5.3
Reconstruction of *Australopithecus robustus* (formerly known as *Zinjanthropus*). *Australopithecus robustus* is considered by some to be a separate species that became extinct, and by others to be a subspecies or variant within *Australopithecus africanus,* and thus possibly a contributor to the human line. (Reproduced by permission of the Trustees of the British Museum, Natural History.)

A fossil form called *Ramapithecus,* named for the Indian god Rama, has been found in present-day India and Pakistan as well as in Kenya. These specimens have been dated as having lived about fourteen (maybe fifteen) million years ago to as recently as nine (maybe eight) million years ago. Their teeth are more humanlike and less like those of apes, and the more recent fossils have jaws more like those of humans than do the older fossils. Whether the ancestors of humans, of apes, of both, or of neither, have indeed been recovered in one or more of the ramapithecine fossils isn't certain, but it seems likely we have at least recovered some relatives of those ancestors.

Big events were occurring during the Miocene. The sea separating Africa and Eurasia retreated, and many animals passed through the tropical forest corridors of present-day Gibraltar and Saudi Arabia. In Eurasia, the Alps, Urals, and Himalayas were uplifted, and there was rifting and volcanic activity in Africa. A cooling and drying during the Pliocene progressively changed much of the tropical forest to extensive grassland. Among the animals that entered this expanding niche were the australopithecines. Namers of fossils have distinguished a dozen or so species and five or so genera. There seems to be some evidence that at least two of them were indeed separate species, and that one of these lived during the period from about 3 to 1.2 million years ago and then became extinct. (Having lasted that long, it was much more successful in terms of longevity than *Homo sapiens* has yet proved to be.) Both species probably made and used tools. The second, *Australopithecus africanus,* whose fossils have been dated from before 4 to 1.5 million years ago, appears to have been the direct but not immediate ancestor of *H. sapiens* (Figure 5.3). Fossils of this species have been found throughout present-day Africa (showing some geographical variation) and into Eurasia as far as Java. In the Pliocene, the baboons, the now-extinct terrestrial ape *Gigantopithecus,* and then predatory cats and dogs, shared the expanding savanna with *Australopithecus.*

It is not clear at present whether the great apes and the human line diverged as recently as five million years ago (Figure 4.4) or as long as fifteen million years ago. The uncertainty lies in the interpretation of the ramapithecine and australopithecine fossils, and of such other lines of evidence as serum albumin resemblances. The apes in turn evolved into different groups (Figure 5.1). What is not clear from Figure 5.1 is that the present-day great apes have not been as successful as the monkeys, in terms of either number of living species or number of living individuals. It is possible that the relative success of these two groups has been influenced by the more severe competition imposed on the great apes by their increasingly successful relative, *Homo.*

During the Pleistocene, as ice sheets advanced and then retreated, the successors of *Australopithecus africanus* devised ways to adapt to the changing climate. They lived in caves, built shelters, and domesticated fire. By 1.3 million years ago, they had changed sufficiently from *A. africanus* so that their fossil remains are given a new name and placed in a new genus, *Homo.* Some consider *Homo erectus* to be the next species in the human line. Others consider *H. habilis* to have evolved first, which in turn evolved further to become *H. erectus.*

At various times during the next million years, *H. erectus* occupied much of Africa, Europe northward to at least present-day Germany, and Asia northward into China. A large cave near Peking was occupied for perhaps 70,000 years by hunters who left hearths, tools, remains of food, and their own bones. The different, named fossils such as Peking man, Heidelberg man, and Java man probably constitute a sample of the geographic and individual variability within the species *H. erectus* over the great area and great time that it existed.

After a million years, the changes in *H. erectus* had been sufficient so that fossils of their descendents (Swanscombe man and Steinheim man, who lived 150–250 thousand years ago in what is now England and Germany) have been admitted—perhaps somewhat arbitrarily—into *H. sapiens*. By 70,000 years ago, there were Neanderthal people in what is France today. About 30,000 years ago, the large jaw and heavy brow structures characteristic of Neanderthal fossils were largely succeeded in the fossil record by fossils with smaller jaw and brow structures, which typify Cro-Magnon fossils and modern people (Figure 5.4). Whether this rapid change was accomplished by a competitive displacement of one race by another, or by a general change throughout *H. sapiens* in response to the new selective opportunities and demands of more extensive tool using, is not clear.

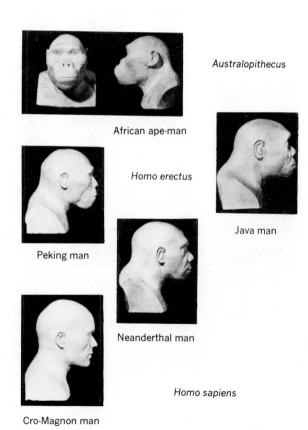

Australopithecus

African ape-man

Homo erectus

Java man

Peking man

Neanderthal man

FIGURE 5.4
The probable appearance of early hominids reconstructed by Maurice Wilson. (Reproduced, with rearrangement, by permission of the Trustees of the British Museum, Natural History.)

Homo sapiens

Cro-Magnon man

5.2 BECOMING HUMAN

The transformation from early primates through many intermediate forms to *Homo sapiens* includes various anatomical, physiological, and behavioral changes. Three important ones are descent from the trees, invention of tools, and increase in size and complexity of the brain. Box 5.A and Figure 5.5 give some information on the cranial capacity of our species and a number of our relatives.

The evolution of mind is a topic of some controversy. Perhaps it can be explained entirely on the basis of physics and chemistry, as now life can be. But some philosophers and biologists, including the German evolutionist **Bernard Rensch,** argue that there is some protopsychic property in all matter that has evolved into mind just as chemicals have evolved into life. Thus it may be present in rocks, and water, and other nonliving things. Clearly, there are isolated sensations in protists and, perhaps, mental images or recollections in the lowest invertebrates. They are more complex in such animals as worms, and there is lasting coherent awareness in higher forms. Plants may respond to thoughts and vibrations such as music. Insects and fishes are capable of memory storage, as judged from delayed reactions to visual stimuli. Still higher forms are capable of some

---

BOX 5.A CRANIAL CAPACITIES OF HUMANS AND SOME RELATIVES

Species	Range[1] of brain size in cubic centimeters
Modern chimpanzee	350–450
Modern gorilla	?–700
Fossil australopithecines	425–775
Fossil *Homo erectus*	815–1067
Fossil Neanderthal	1200–1500
Modern human	<1000–2102

[1]Excluding, where possible, pathologically abnormal individuals

The gorilla is the largest of the primates, but a gorilla brain is only about one-third the size of a human brain. In macroevolution, a larger brain permits more complex neurophysiological organization, which in the primate line is associated with the development of human intelligence ("larger," in this context, may be taken to refer both to absolute size and to size relative to the body it controls).

The changes in organization of the brain were probably more important than changes in its size during primate evolution. Some evidence for this can be inferred from differences among primate brains in the proportions of the major cortical subdivisions. For instance, the chimpanzee has a relatively large occipital lobe, which includes the visual centers, while the hominid line has relatively large parietal and temporal lobes, which include centers of sensory integration and association, and of memory. Thus brain size alone is a tenuous indicator of intellectual capacity. The brain of Ivan Turgenev was the largest of those measured to date (2,102 cc), Jonathan Swift's exceeded 2,000 cc, Thackeray's and Schubert's were near average (1,664 and 1,420 cc) and Anatole France did his considerable thinking with a brain of 1,017 cc. The brain of a human with microcephaly may be only 600 cc, well within the size range of gorillas. Microcephalic persons are clearly subnormal in intelligence, but they are human in behavior, including having the capacity to learn and utilize language.

84

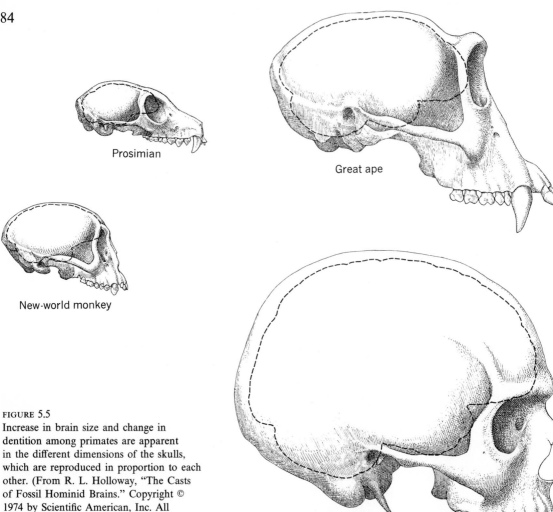

FIGURE 5.5
Increase in brain size and change in dentition among primates are apparent in the different dimensions of the skulls, which are reproduced in proportion to each other. (From R. L. Holloway, "The Casts of Fossil Hominid Brains." Copyright © 1974 by Scientific American, Inc. All rights reserved.)

Prosimian

New-world monkey

Great ape

Modern human

generalization and of transfer of information learned in one situation to another. The ability to form abstract concepts, to understand symbolic gestures and the idea of numbers and values, is found in many mammals. Comprehension of causal relations appears next on the evolutionary ladder. Finally, the use of reason, foresight, and displacement (see Section 2.5) developed. These last achievements place a selective advantage on complexity of brain structure, especially of the cerebral hemispheres.

The origin of tools dates back to prehuman times. Today, chimpanzees are known to make simple tools for obtaining food and water. They have been observed to fashion sticks to dislodge insects (which they eat) from crevices in trees, and to make sponges out of leaves to soak up water for drinking. When more sophisticated tools were constructed by the transition forms between apes and humans, they profoundly influenced the course of future evolution.

Many of the important tools were in fact weapons. These were used to kill or intimidate not only large, dangerous, or fast animals, but also that most dangerous of game, other weapon-bearing hominids. Advanced individuals or groups would have had enormous selective advantage as intellectual and social innovation furthered the evolution of weapons. The biological and cultural evolution of the weapon makers followed as a consequence.

Tools also led to record keeping (which included inscriptions on stone and bone, and perhaps other less easily preserved material) and even to the beginnings of art. One such inscribed bone, an ox rib, has been dated as being 135,000 years old.

The general logic in the use of information from contemporary primates is that if a structure or behavior is widespread in a group of related living primates, it was probably present in common ancestors of the group.

Human and gorilla infants share an inability to hold onto their mothers. Although most mammals develop to maturity in a few years or less, maturation is much longer among the apes (chimpanzees required 8–10 years to reach maturity). Selection in the human line has favored an even longer period of development, during which the young can form family groups with the adults, and learn from them. Human, monkey, and ape females share a menstrual cycle of approximately one month, with the sexual activity of monkeys and apes concentrated in a period of estrus close to ovulation. The sexual activity of human females is not restricted to the short period during each month when the likelihood of conception is the greatest, a behavioral difference with important implications for pair-bonding and other social organization.

The foraging or hunting ranges of the ancestors of *H. sapiens* were probably greater than those of present-day baboons and gorillas, which in turn are generally greater than those of monkeys. The greater a species' range, the more selective advantage would be conferred for memory and for means of transport and storage of food. Having a wide range would also increase contact and exchange between different neighboring groups, and might increase the need for defense of home territory.

Many primates manipulate objects (see also Figure 3.2), frequently as a part of feeding behaviors. But the ability to make substantial use of many different kinds of objects evolved only once. Next to humans, the chimpanzee performs the greatest amount of object manipulation. Chimpanzees build nests; use sticks to get insects and honey; use stones to break nuts; use both sticks and stones in aggressive or defensive display; and throw both underhand and overhand.

Bipedal locomotion (Figure 5.6) freed the arms and hands for fashioning and using tools in gathering food and in hunting. Memory, foresight, and originality were favored. This, in turn, led to a reform in social organization, because hunting large and dangerous game was likely to require cooperation with others. Monkeys and apes are organized in social groups, and these social groups are

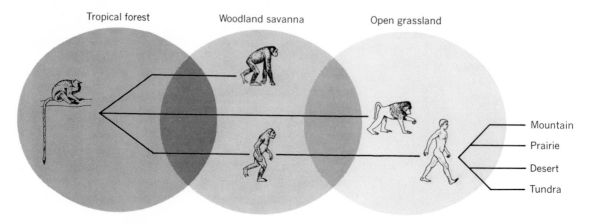

Tropical forest Woodland savanna Open grassland

Mountain
Prairie
Desert
Tundra

FIGURE 5.6
Evolution of bipedal locomotion. (From J. Napier, "The Antiquity of Human Walking."
Copyright © 1967 by Scientific American, Inc. All rights reserved.)

reinforced by various behavior patterns. Defense against predators, particularly big cats, is an important function of the primate social group. Whether a behavior pattern is inherent or learned is not always clear. But if learned, there must have been strong selection for genetic changes that increased the ease and even pleasurableness of such learning, with great survival advantage accruing to those who learned quickly. The implications of this for human evolution are profound.

Teeth and jaws, being hard and not easily damaged by predators or carrion eaters, are the most commonly fossilized parts of our ancestors' remains. There was a progressive reduction in the size of the canine teeth during the transition from primitive hominid to human (Figure 5.5). This is possibly associated with a replacement of their biting and tearing functions by weapons and tools. Since there is evidence that some change in dentition may have preceded significant weapon and tool use, it may have been a selective response to changes in social organization and behavior. For instance, there would have been advantages to intragroup cooperation, and thus to a reduction in intragroup aggression and display. Perhaps the fact that the females tended to be sexually receptive much of the time reduced the ability of a dominant male to service an entire social group, and thus the importance of the canines in display and intragroup aggression by the dominant males was lessened. Whatever the historical combination of selective pressures, the change in dentition (see Figure 5.5) clearly makes human mouth-to-mouth kissing a more pleasant event.

It is also suggested that bipedal walking may have freed the mouth and teeth, formerly much used for carrying things, for the development of human speech. This suggestion is perhaps overly facile. Many animals find time to chatter, and even vocally communicate in effective ways. But they have not developed a complicated symbolic language. One of the most important features of being human is that we have. At this moment we are using a way of communicating not available to any other of Earth's species.

Tropical forest
woodland
grassland

TWG

87

5.3 CULTURAL
EVOLUTION

TWG

5.3 CULTURAL EVOLUTION

In a genetically variable species, evolution in response to powerful selection can be rapid, as users of insecticides and antibiotics have been dismayed to learn. The genetic changes that occurred in the various primate lines 3–60 million years ago were generally quite adequate for adapting to the slow environmental changes of those times. However, during the late Pliocene and the Pleistocene, the rate of environmental change appears to have accelerated, placing a premium on adaptability. Most organisms continued to rely on biological evolution, but *Homo sapiens* invented a new response, that of cultural evolution. By 30,000 years ago, there were humans of modern form on Earth. Biological evolution has undoubtedly continued in the human line since then, but the main events of human biological evolution had already taken place. Cultural evolution became increasingly important in human adaptation, and gave us a tremendous competitive advantage over other mammals of comparable generation time, which were depending on genetic adaptation.

Cultural evolution is Lamarckian (see Box 3.A). Since change could be communicated to individuals throughout a population, rather than only to a few offspring, the time necessary to incorporate a change was greatly reduced. Time intervals between major changes contracted from millions, to thousands, to dozens of years. For example, it took millions of years of organic evolution before birds could fly, but we acquired the same skill within a few years of our discovery of appropriate power sources and the principles of flight. The cultural evolution of our ability to fly is such that we have traveled to and from Earth's Moon and have landed equipment on Venus.

The same interaction of cultural and organic evolution that has produced the great advances in our understanding of physical and biological phenomena is responsible for endowing us with tremendous control over nature. Our mastery of Earth is not complete by any means: the polar cap may melt, another ice age may make its appearance, contaminants from outer space may be uncontrollable, or still unimagined catastrophes may confront us. Barring such disasters, our species is in command, and it is not likely to allow any other that might evolve from other existing forms of life to displace it as the master of this planet.

An important debate is in progress concerning the interaction between biological and cultural evolution. It includes such questions as: To what degree does and can culture override biological potentials? To what degree is a population's culture dependent on, and a result of, its genes? Will culture accelerate or decelerate biological evolution?

Before long, the question of controlling further human biological evolution will face us. One of the favorite subjects of science fiction is the emergence of *Homo superior*. Artificial selection is, of course, technically as applicable to human beings as to chickens or corn. But the exercise of such selection is outside the realm of biology. Religious, ethical, political, psychological, and social considerations are paramount in the possibility of humans directing their own evolution, with or without the consent of the selected (see Chapter 22).

BOX 5.B LANDMARKS IN HUMAN CULTURAL EVOLUTION

The Neanderthals were succeeded by Cro-Magnons, who resembled present-day humans (Figure 5.4). The Cro-Magnons lived during the Old Stone Age (which is also called the Paleolithic), and developed a culture that included notable art work. Their cave paintings, which have been found in central and southern France and in Spain, are remarkable in their sophistication. Most of their subjects were the animals that they hunted. The pictures were probably painted as a part of the practice of compulsive or sympathetic magic, intended to secure success in hunting. As the Paleolithic was a glacial period, it is not surprising that one of the animals depicted is the reindeer, which ranged as far south as Spain some 15–20 thousand years ago. The paintings reproduced here (losing much by not being in color) are from caves in Altamira, Spain, and were made towards the end of the Paleolithic by members of the so-called Magdalenian culture. They represent a reindeer, a bison, and a wild boar.

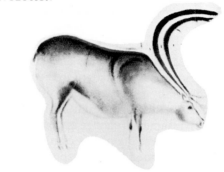

The Old Stone Age began more than 2 million years ago and was succeeded by the Middle (Mesolithic) and then the New (Neolithic) Stone Ages, during which human development extended as noted in the table below.

The landmarks of human cultural evolution listed do not necessarily represent change in human intellectual development. Indeed, among others, Claude Levi-Strauss, the French anthropologist, maintains that the human intellect has operated in the same basic patterns since human society appeared.

The transitions from the social organizations of dispersed hunters successively through those of larger hunting groups, agriculturists, preindustrial cities, and the feudal system, to current industrial societies, were each accompanied by a burst of population growth, discussed in Section 15.1.

CAVE PAINTING REPRODUCTIONS COURTESY OF THE AMERICAN MUSEUM OF NATURAL HISTORY.

Age	Years ago	Events of cultural evolution
Paleolithic	>2,000,000	Production of tools and cave drawings
Mesolithic	12,000	Domestication of animals (dogs first), development of water transport
Neolithic	7,000	Change from hunting and gathering to agriculture
Bronze	5,000	Use of copper and tin, origin of urbanization
Iron	3,500	Beginnings of recorded history
Industrial	150	Industrial revolution
Atomic	30	Use of nuclear energy

Biological evolution has been around for a long time, while cultural evolution is a new experiment. Some expect this experiment to fail, and the survivors to fall back on tried-and-true biological evolution. A passage from Ian McHarg's *Design with Nature* summarizes this view: "The atomic cataclysm has occurred. The earth is silent, covered with a gray pall. All life has been extinguished save in one deep leaden slit, where, long inured to radiation, persists a small colony of algae. They perceive that all life save theirs has been extinguished and that the entire task of evolution must begin again—some billions of years of life and death, mutation and adaptation, cooperation and competition, all to recover yesterday. They come to an immediate, spontaneous and unanimous conclusion: 'Next time, no brains'."

Even if the experiment survives and continues, there are thoughtful suggestions that human culture is now changing our environment faster than even cultural evolution can respond. Perhaps cultural evolution has run amok, overriding the controls that used to keep the human species in tune with available resources. Perhaps the modern human species has no such set of built-in controls.

Human biology evolved in a setting characterized by small populations, small social groups, great hazards, and need for a broad range of personal skills. The human in modern society appears to be too aggressive, too eager to be dominant, and too acquisitive. Organizing a modern society is difficult when the organism to be organized has evolved along the lines of *Homo sapiens.* Perhaps many anti-social acts, such as selfishness, cruelty, and war, are easily learned by the members of *H. sapiens* because such acts are in accord with fundamental human biology. If indeed such things are a part of the human evolutionary baggage brought to the twentieth century, it is important to recognize and understand them in planning how to organize a safe but fulfilling society for this difficult species. One of the pressing problems, that of large population size, will cause untold misery if not controlled. We will return to this in Chapter 15, after discussing some of the details of hereditary mechanisms.

6

INFORMATION
MECHANISMS
OF THE CELL

Among the most spectacular achievements of modern biology are the deciphering of the language in which genetic information is transmitted, and the construction of plausible models, some of which have been verified, of how the language is translated into action by the cell. An international group of geneticists, microbiologists, biochemists, and crystallographers participated in these discoveries. Nobel prizes have thus far been awarded to American (G. W. Beadle, M. Delbrück, A. D. Hershey, R. W. Holley, F. G. Khorana, A. Kornberg, J. Lederberg, S. E. Luria, M. W. Nirenberg, S. Ochoa, E. L. Tatum, J. D. Watson) British (F. H. C. Crick, M. H. F. Wilkins), and French (F. Jacob, A. Lwoff, J. Monod) scientists for their contributions to the understanding of the information mechanisms of the cell. Many others have played important roles in elucidating the complicated and beautifully elegant mechanism by which a cell receives instructions, carries them out, and passes them on to its descendents.

No full historical account of these recent discoveries will be given here. Rather, a capsule description of current theory, very much simplified, will be given. Many details unessential to the understanding of the remaining chapters of this book will be omitted. But it is strongly urged that those for whom the beauty of the genetic apparatus presents an intellectual or esthetic fascination become acquainted with further particulars of molecular genetics by additional reading.

6.1 CHEMICAL STRUCTURE OF THE HEREDITARY MATERIAL

It may be recalled from Section 2.3 that the bearers of the genetic message are nucleic acids. DNA, which is found in cell nuclei, contains the specifications for the manufacture of proteins by the cell. Its structure is described in Box 6.A. The fact that DNA is the vehicle of hereditary information was first proved by the experiments of the American biologists O. T. Avery, C. M. MacLeod, and M. McCarty (Box 6.B). The process of genetic transfer demonstrated in these experiments, **transformation,** has been unequivocally demonstrated only in protists. Nevertheless, DNA appears to be the hereditary language on Earth and to have a common structure in all forms of life in which it is found (some viruses have only RNA).

The language of DNA might be described as having three-letter words, each written as a triplet sequence of bases. Since there are four different bases in DNA, there can be 64 (4^3) different "words," such as ACC, CTG, and GCA. Through the brilliant investigations of Marshall W. Nirenberg and others, the meaning of each of the 64 different triplets, or **codons,** has been deciphered. Each one, except for three, ATT, ATC, and ACT, specifies one of the twenty common amino acids that make up the polypeptide chains of proteins. Figure 6.1 shows the genetic code for both DNA and RNA. It may be noted that although some of the amino acids are coded for by only one triplet (for example, methionine by TAC), others are specified by as many as six codons (e.g., serine, AGA, AGG, AGT, AGC, TCA, TCG).

The sequence of codons in the DNA specifies the sequence of the amino acids in the polypeptide chain manufactured by the cell. When UAA, UAG, or especially UGA (the RNA equivalents of the three exceptional DNA codons mentioned in the previous paragraph) appear in the chain, a halt to protein synthesis is called for: they are the stops of the molecular language.

We have provisionally defined the gene as a unit of heredity. We can now redefine it in terms of molecular structure, although in later discussion of Mendelian and population genetics we may at times revert to the original definition. More strictly specified, *the gene is a stretch of DNA coding for a particular polypeptide.* The number of genes in different organisms varies from no more than three in small viruses, which use RNA for coding, to several thousand, or enough

BOX 6.A THE STRUCTURE OF NUCLEIC ACIDS

The nucleic acids, DNA and RNA, are chains made up of subunits called **nucleotides,** each consisting of phosphoric acid, a nitrogenous ring compound, and sugar. In DNA the sugar is deoxyribose, and in RNA it is ribose. The nitrogenous compounds, or bases, are five in number, two purines, **adenine (A)** and **guanine (G),** and three pyrimidines, **cytosine (C),** common to both DNA and RNA, **thymine (T)** present only in DNA, and **uracil (U)** present only in RNA.

The structure of a nucleic acid may be diagrammed thus:

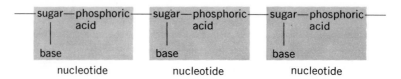

Different kinds of nucleic acid molecules are of different lengths, each highly precise. Some are known to contain as many as 200,000 nucleotides. In the base position at any place there may be either one of the purines or one of the pyrimidines. The following are examples of possible sequences of bases

DNA: AACGTAGCTGGT

RNA: UCUGGUCACAUG

The number of different possible sequences is enormous. Each molecule of DNA consists of two strands of nucleotides joined by hydrogen bonds and intertwined in a helix (excepting some single-stranded DNAs found in viruses). The bonding, because of the structural properties of the bases, is always between A and T, and between G and C. A segment of a DNA molecule may be illustrated as if it had only two dimensions (although, of course, it has three) to give us an idea of how the parts of nucleotides fit together to make a double-stranded chain:

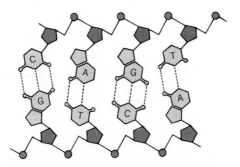

The dark circles are phosphoric acids, the dark pentagons are sugars, the lighter one-part and two-part shapes are pyrimidines and purines, respectively, and the dashed lines are hydrogen bonds.

A three-dimensional view of a stretch of DNA might be represented in the following way according to the model first proposed by Watson and Crick:

Adenine ⊏▭⊐

Thymine ⊏▥⊐

Cytosine ▬▬▬

Guanine ⊏⋯⋯⊐

The specificity of the bonding (A only with T, and G only with C) leads us to say that there is *complementarity* between the two strands of DNA; that is, each base of one strand specifies the complementary base of the other. DNA replicates itself by first separating into two single strands, each of which then has a complementary strand synthesized along its length, resulting in two identical double-stranded molecules:

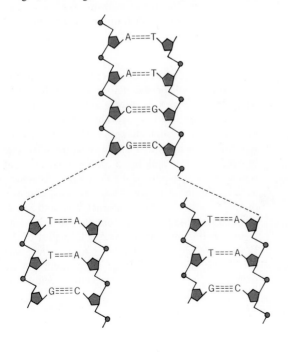

BOX 6.B TRANSFORMATION

In 1944, experiments conducted with pneumococcus bacteria (an organism that causes pneumonia in mammals) clearly showed the importance of DNA in heredity. At that time, scientific opinion on how genetic information was stored and used by organisms was divided—that is, there was uncertainty over what type of molecule carried the genetic information. It was generally accepted that chromosomes contain genetic information, and that chromosomes include DNA and protein in their structure. Most geneticists discounted DNA as a candidate because of its apparently simple structure. After all, DNA is uniform in its sugar and phosphate components, and has only four bases. The weight of opinion was that proteins (with their 20 amino acids, their ability to join with other organic and inorganic molecules, and their great diversity in size and shape) were much more likely to possess the complexity necessary to store the genetic information for building and running an organism.

The experiments used two alternative variants of the bacteria, R and S, which differ in the following characteristics:

1. S cells form a polysaccharide capsule.
    R cells do not form such a capsule.

2. S colonies appear smooth when grown on culture plates.
    R colonies appear rough when grown on culture plates.

3. S is virulent.
    R is relatively harmless.

4. R and S evoke formation of different antibodies in rabbits (see Box 3.C and Chapter 20).

The first and third differences are interesting in other contexts, but only (2) and (4) are necessary for understanding the experiment:

 (a) When R cells were grown on culture plates, they formed *only* rough (non-virulent) colonies.

(b) Rabbits were injected with R cells. When their blood serum, containing anti-R antibodies, was added to plates inoculated with growing R cells, growth of rough colonies (or any colonies) as in (a) did not occur. The antibodies had killed the R cells.

(c) S cells were killed in such a way that the cells were destroyed, but the DNA from them remained chemically active. When the DNA extract from such cells was placed on a culture plate, predictably, nothing grew.

(d) Given this background, the crucial experiment entailed mixing live R cells and DNA extract from killed S cells—i.e., a combination from steps (a) and (c)—and spreading the mixture on culture plates. After a short period of growth, serum containing anti-R antibodies was added. As expected, no rough colonies grew, but the important result was that several smooth (virulent) colonies grew.

The explanation of the result in (d) is that during the short growth period before the anti-R serum was added, a few R cells were able to incorporate DNA from the S-cell extract. This DNA was active, and produced enough changes in some of the cells that they were not killed by anti-R antibodies. The information for making a polysaccharide coat was apparently a part of the necessary protection, as all colonies that grew had this characteristic of the S cells. The S-cell extract was sufficiently purified in subsequent experiments that it became clear that this information *had* to be a part of the DNA molecule. Interestingly, it took eight years for the significance of these experiments to be generally appreciated by molecular geneticists (see Section 3.2).

First nucleotide	A or U		G or C		T or A		C or G		Third nucleotide
A or U	**AAA** *UUU*⎫ Phe **AAG** *UUC*⎭ **AAT** *UUA*⎱ Leu **AAC** *UUG*⎰		**AGA** *UCU*⎫ **AGG** *UCC*⎪ Ser **AGT** *UCA*⎬ **AGC** *UCG*⎭		**ATA** *UAU*⎱ Tyr **ATG** *UAC*⎰ **ATT** *UAA*⎱ Stop **ATC** *UAG*⎰		**ACA** *UGU*⎱ Cys **ACG** *UGC*⎰ **ACT** *UGA* Stop **ACC** *UGG* Trp		A or U G or C T or A C or G
G or C	**GAA** *CUU*⎫ **GAG** *CUC*⎪ Leu **GAT** *CUA*⎬ **GAC** *CUG*⎭		**GGA** *CCU*⎫ **GGG** *CCC*⎪ Pro **GGT** *CCA*⎬ **GGC** *CCG*⎭		**GTA** *CAU*⎱ His **GTG** *CAC*⎰ **GTT** *CAA*⎱ Gln **GTC** *CAG*⎰		**GCA** *CGU*⎫ **GCG** *CGC*⎪ Arg **GCT** *CGA*⎬ **GCC** *CGG*⎭		A or U G or C T or A C or G
T or A	**TAA** *AUU*⎫ **TAG** *AUC*⎬ Ile **TAT** *AUA*⎭ **TAC** *AUG* Met		**TGA** *ACU*⎫ **TGG** *ACC*⎪ Thr **TGT** *ACA*⎬ **TGC** *ACG*⎭		**TTA** *AAU*⎱ Asn **TTG** *AAC*⎰ **TTT** *AAA*⎱ Lys **TTC** *AAG*⎰		**TCA** *AGU*⎱ Ser **TCG** *AGC*⎰ **TCT** *AGA*⎱ Arg **TCC** *AGG*⎰		A or U G or C T or A C or G
C or G	**CAA** *GUU*⎫ **CAG** *GUC*⎪ Val **CAT** *GUA*⎬ **CAC** *GUG*⎭		**CGA** *GCU*⎫ **CGG** *GCC*⎪ Ala **CGT** *GCA*⎬ **CGC** *GCG*⎭		**CTA** *GAU*⎱ Asp **CTG** *GAC*⎰ **CTT** *GAA*⎱ Glu **CTC** *GAG*⎰		**CCA** *GGU*⎫ **CCG** *GGC*⎪ Gly **CCT** *GGA*⎬ **CCC** *GGG*⎭		A or U G or C T or A C or G

FIGURE 6.1

The genetic code. The DNA codons appear in boldface type; the corresponding RNA codons are in italics. The names of the amino acids have been abbreviated: **A**lanine, **Arg**inine, **Asn**paragine, **Asp**artic acid, **Cys**teine, **Glu**tamic acid, **Gln**utamine, **Gly**cine, **His**tidine, **Ile**soleucine, **Leu**cine, **Lys**ine, **Met**hionine, **Phe**nylalanine, **Pro**line, **Ser**ine, **Thr**eonine, **Trp**ryptophan, **Tyr**rosine, and **Val**ine. The RNA codons UAG, UAA, and UGA are "nonsense" words, which normally signal the termination of polypeptide synthesis, and thus serve as punctuation. Molecular biologists have named UAG "amber," the translation of the German "Bernstein," the name of its discoverer. They have furthered the nonsense terminology with "ochre" for UAA, but are apparently too ashamed of themselves to extend this nonsense jargon to the third nonsense codon, UGA. An AUG codon near or possibly at the beginning of an mRNA (see Box 6.C) begins polypeptide synthesis with a special tRNA bearing N-formyl methionine. The final form of many proteins has this initial amino acid chopped off.

to code for 2,000–3,000 different proteins, in the bacterium *E. coli*, and, perhaps, to some hundreds of thousands, in humans. In the egg of the frog *Xenopus* it has been estimated that there is enough DNA to constitute 756,000 genes exclusive of those coding for ribosomal RNA (see Box 6.C). In cattle sperm there may be enough DNA for 6–7 million genes, although this may, in fact, be a hundredfold overestimate of the actual number. Whereas a quarter to a third of the genes in *E. coli* are known, only a small fraction have been identified in higher organisms. It is highly probable that the number of proteins coded for in humans is considerably lower than the number of genes, some of which may code for the same protein (the property of *redundancy*) or have other functions than directing the order of amino acids in polypeptides. The general properties of the code can be summarized in this way:

1. It is organized as three-base (triplet) codons.
2. Most of the amino acids are coded for by more than one codon.

3. It is nonoverlapping—that is, the six consecutive bases AUGCAC spell out only two consecutive codons, AUG and CAC. (In an overlapping code these six bases would contribute to the spelling of eight consecutive codons: --A, -AU, AUG, UGC, GCA, CAC, AC-, and C--.)

4. It is commaless—that is, the last base of one codon is not separated from the first base of the next codon by extra space or punctuation. For example, AUGCAC is distinguished as two codons, AUG and CAC, without being marked by anything special between the G and C.

5. Only one strand of DNA is read, and it is read in a given direction. As an analogy, the sentence "THE CAT SAT" provides intelligible information. But read backwards, "TAS TAC EHT" means little.

6. The code appears to be universal, which may be a bit of an overstatement. But all forms of life on Earth thus far investigated use the same codons for the same amino acids and for signalling termination.

6.2 READING THE HEREDITARY MATERIAL

The DNA performs two operations in the cell. The first, called **transmission** because the same message is forwarded, begins prior to cell division when the DNA replicates itself (Box 6.A), and is completed during cell division when the replicated DNA is equally divided between daughter cells (Box 7.A). This operation is also called **replication.**

The second operation is called **transcription** and is the first step of protein synthesis. DNA is transcribed into the related but different language of RNA, using words that contain U's and not T's. In this manner, DNA directs protein manufacture by synthesizing **messenger RNA (mRNA)** complementary to itself, just as it produces complementary DNA strands during replication.

Messenger RNA passes through the nuclear membrane (Box 2.C and Figure 2.3) into the cytoplasm, where ribosomes attach to it and initiate protein, or polypeptide, synthesis. Another form of RNA (which is also made in the nucleus), **transfer RNA (tRNA),** is present in the cytoplasm. Molecules of tRNA are essentially small adapters, each having a three-base anticodon specific for a complementary site on mRNA, and a site that can attract and later detach a particular kind of amino acid. The tRNA molecules transport the appropriate amino acids to the ribosomes. Then the amino acids are fitted together in the sequence originally dictated by the DNA, and eventually are released as a completed polypeptide. This final operation is known as **translation** because the triplet words of the mRNA and tRNA are translated into the very different language of chains of amino acids.

The ribosome is largely constituted of RNA (of a still different type—**ribosomal RNA** or **rRNA**). While the ribosome provides the site of polypeptide synthesis, its rRNA does not appear to have translatable genetic information.

The major steps of protein synthesis are presented in Box 6.C. The relationships and operations of the nucleic acids are presented diagrammatically as follows:

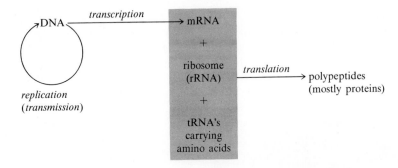

The details of protein synthesis vary from species to species. Thus, for example, the longevity of RNA molecules differs among various organisms. In some bacteria, mRNA molecules are functional for only one or two minutes, which is a sensible strategy for organisms whose needs for one or another protein can be drastically and suddenly changed by changes in their environment. On the other hand, in the homeostatic environment of vertebrate cells that are manufacturing, for instance, hemoglobin, mRNA can be very long-lived and carry out its functions for a considerable period of time.

The whole operation of synthesizing proteins does not proceed in a haphazard fashion, but rather is under the delicate control of regulatory systems. A model proposed for a regulatory system in bacteria is described in Box 6.D. An important feature of the system is that the turning on and off of protein synthesis is based on feedback. Accumulation of the final (or an intermediate) product of the synthesis in the cell, or alternatively, exhaustion of some building material, turns the process off. In lower organisms, an **operon** is defined as a stretch of DNA containing genes that code for mRNAs—the **structural genes**—and **regulatory genes** that control when and to what extent the structural genes are transcribed. Thus, in the bacterium *Salmonella*, the histidine operon makes ten different enzymes and has around 11,000 nucleotides. In one virus some 75 genes, organized in blocks controlling DNA synthesis, its timing, and the assembly of the virus from its component parts, have been identified. Such operons are not known in higher forms, but there is evidence that some higher organisms have gene-complexes that comprise two or more contiguous genes affecting the same developmental process. Perhaps, in the course of evolution, operons became fragmented and their constituent parts distributed among different chromosomes.

Most of the mechanisms described have been worked out in viruses and bacteria. As we move from protists to multicellular forms of life, many complexities arise. For instance, a virus consists merely of a stretch of nucleic acid and a protein coat; a bacterium can have a single circular chromosome carrying the DNA. Chromosomes of plants and animals are structurally much more complicated.

BOX 6.C PROTEIN SYNTHESIS

The ribosomes, constructed mostly of rRNA, are the sites of protein (and other poly-peptide) synthesis. Ribosomes normally operate as a series of independent assembly sites spaced along a linear mRNA molecule. Each time a ribosome passes along the length of an mRNA, one polypeptide molecule is produced. The following drawings illustrate this process in eukaryotic (with a nucleus) and prokaryotic (without a true nucleus) cells.

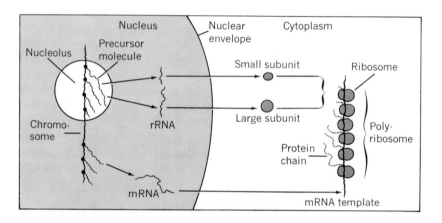

In eukaryotic cells, most transcription takes place in the nucleus and translation takes place in the cytoplasm. Ribosomal RNA transcribed from genes in the nucleolus forms the two large RNA molecules found in the two units of a ribosome. (Not all chromosomes have nucleoli.) Messenger RNA's transcribed from other genes serve as templates on which polypeptides are assembled.

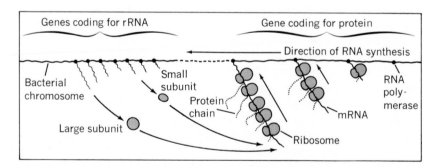

Simultaneous transcription and translation take place in prokaryotic cells such as bacteria. In the bacterial system diagrammed above, rRNA is transcribed from two genes, apparently contiguous, on certain segments of the single bacterial chromosome. Messenger RNA is transcribed from genes at other sites and immediately translated into protein by ribosomes.

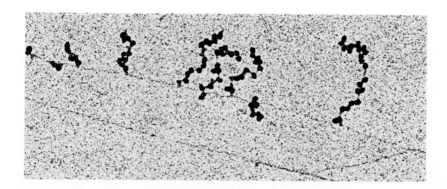

Shown above is an electron micrograph of an unidentified operon of the bacterium *Escherichia coli*. The different sized groups of attached ribosomes are formed by mRNA's in successive stages of transcription from DNA. The images in this micrograph are magnified about 70,500 times.

Details of the transcription and translation events are given below. Note that the mRNA is complementary to the righthand DNA strand. Polypeptides are formed by linkage of amino acids in the sequence determined by the triplets of mRNA, which pair with appropriate triplet anticodons of tRNA at the ribosome.

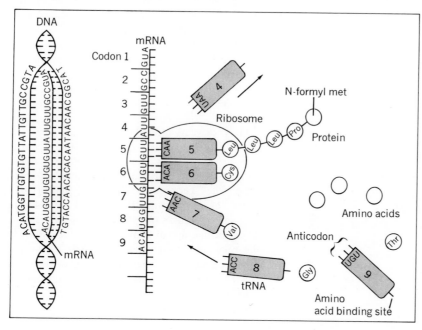

THE FIRST TWO ILLUSTRATIONS ARE ADAPTED FROM O. L. MILLER, JR., "THE VISUALIZATION OF GENES IN ACTION" COPYRIGHT © 1973 BY SCIENTIFIC AMERICAN, INC. ALL RIGHTS RESERVED. THE ELECTRON MICROGRAPH IS FROM O. L. MILLER, JR., B. A. HAMKALO, AND C. A. THOMAS, JR., "VISUALIZATION OF BACTERIAL GENES IN ACTION." SCIENCE 169:392–395. COPYRIGHT 1970 BY THE AMERICAN ASSOCIATION FOR THE ADVANCEMENT OF SCIENCE. THE FINAL ILLUSTRATION IS ADAPTED FROM M. NOMURA, "RIBOSOMES." COPYRIGHT © 1969 BY SCIENTIFIC AMERICAN, INC. ALL RIGHTS RESERVED.

BOX 6.D A BACTERIAL REGULATORY SYSTEM

The diagram below is based on a model of regulation of protein synthesis in bacteria proposed by François Jacob and Jacques Monod. Three structural genes are contiguous to each other in the lac operon, so named because their DNA specifies the amino acid sequences of three proteins that provide for the metabolism of lactose in the bacterium *Escherichia coli.* Transcription of mRNA from the *z, y,* and *a* structural genes is under the

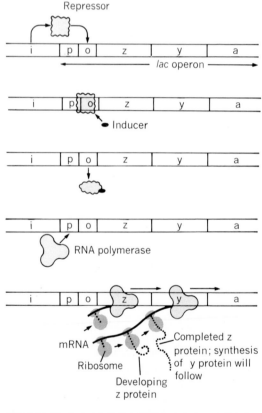

THE ILLUSTRATION IS ADAPTED FROM M. PTASHNE AND W. GILBERT, "GENETIC REPRESSORS." COPYRIGHT © 1970 BY SCIENTIFIC AMERICAN, INC. ALL RIGHTS RESERVED.

control of a contiguous operator gene (*o*). The operator gene is controlled by a regulator gene (*i*), which may be located some distance away. The model may work in different ways. In this system, the regulator produces a substance, the *repressor.* When the repressor is freed from the operator by an inducer (in this case, a lactose derivative that attaches to and changes the shape of the repressor), it allows molecules of the enzyme *RNA polymerase* to attach at the promoter (*p*) gene and to transcribe the appropriate mRNA; protein synthesis is then carried out. When repressor again attaches to the operator gene, it switches off the production of mRNA. If mutation alters the operator gene or renders the repressor ineffective, protein synthesis is not regulated and goes on continuously.

In other systems, the end products of protein synthesis may regulate further synthesis by interfering with transcription of new mRNA. Production of the end product then decreases and stops as available mRNA is broken down. The process in either system is a feedback process, and is end-directed. This model is based on experiments with bacteria, and although some attempt has been made to apply it to higher organisms, it is likely that the regulatory systems in them are more complex.

Only 26 years after the transformation experiments that first showed DNA to be the carrier of genetic information (Box 6.B), Jonathan Beckwith's group at Harvard isolated the first intact gene. They chose the *z* structural gene (which codes for the enzyme *β*-galactosidase) and the adjacent promoter (*p*) and operator (*o*) regions. Their technique will allow more detailed study of the regulatory system, and also tells us the size of these genes. The *β*-galactosidase gene measures 3,700 base pairs, and the (*o*) and (*p*) regions are a little more than 400 base pairs long.

They contain, in addition to DNA, several types of protein, certain of which are apparently involved in regulatory functions of RNA production. The processes whereby chromosomes replicate and serve as directors of ribosomal activity are surely more complicated than described here, and much is to be learned about them. But the basic principles that have evolved by natural selection to ensure the transmission of genetic instructions from cell to cell (intraorganismal heredity) are, no doubt, general throughout all forms of living matter.

6.3 DEVELOPMENTAL GENETICS

Another important feature of multicellular organisms that does not have a counterpart in bacteria is differentiation of parts and supercellular organs. The cells that make up a specialized part or organ are different and do different things from other cells in the same organism because only a small and characteristic fraction of the organism's set of genes is expressed in each cell. In an organ such as the kidney or liver, perhaps only a few percent of the genes are actively directing protein synthesis, the rest being more-or-less permanently muzzled. The questions of how only certain genes are allowed to function, and how the regulatory decisions are made, when a cell divides, that cause the daughter cells to take different developmental pathways, are among the highest priority questions in biology today.

The chromosomes in the salivary-gland cells of drosophila, the midges, and other insects show evidence of such regulation. Enlargements and puffs appear at different places along the lengths of the chromosomes at different times in the course of development. By introducing into the cells radioactive uracil, which is incorporated into the RNA manufactured, it is possible to see that these puffs are, indeed, associated with RNA synthesis (Figure 6.2). In other words, the instructions a cell receives from the nucleus are under regulatory control, which directs the particular activity the cell is to undertake at a particular time.

Various types of controls, directing the development of cells and then what they do once developed, must exist. For instance, the cytoplasm of different cells may contain different materials. Although the total genetic message is passed from mother to daughter cells in the chromosomes, the cytoplasm may be unequally distributed at some cell divisions. Experiments with sea urchins and other organisms have shown that differentiation of cytoplasmic contents does occur. In part, such cytoplasmic inequalities might be directed by the position of the cells in the developing organism. Thus, outer cells may acquire different potentialities for development than those surrounded by other cells.

It has also been demonstrated that nuclei themselves may become differentiated as development proceeds. Embryologists have destroyed the nucleus of an amphibian egg and by a kind of microsurgery introduced a nucleus from another

FIGURE 6.2
RNA synthesis associated with puffing in
the salivary-gland chromosome of a midge.
The black spots indicate the presence of
newly produced RNA, which was radioactive (and thus visible on film sensitive
to radioactivity) because radioactive uracil
had been added to the organism's food.
(Uracil, rather than thymine, is incorporated
into RNA, and thus can be used to
distinguish RNA from DNA—see Box 6.A.)
The radioactivity concentrates at those
places where RNA synthesis is most actively
proceeding. The upper picture shows a
chromosome with an active puff; the lower,
for comparison, a chromosome in which
RNA synthesis has been inhibited. (By
Claus Pelling.)

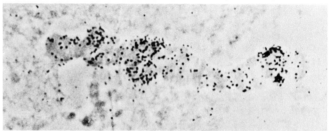

cell. In a series of such experiments, if the age of the replacement nucleus is
varied, it is found in some experiments that progressively older nuclei have an
increasingly reduced repertory of activities. That is, the differentiated nuclei
have become unable to direct the manufacture of all the substances specified in
the total genetic message and now instruct cells to perform only certain activities.
The activities, of course, are different in different tissues, and eventually lead
to the distinct shapes and functions of various body organs. In some cases, nuclei
transplanted to a different kind of cell do not adequately direct the metabolism
and development of the new kind of cell. But, if transplanted a second, third,
or more times to this new cytoplasmic environment, they respond by increasingly
directing the appropriate activities of the new host cells. Thus, the apparent differentiation of nuclei may be partly or even wholly altered or reversed. These observations have important implications for studies of aging (Sections 6.6 and 7.1)
and the technique of cloning (Sections 11.3 and 22.2).

Many other considerations of developmental genetics might be mentioned,
but enough has been said to indicate existing problems and the type of answers
that are being sought (see also Section 19.2). Another role of RNA in cellular
activities, although it does not technically belong to the subject of developmental
genetics, may be discussed here.

The Swedish neurologist Holger Hyden has developed a theory that memory
storage in the brain occurs through the production of specific proteins. Nucleic
acids would, according to this theory, be expected to control the accumulation
of information in the brain. Significantly enough, Hyden has found that the RNA
content of human nerve cells rises progressively from the time a person is three

years old until age 40 and declines precipitately at about age 60, thus showing a correlation with the temporal pattern of the efficiency of brain function.

Other investigators, in particular, the American J. V. McConnell, reported that the simple cannibalistic planarian flatworm, which has a very rudimentary central nervous system, can be trained to respond to light and electrical shock stimuli. When trained worms were fed to untrained ones, the latter seemed to acquire a memory of the training. The question immediately arose whether it was RNA that incorporated the learned responses into the primitive brains of the untrained worms.

Some of the most provocative experiments used rats trained to avoid darkness, contrary to their normal behavior of seeking out dark places. The brains of trained rats, and also some from untrained rats (the controls), were ground up and injected into animals that had received no such training. Those receiving the control brain extract continued to seek out darkness, while those receiving brain extract from the trained rats largely avoided the dark. A chemical consisting of a chain of only 15 amino acids was isolated by Georges Ungar, the investigator, who named it "scotophobin" (from the Greek words for "dark" and "fear") and verified in additional experiments that it was the substance which altered the behavior. Later, Wolfgang Parr synthesized this substance in the laboratory. The synthetic scotophobin worked nearly as well as the scotophobin isolated from the trained rats in changing the behavior of untrained animals that received injections of it.

The whole issue is still in doubt. If it is confirmed that learning can be transferred by injection, classroom instruction may receive a brand new dimension, in which brain extracts of learned professors could be served to students instead of lectures. A more serious prospect is that target populations of some animals, including defending armies or subject populations of humans, might have their behavior altered by receiving some appropriate chemical.

6.4 POINT MUTATION

Just as the genetic code provides for heredity, so it also provides for the production of variation. In Chapter 16, we shall discuss the various kinds of mutation that can occur. Here we shall consider only one kind, which, perhaps, is the most important supplier of variation for selection to work upon: the class of mutations known as **point mutations,** which are intragenic.

Errors can occur in the course of transmission of the genetic message from cell to cell (somatic mutations) or from parent to offspring (germinal mutations). A different base may be substituted at any position in the original DNA message. A base may be deleted from any part of the DNA message, or one may be inserted. In all three situations, the meaning of the message may be changed, as shown in Figure 6.3. Once a change has been made, it will be transmitted to the

104

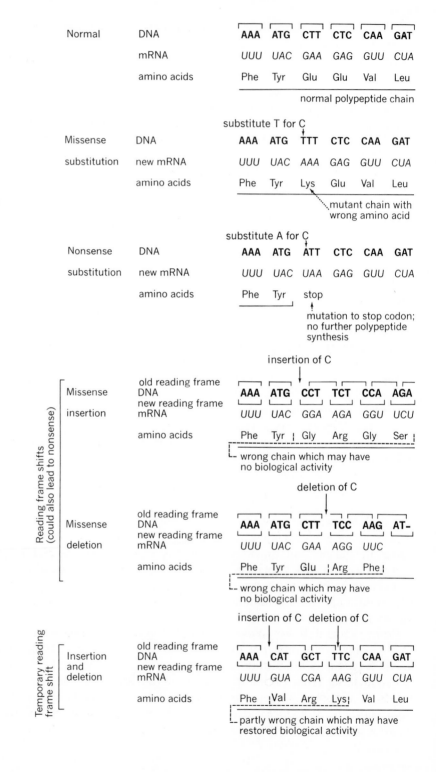

FIGURE 6.3
The molecular basis of point mutation. Each substitution, deletion, or insertion is a mutational event that alters the original (normal) DNA, and thus, in turn, the primary product (mRNA) and the secondary product (a polypeptide).

following generations of cells or of organisms, unless the distortion is so severe as to prove **lethal** before any descendents are produced.

The severity of the effect of a mutation depends on many factors. If, for instance, the DNA codon AAA mutates to AAG, no observable effects will ensue, since both code for the same amino acid, phenylalanine (Figure 6.1). Slight changes involving only a few wrong amino acids in a polypeptide, such as the missense substitution or the neighboring insertion and deletion shown in Figure 6.3, may have only a small effect. This is most likely to be true if the changes do not affect the configuration of the polypeptide, or are not at an active site of the molecule. Other errors, for instance the deletion of a whole codon, such as appears to account for one of the abnormal hemoglobins, may have trivial genetic consequences.

Or, as will be shown later in this section, a change of even a single nucleotide may have spectacular phenotypic effects.

But some mutations, like the missense examples in Figure 6.3, may produce proteins that have no biological activity, or that cause significant abnormalities. These, and mutations in which the stop codon appears in the wrong place in the instructions, may prevent the cell from manufacturing a necessary enzyme or protein, thereby creating a **metabolic block.** It should be apparent that the position of a mutation in the DNA sequence is an important determinant of the severity of its consequences. A changed codon at the end of a DNA sequence is likely to have small effects, because most of the protein will continue to be made according to the unchanged instructions that precede it. A stop introduced by a mutation at the beginning of a chain will prevent further synthesis of mRNA, and hence, of the protein.

Clearly, the potentialities for the number of mutations that a single gene may undergo are enormous, since each of its 1,000 or so nucleotides theoretically can mutate three different ways. Perhaps some of these possibilities cannot occur because of chemical considerations, but even so, a huge variety *is* possible. Indeed, wherever a search was feasible and undertaken, a large number of different forms of the same gene, as judged by their effects on the phenotype, were found. As noted in Chapter 4, alternative forms of the same gene in a given species, presumably due to one or more changes in codons, are known as alleles; the existence of many forms of a gene is known as **multiple allelism.** We shall now examine two instances of changed instructions in human genes. The first relates to the effects of a base substitution in one of the genes controlling the manufacture of hemoglobin. The other deals with blocks in the metabolic pathway of the amino acid phenylalanine.

Box 6.E describes two mutations that so change instructions to red blood cells that they produce abnormal, instead of normal, hemoglobins. In both instances, only a single codon is affected, and probably only one base (out of the 438 coding for the 146 amino acids of the beta chain) has been changed. The phenotypic effects of the substitution are dramatic. Persons whose cells make hemoglobin S instead of hemoglobin A suffer from a disease called **sickle-cell anemia,** so named because, when the supply of oxygen is reduced as it is under the coverslip of a

BOX 6.E HEMOGLOBIN

The molecules of all vertebrate hemoglobins, except for those of the lamprey and hagfish, are composed of amino acids assembled into two pairs of chains. There are five different chains in normal humans, and they differ from each other in their amino acid sequences. The production of each is controlled by a separate gene; these genes probably arose in the course of evolutionary history from a single gene that carried instructions for the alpha chain. A normal human will, at one time or another, have four different kinds of hemoglobin. The subscripts indicate that two chains of each kind make up the hemoglobin molecule.

Adult hemoglobin	α_2	β_2
Fetal hemoglobin	α_2	γ_2
Kunkel's component (about 10 percent of the total hemoglobin in the adult)	α_2	δ_2
Embryonic hemoglobin (disappears after 12 weeks of gestation)	α_2	ε_2

Each chain has approximately 140 amino acids, and the exact sequence of amino acids is known for four of the chains.

The alpha and beta chains differ in 21 amino acids, the beta and gamma in 23, and the beta and delta in not more than 10. The reconstruction of the evolution of genes determining hemoglobin production from a gene controlling production of muscle protein (myoglobin) can be diagrammed as follows:

myoglobin

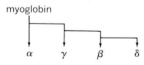

α γ β δ

By contrast, human alpha and gorilla alpha chains differ in only two amino acids, and human and gorilla beta chains differ in only one. Thus it appears that the different hemoglobin genes duplicated and began separate evolution much earlier in the history of the vertebrates than the recent divergence of humans and gorillas.

Numerous mutations in the genes specifying the different chains are known. Only four, including those that result in the production of hemoglobins C and S, are common in human populations. Nearly 150 other abnormal hemoglobins have been discovered, but most of these have been found in only one family, and none has been found in more than a few families, which may be in widely separated places. We shall consider in particular hemoglobins S and C, which both result from mutations in the gene responsible for the beta chain. The differences between the normal and two mutant amino acid sequences in a short stretch of the beta chain are shown below:

Position in chain	Normal adult Hemoglobin A	Hemoglobin S	Hemoglobin C
4	Threonine	Threonine	Threonine
5	Proline	Proline	Proline
6	**Glutamic acid**	**Valine**	**Lysine**
7	Glutamic acid	Glutamic acid	Glutamic acid
8	Lysine	Lysine	Lysine

In both abnormal sequences an amino acid in the sixth position in the chain has been substituted. Although it is not known which of the glutamic acid codons is actually present in the corresponding position in the DNA, it is possible to see that each change may be only the substitution of a single nucleotide. Thus CTC, which codes for glutamic acid, could have been changed to CAC (valine) and to TTC (lysine) to produce the abnormal types.

We shall return to defective hemoglobins in Section 7.4.

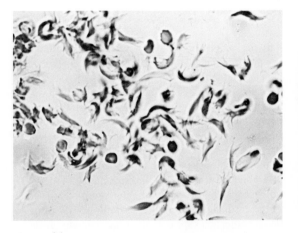 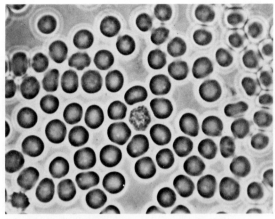

FIGURE 6.4

Left: Red blood cells from a person who has sickle-cell anemia. These cells have collapsed in the oxygen-deficient environment that develops under the coverslip of a microscope slide. *Right:* Normal red blood cells. (Photomicrographs by Anthony C. Allison.)

microscope slide (Figure 6.4), or in the blood stream at high altitudes, the defective hemoglobin molecules clump together in rods and cause the cells to assume the shape of sickles. The disease is severe, and has serious secondary consequences in addition to its primary effects (Figure 6.5). The sickling phenomenon is probably caused by a difference in the structures of valine and glutamic acid. This in turn gives rise to a different shape of the hemoglobin molecule, including a projection from the beta chain that can fit into an adjacent alpha chain and cause a sort of stacking of the molecules.

Hemoglobin C (produced by the second mutation described in Box 6.E) does not cause sickling but, like hemoglobin S, produces anemia. The evolutionary interplay of these and related hemoglobin variants will be considered in Section 7.4.

Our second example deals with complete metabolic blocks, which appear if there is a failure of production of specific enzymes essential to the normal functioning of the human body. Several are described in Box 6.F. The block in the pathway between tyrosine and melanin produces albinism (Figure 6.6). The inability of the liver to manufacture the enzyme phenylalanine hydroxylase has the more deleterious effect of causing the disease **phenylketonuria (PKU).** PKU deserves discussion at some length.

The accumulation of phenylalanine and of phenylpyruvic acid in the bloodstream of PKU individuals causes, in a manner unknown, severe mental retardation. Indeed, the disease was discovered by observing a high concentration of phenylpyruvic acid in the urine of mentally retarded inmates of a Norwegian institution. The incidence of the disease in populations varies, and no exact figures are available. Estimated numbers of babies born with the defect run from one

BOX 6.F PHENYLKETONURIA

The first step in the metabolic breakdown of the essential amino acid phenylalanine is mediated by a liver-produced enzyme, phenylalanine hydroxylase. This enzyme is responsible for the substitution of an OH group for an H atom in phenylalanine, converting it to the amino acid tyrosine. Tyrosine, in turn, through a series of intermediate steps is converted into **melanin,** the skin pigment, and other substances. It is also broken down further along the pathway illustrated, in which the existence of intermediary steps is indicated by dotted arrows. If phenylalanine hydroxylase is absent, phenylalanine is in part converted into phenylpyruvic acid, which accumulates, together with phenylalanine, in the bloodstream. These substances are toxic to the central nervous system and produce some of the symptoms of the genetic disease phenylketonuria. Other genetic metabolic defects in the tyrosine pathway are also known. As indicated in the diagram, absence of enzymes operating between tyrosine and melanin is a cause of **albinism.** Two other blocks illustrated produce tyrosinosis, a rare defect that causes hydroxyphenylpyruvic acid to accumulate in the urine, but requires no treatment; and **alkaptonuria,** which makes urine turn black on exposure to air, causes pigmentation to appear in the cartilage, and produces symptoms of arthritis. Another block in a different pathway from tyrosine, somewhat more complex, produces thyroid deficiency leading to goiterous cretinism.

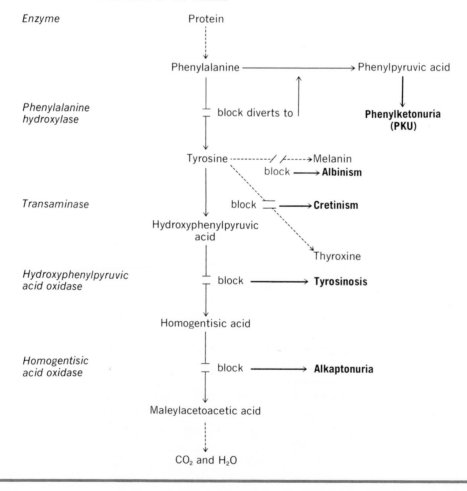

in 10,000 to one in 20,000 among the white U.S. population; the estimated frequencies are considerably lower for blacks. As a reasonable estimate, more than 500 white PKU babies are born per year in the United States, and another 1,000 or more born per year carry a newly arisen mutant gene for the disease without its being phenotypically expressed.

In recent years, tests have been devised to identify high phenylalanine blood levels in newborn babies. The *diaper test* depends on change in the color of the urine when ferric chloride is added to it. The more efficient bacteriological *Guthrie test* is based on the ability of certain strains of bacteria to grow only when high levels of phenylalanine are present in their diet. To perform this test, blood from babies is added to bacterial cultures lacking phenylalanine. If they grow, it is an indication that the phenylalanine level of blood is high. Some 37 states now make the Guthrie test compulsory for all newborn babies in order to identify PKU

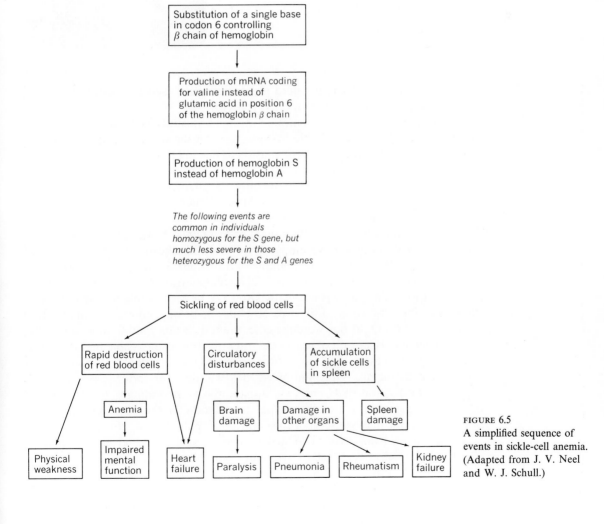

FIGURE 6.5
A simplified sequence of events in sickle-cell anemia. (Adapted from J. V. Neel and W. J. Schull.)

FIGURE 6.6
An albino. (Courtesy James Walsh.)

infants. The test is not one hundred percent effective, and errors in both directions, false positives and false negatives, occur.

Phenylalanine is highly concentrated in such foodstuffs as bread, cheese, eggs, fish, and milk. Since PKU babies cannot perform the normal metabolic breakdown of this amino acid, treatment consists of placing them on a regime of vegetables and fruit and a special protein mixture, low in phenylalanine. It is not a particularly palatable diet, but apparently an adequate one for these babies.

Success has been reported for this treatment when it is initiated early in life. In one study, 14 out of 17 children who had been placed on the diet soon after birth were within the normal I.Q. range; children who began eating this special diet after six years of age were already severely handicapped, and showed no improvement. We have here an example of nutritional management of a genetically caused effect. However, there are still a great many unsolved PKU problems.

To start with, diet alone is not really known to be responsible for the improvement. Possibly, the successfully treated children belong inherently among the higher range of I.Q.; if this were the case, the test results would not be an index of improvement as a result of nutritional therapy. It is also possible that the extra attention paid to them is a cause of the rise in I.Q. Other nongenetic complications enter the picture. Thus, untreated PKU mothers can induce the disease in their genetically non-PKU children, apparently by providing a toxic environment during gestation. Besides mental retardation in the offspring, maternal phenylketonuria may cause abortions and other pathological effects.

Furthermore, if an error of diagnosis is made and a normal baby is placed on the PKU diet, mental retardation can occur because of phenylalanine deficiency. Other forms of protein malnutrition can have similar effects. We shall discuss the population genetics of PKU in Section 13.3.

6.5 NEUTRAL MUTATIONS AND NON-DARWINIAN
EVOLUTION

Darwin's name has become associated with evolution in response to natural selection. As it became clear that some evolutionary change in organisms might also occur in response to completely random processes, the term "non-Darwinian evolution" was applied to cases where some of the differences between taxa did not seem to be particularly advantageous, but seemed selectively equivalent. How much of the genetic variability within and between populations is selectively neutral is an interesting and important issue, and is being actively investigated.

Organisms extract energy from organic molecules by combining parts of them with oxygen (a process called respiration). Cytochrome c is essential for this process. The ability—and the requisite molecules—to carry out this metabolic process probably evolved in single-celled organisms 1.5–2 billion years ago, and was a necessary preliminary to the evolution of complex, many-celled organisms. Cytochrome c is also one of the molecules that seems useful as an evolutionary clock (Section 3.5), which is based on the accumulation of selectively neutral mutations. It may seem strange that such an important molecule would be the repository of selectively neutral mutations, but an understanding of its structure perhaps provides some of the answer to this dilemma (Box 6.G).

Changes in cytochrome c over time have been less frequent than changes in such molecules as fibrinopeptides and hemoglobins. We have no reason to think the gene for cytochrome c mutates more slowly than the genes for these more rapidly changing molecules, or that there are differences in mutation rates within the different regions of the cytochrome c gene, which we may refer to as invariant, conservative, and radical regions. The mutations are probably random, and what we observe when we compare the sequence of amino acids from species to species are the molecules that have survived the rigid tests of function and successful reproduction. Invariant regions evidently are invariant because any mutational changes are, without exception, selected against. Conservative changes can be tolerated because they preserve the essential chemical and configurational properties of the molecule. Radical changes presumably identify regions of the molecule that do not much matter, other than adding correct distance, in the function of the molecule. It is these tolerable conservative and radical changes that are the "neutral" mutations.

Such neutral mutations, by definition, are unaffected by selection, but are subject to the forces of drift, migration, and possibly segregation distortion (Section 4.3). Drift and migration are amenable to mathematical treatment, and the predictions arising from such treatment are reasonably consistent with observations on rates of substitution in the "neutral" parts of such molecules as cytochrome c.

It has been further suggested that much of the enormous genetic variability observed and experimentally demonstrated in many present-day species, including humans, may similarly be the result of neutral mutations that are in the

BOX 6.G CYTOCHROME C

The composition of cytochrome c has been analyzed in 38 species, including 11 mammals, 4 birds, 2 reptiles, 1 amphibian, 2 fishes, 4 insects, 2 yeasts, 1 mold, and 10 angiosperm plants. The cytochrome c molecules of human and chimpanzee consist of 104 amino acids in exactly the same order and folded into the same three-dimensional structure. By contrast, human cytochrome c differs from that of the red bread mold (*Neurospora*) in both length of molecule and its internal detail, having different amino acids at 44 of the 103 positions that they have in common. Yet the three-dimensional structures of these two molecules are still very similar, and human cytochrome c can serve as an electron acceptor and donor in the respiration system of the bread mold, and vice versa.

The illustration presents complete amino acid sequences for humans and chimpanzees, rattlesnakes, and red bread mold, aligned for easy comparison. A summary of the amount of variation in cytochrome c, position by position, of 38 species is also presented as a histogram. Note that some parts of the amino acid sequence do not vary. Thirty-five of the 104 positions are completely invariant among the 38 species, including a long sequence from position 70 through 80. (The bacterium *Rhodospirillum* has a different amino acid at 15 of the 35 "invariant" positions. Thus the number of cytochrome c invariant positions has been reduced to 20, now that the cytochrome c of this bacterium has been sequenced.) Another 23 positions are occupied by only two different (but very similar) amino acids. An additional 17 positions are occupied by only three, which are also similar to each other in structure and chemical properties. Such interchanges among similar alternatives are called *conservative* substitutions, because they conserve the overall chemical nature of that part of the molecule. In some places along the chain, *radical* changes can be tolerated. Position 89, for example, may have aspartic acid or glutamic acid (acidic), lysine (basic), serine, threonine, asparagine or glutamine (polar but uncharged), alanine (weakly hydrophobic) or glycine (no side chain). Only large hydrophobic amino acids seem to be forbidden at position 89.

Amino acid key (see Figure 6.1)

Hydrophobic, aromatic rings: F (Phe), W (Trp), Y (Tyr)
Hydrophobic, not aromatic: I (Ile), L (Leu), M (Met), V (Val)
Hydrophilic, basic: H (His), K (Lys), R (Arg), X (methylated Lys)
Hydrophilic, acidic: D (Asp), E (Glu)
Ambivalent: A (Ala), C (Cys), N (Asn), P (Pro), Q (Gln), S (Ser), T (Thr)
No side chain: G (Gly)

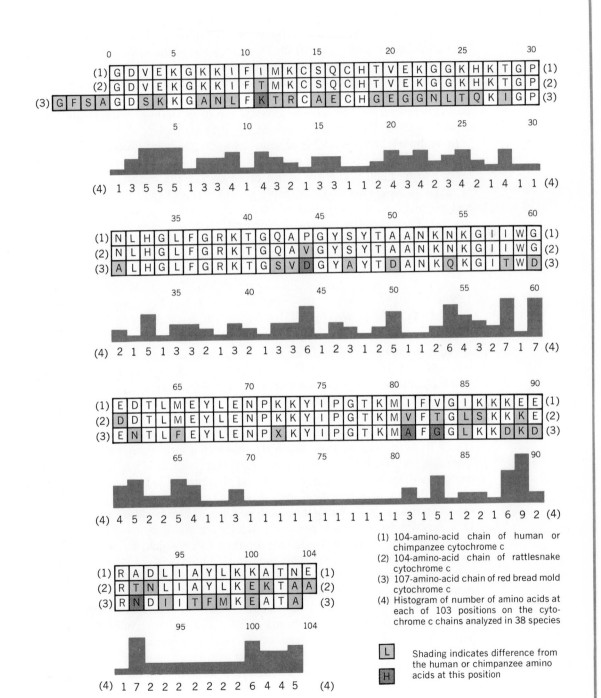

(1) 104-amino-acid chain of human or chimpanzee cytochrome c

(2) 104-amino-acid chain of rattlesnake cytochrome c

(3) 107-amino-acid chain of red bread mold cytochrome c

(4) Histogram of number of amino acids at each of 103 positions on the cytochrome c chains analyzed in 38 species

L — H
Shading indicates difference from the human or chimpanzee amino acids at this position

process of being substituted by the forces of drift, migration, and segregation distortion. This idea remains controversial. Only indirect arguments can be made, because small selective differences are exceedingly difficult to demonstrate. Yet small selective differences over large periods of time could be actively maintaining variability. One of the problems for proponents of the idea that neutral alleles contribute much of the genetic variability within populations is that there is little current variation in molecules such as cytochrome c within a species, even though the between-species variation pattern identifies a number of apparently neutral sites in the molecule that can tolerate such variation. Perhaps, within a well-tuned organism, these sites are not as neutral as they look.

In summary, there is a class of mutations, mostly point mutations affecting a single codon to create a one-amino-acid missense substitution (Figure 6.3), that appear to be neutral and are useful in measuring the evolutionary divergence of major groups. Of the considerable genetic variability in present-day species and populations, it is not clear whether only a little or a lot of it is selectively neutral. That which is present and neutral may be viewed as interesting (even esthetic) genetic variation, or irritating genetic noise, depending on one's point of view.

6.6 INDUCTION OF GENETIC CHANGE

A fuller discussion of mutations and the factors involved in **mutagenesis** is to be found in Chapters 16 and 17. Here, only some few comments on the role of nucleic acids in the production of variation will be made.

If mutations are viewed as changes in the genetic instructions, it is clear that they have to be incorporated in the vehicle that carries the message, the DNA of the chromosomes. In bacteria, this can be accomplished by transformation (Box 6.B) or by **transduction** (Box 6.H), in which a virus carries a piece of DNA from one bacterium to another, which then incorporates it into its own chromosome. Many attempts to demonstrate these phenomena in higher organisms have been made, but no unequivocal successes have been obtained, although a Russian report on success in changing the genotype for egg color in silkworms is on record. One experiment for which positive results were claimed involved injection of DNA from one breed of ducks into another, after which the recipients began producing ducklings of a different type from either original strain. But the uncertainty of origin of the experimental birds and the failure of follow-up experiments to duplicate the positive results throw doubt on this claim.

There is still another form of genetic transfer of information between cells that does not involve mating (mating will be covered more fully in Chapter 7). This involves the poorly understood intracellular particles called **episomes,** which are transferred via cell-to-cell contact without accompanying chromosomal transfer. Among the most troublesome are a class of episomes called *resistance transfer factors,* which apparently carry genes for resistance to a whole array of antibiotics. Episomic transfer is far less restricted than transduction, since genetic integration is not necessary for maintenance and expression of episomic genes. The RTF elements can be passed freely between bacteria belonging to different genera (*Escherichia*, *Salmonella*, *Shigella*), and a single episomic transfer may confer simultaneous resistance to as many as five completely different classes of antibiotics. This clearly has enormous implications for medicine as well as genetics.

In general, believers in the inheritance of acquired characters suggest that transformation of the genetic message can be obtained by means not involving DNA. One such alleged method is blood transfusions, not only in organisms such as chickens, whose red blood cells have nuclei and therefore DNA, but also in mammals, which have no nuclei in the mature red blood cells. The official party line in the USSR regarding genetics upheld the idea of inheritance of acquired characters for nearly twenty years (until 1964), but it is now apparent that the successful experiments reported there were fraudulent (see Chapter 23). No carefully controlled blood-transfusion experiments, including the one by Galton (Section 3.6), may be said to have given positive results, although there are a couple of doubtful reports from France and Switzerland. Indeed, in the light of what we now know about the chemical nature of hereditary instructions, it seems highly improbable that acquired characters can imprint themselves on the DNA transmitted to the next generations. And the fact that experiments with cutting off the tails of successive generations of mice failed to produce a tailless strain should not surprise us. After all, circumcision has been practiced by Semitic people for countless generations, yet Semitic males continue to be born with foreskins.

This is not to say that artificially produced changes in genetic instructions are not possible. Many mutagens of a physical nature (such as ionizing radiation) or a chemical nature (such as nitrous acid, among numerous other substances) are known. But all of these produce their effect by changing, in some way, the DNA of the cells, and thus the DNA of the descendents of these cells.

The genetic aspects and consequences of spontaneously or artificially induced mutations are discussed in various succeeding chapters. Important somatic considerations, however, are also involved.

A possible instance of a spontaneous change in the genetic message of somatic cells in humans has been reported. There is a virus causing tumors in rabbits that is not pathogenic to humans. Infected rabbits were found to have acquired the ability to synthesize an enzyme, a form of arginase that operates in the metabolism of the amino acid arginine, which they could not do before. Probably, genes carried by the virus somehow became incorporated into rabbit liver cells. Support for this explanation is provided by synthesis of the new kind of arginase in tissue culture infected with the virus. When blood serum of scientists who had worked with the virus for a long time was examined, indications were obtained that the genetic information coding for the new arginase had also been incorporated into their cells. This property would be transmitted to offspring only if the new DNA could find its way into the gonads, where germ cells are produced. But, if the

BOX 6.H TRANSDUCTION

Transduction was first shown with the mouse-typhoid bacterium, *Salmonella typhimurium*, in a series of experiments by N. Zinder and J. Lederberg commencing in 1949. Transduction is similar to transformation (Box 6.B) in that genetic information of one bacterial cell is incorporated into a second without direct contact between cells. Unlike transformation, transduction requires the activity of a virus to effect the transfer. An experiment demonstrating transduction is outlined as pictured and described below:

(A) The experiment is conducted in a U-tube, separated into two compartments by a filter. The filter allows free passage of nutrient medium and virus-sized particles, but does not allow bacteria to pass. A "recipient" strain of *Salmonella* is on the left, a "donor" strain on the right. One of the recipient bacteria has undergone destruction by viruses, and some of the released viruses are passing through the filter and beginning to attack the highly susceptible donor strain.

(B) Detail of a recipient-strain bacterium. The DNA of a virus (called a prophage) is passive within the cell, generally reproducing with the bacterial DNA when the cell divides, but not actively directing cell activity.

(C) However, at low frequency, the prophage can be "induced" (for instance, by ultraviolet light) to begin directing the cell to make virus coat protein. The virus DNA then quickly replicates itself many times, and each DNA strand is coated with virus protein manufactured by the cell. The bacterium is destroyed in the process, and the complete viruses are released.

(D) A few viruses pass through the glass filter and encounter bacteria of the donor strain. These bacteria are highly susceptible to them, and the process described in (C) proceeds rapidly, without induction, in the donor colony.

(E) At a low frequency, perhaps one in a million, the new viruses incorporate a bit of bacterial chromosome (the dark gray square in one virus) inside their protein coats.

(F) If one of these viruses carrying bacterial DNA makes its way back across the filter to the recipient strain it most likely will inject its DNA into a recipient bacterial cell, but the DNA will not replicate and destroy its host (i.e., it will become a prophage).

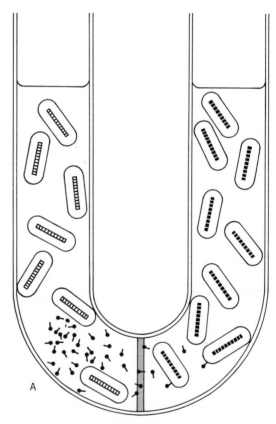

A

(G,H) Instead, the donor bit of bacterial DNA may pair with, and then become incorporated in, the recipient chromosome. If the gene from the donor is different from that in the recipient, new variability is introduced into the recipient strain, and can be selected for to demonstrate that transduction has occurred. (This is really a migration event, but tends to mimic mutation.)

A clever way to distinguish transformation and transduction is to remember that transduction is analogous to sending the DNA message in an envelope (the virus protein coat), while transformation is more like sending a postcard, the card itself (the DNA) bearing the message.

Much work on transduction has made the following generalizations available:

1. Some viruses transduce any bacterial genes, apparently incorporating any fragment of host-cell DNA of the right size into their developing protein coat.

2. Certain viruses transduce only particular sections of host-cell chromosome, and thus only certain genes. Such were used in the isolation of three genes of the *lac* operon (see Box 6.D).

3. Transduction generally occurs at a very low frequency. In certain circumstances, its frequency can be greatly increased, leading to possibilities for genetic engineering by means of virus-mediated insertion of specific genes into host organisms (Section 22.2).

4. So far, most cases of transduction and transformation have been limited to bacteria. There are recurring reports of these processes being demonstrated in higher organisms, including human cell cultures, but most of the experiments have not been satisfactorily repeated.

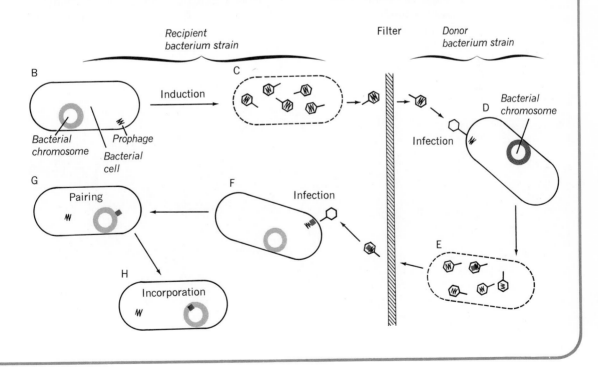

report is substantiated, this work will provide the first step towards the eventual potentiality of changing the genetic messages in a human body by what has been called euphenics and **genetical engineering** (Section 22.2).

Viruses are known to interfere with the hereditary information of cells in other ways too. For instance, at least 40 different viruses that induce formation of tumors in animals have been identified. Hereditary information normally regulates cell division. Cancer, in general, is the result of uncontrolled cell division, and any factors that interfere with the control of normal cellular growth could be carcinogenic.

To move to an even broader topic: Nucleic acids must play a role in the process of **aging.** Although the exact nature of senescence is not understood, it is clear that aging and eventual death are somehow programmed into the hereditary apparatus. One possibility that has been suggested is that somatic mutations producing metabolic blocks accumulate gradually throughout the lifetime of an individual until the loss of many vital functions causes death. Another concerns the correlation of gene redundancy (i.e., genes duplicated more than once within a cell) with life-span. It suggests that in short-lived species, a few mutations or even a single mutation could lead to cell senescence or death, while in long-lived species, many more of the important redundant genes have to be damaged or inactivated before serious cell damage occurs. These possibilities do not seem reasonable, as will be discussed further at the end of Section 7.1. However, if aging is somehow a programmed developmental process, it is easy to speculate that it might be eventually possible to replace genetic messages as the body ages, and thus produce immortal people. *Homo continuus,* as we might then rename our species.

7

BIOLOGICAL COMMUNICATION BETWEEN GENERATIONS

Before the genetic principles mathematically described by Mendel could be understood biologically, it was necessary to understand what cells are and how they divide. The conceptual separation of the genetic constitution of an organism from the particular traits and characteristics that the organism develops in its environment was a major insight. This chapter concludes with an account of the interaction of genotype and environment in influencing both the expression and evolution of several blood abnormalities in humans, with particular attention being given to sickle-cell anemia.

7.1 CELL DIVISION

Most bacteria reproduce by simple fission, although some strains can reproduce sexually as well as asexually. When reproduction is asexual, each daughter bacterial cell receives a set of genetic instructions from the mother cell. The instructions are coded in the DNA of the bacterial chromosome (see Box 2.C), which replicates itself prior to cell division. In somatic cell division of higher organisms the process is more complex, but the principle is the same: the hereditary apparatus replicates itself and each daughter cell receives the full set of the genetic information contained in the chromosomes of the parent cell. The number of chromosomes of normal somatic cells remains constant. The process of replication and distribution of chromosomes to daughter cells is called **mitosis.**

In the sexual reproduction of higher organisms, however, this process for the transmittal of genetic messages would lead to an absurd situation. If both of the germ cells (**gametes,** in higher organisms called egg and sperm) supplied by the parents contained all of their chromosomes, the resulting zygote would have a double number of them, the next generation four times as many, and so on *ad infinitum.* This does not happen. Instead, during the special cell divisions called **meiosis,** which lead to gamete formation, there is a reduction division that halves the number of chromosomes in a highly regular fashion, so that each gamete is provided a single set of chromosomes. The original number is reconstituted at fertilization, giving the zygote two full sets of chromosomes. Thus, both the parents and offspring are **diploid** (two sets of chromosomes) and the gametes are **haploid** (one set). Mitosis and meiosis are illustrated in Box 7.A.

It is important that in meiosis each pair of chromosomes behaves independently of the other pairs at the time of aligning in the center of the nucleus. Thus, if we label four chromosomes of some animal **A, B, C,** and **D,** and assign subscripts **s** and **d** to indicate their origin from the sire (by way of his sperm) and the dam (by way of her egg), a zygote would have a diploid ($2N$) chromosomal formula of $\mathbf{A_s A_d B_s B_d C_s C_d D_s D_d}$. When this zygote becomes an adult and starts producing gametes, only one member of each pair of chromosomes is distributed to each germ cell. The number of chromosomes in each germ cell is haploid (N).

Because different chromosomes **segregate independently** of each other, different kinds of gametes are produced. In our example, sixteen kinds are possible. Note that the immediate parent of these gametes is the zygote grown to adulthood, and the sire and dam of that adult are the grandsire and granddam of the gametes:

Grandsire's sperm	Granddam's egg	Fertilization	Zygote
$A_s B_s C_s D_s$ +	$A_d B_d C_d D_d$	$\longrightarrow$	$A_s A_d B_s B_d C_s C_d D_s D_d.$

Then, when the zygote has developed into an adult, many meioses produce the following kinds of gametes:

All chromosomes from the grandsire	One from granddam three from grandsire	Two from each grandparent	Three from granddam one from grandsire	All chromosomes from the granddam
$A_s B_s C_s D_s$	$A_d B_s C_s D_s$	$A_d B_d C_s D_s$	$A_s B_d C_d D_d$	$A_d B_d C_d D_d$
	$A_s B_d C_s D_s$	$A_d B_s C_d D_s$	$A_d B_s C_d D_d$	
	$A_s B_s C_d D_s$	$A_d B_s C_s D_d$	$A_d B_d C_s D_D$	
	$A_s B_s C_s D_d$	$A_s B_d C_d D_s$	$A_d B_d C_d D_s$	
		$A_s B_d C_s D_d$		
		$A_s B_s C_d D_d$		

Each of these sixteen gametic chromosomal complements is equally likely among the many eggs produced by a female (or the many many sperms produced by a male) having the chromosomes $A_sA_dB_sB_dC_sC_dD_sD_d$. The number of different kinds of gametes possible may be predicted by the expression 2^N, with the exponent (N) being the haploid number of chromosomes (note that $2^4 = 16$). We shall see later that in reality the potential number of kinds of gametes is tremendously greater because of crossing-over, an exchange of parts between members of a chromosome pair such as A_s and A_d. If fertilization is at random, that is to say, if chance governs which of the many different gametes produced by an individual unites with a randomly chosen gamete from another parent, it is easy to compute the proportions of the various chromosomal complements expected among the progeny. The number of diploid chromosomal complements possible in our example is 3^N or 81. (Note that this is also the expression used to calculate gene combinations in Box 4.E.)

This example, in which only four pairs of chromosomes are segregating, demonstrates that the independent segregation of chromosomes in different meioses within the same parent produces considerable variability among the gametes of that parent, and the different chance combinations of two such gametes from different parents produces even greater variability among the offspring. If most humans contain variability in all 23 of their chromosome pairs, as indeed we think they do, then the chance that two offspring of the same two parents could receive the same chromosome complement (unless they are identical twins) is only 1 in 3^{23} (a vanishingly small probability). (And we are again ignoring the immense contribution to variability made by crossing-over.)

There is increasing evidence that aging in humans, and in most organisms, is somehow related to cell divisions. True, medicine has made great advances in increasing the average age of the human population, by reducing mortality due to injury or disease. But there seems to be a somewhat elastic upper limit to human life, in the neighborhood of 80–100 years, that medicine has had little success in extending. Research with normal human cell cultures consistently shows that cells from young humans die after about 50 divisions. (Fifty divisions are quite sufficient to produce a human. In fact, 50 doublings of a culture begun with a single human cell, if no cells were discarded, would produce about 20 million tons of cells.) Cells from older humans die after fewer divisions. Various theories of aging, or maturation, have in common the idea that a cell accumulates changes, or errors, or damage, and that this change is genetic. Time *per se* does not automatically produce these effects, as cells in culture that are inhibited from dividing for extended periods by cold storage, upon being thawed, resume dividing. They complete about the same number of divisions that they would have, had division not been interrupted, and then they die. Occasional exceptions are found, such as the widely used HeLa human cell culture, which appears to be immortal. Interestingly, these exceptional cell cultures often behave like cancer cells.

The restoration of juvenility, like the Fountain of Youth, has been sought but not yet found. Maturation is poorly understood. In animals, the cells that produce eggs and sperm are generally set aside early in embryo development, and take little part in the development and functioning of the animal. Perhaps

BOX 7.A MITOSIS AND MEIOSIS

The life cycle of diploid sexually reproducing organisms can be diagrammatically illustrated as shown below. (♂ is the symbol for male; it represents the shield and sword of Mars. ♀ is the symbol for female; it represents the mirror of Venus.)

A simplified summary of the important events of mitosis and meiosis emphasizing similarities and differences in the two types of cell division is presented on the facing page. A single pair of chromosomes is shown in the illustrations. Most organisms have many pairs of chromosomes.

The difference between meiosis in males and in females should be noted: The premeiotic nucleus in the male divides twice, producing four similar-sized sperm cells. In the female there are also two divisions, but the cytoplasm is divided very unequally. Most of it, and one of the four nuclei, form the egg cell. The remainder, plus the other three nuclei, form much smaller **polar bodies** and are abortive (i.e., do not take part in fertilization and zygote formation—see Figure 2.4).

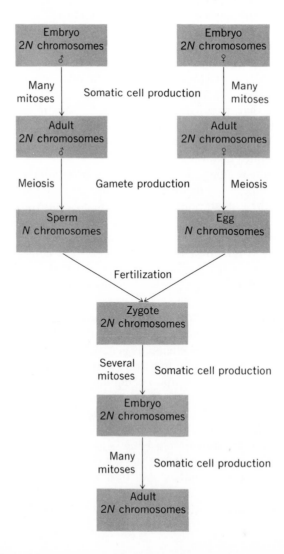

	Mitosis *(two cells are formed from one)*	Meiosis *(gametes are formed)*	

| | One 2*N* nucleus in interphase—DNA doubles | One 2*N* nucleus in interphase—DNA doubles |

Typical chromosome

Arm

Centromere

Chromosomes shorten

Chromosomes shorten

Chromosome arms divide

Chromosome arms divide

Mitotic chromosomes do not pair, but proceed immediately to ⌐

Chromosomes pair, with centromeres and arms of the pairs exactly aligned

All pairs of chromosomes align in the center of the nucleus

Centromeres, and then the chromosome arms, separate

Note—the number of centromeres is being reduced at this point in meiosis

Two nuclei

Unpaired chromosomes align in the center of the nucleus

Unpaired chromosomes align in the center of the nucleus

Centromeres divide

Centromeres divide

Note—the number of chromosome arms is being reduced at this point in meiosis

Centromeres, and then the chromosome arms, separate

Centromeres, and then the chromosome arms, separate

Two nuclei (each 2*N*)

Four nuclei (each *N*)

Interphase nuclei of two somatic cells

Interphase nuclei of four gametes

these germ-line cells remain juvenile while the rest of the animal ages. In perennial plants such as redwood trees, however, the germ line is not kept separate in the developing body of the tree. The cells at the growing point (the apical meristem) produce shoots and leaves that, in each subsequent year, are physiologically more mature. However, these same apical meristem cells produce flowers, and it is within the mature flower tissue that the special cell division, meiosis, takes place. When the meiotic products, the egg and pollen nuclei, are united, the resulting embryo is juvenile. Thus it appears that some process associated with the meiotic division can repair the accumulated genetic damage, or reverse the genetic changes, in the aged premeiotic cells.

7.2 SIMPLE MENDELISM

The elementary statistical consequences of a binary scheme of inheritance (i.e., based on segregating pairs) were described for garden peas by Mendel (Figure 7.1) in 1865. The significance of Mendel's results was not generally appreciated for 35 years (although a Russian botanist, I. F. Schmalhausen, had stressed their importance soon after they were published). It was not until 1900 that independent experiments carried out by the Dutch biologist Hugo deVries, the German botanist Carl Correns, and the Austrian plant breeder Erich Tschermak (who, however, failed to interpret his findings properly) led to general recognition of the fact that Mendel had, indeed, provided the basis of formal transmission genetics. To round out the international character of the rediscovery of Mendelism, it was the French zoologist L. Cuenot and biologists **William Bateson** (British, 1861–1926) and W. E. Castle (American) who, within a few years, accumulated data to show that animals as well as plants follow Mendel's laws. Bateson (Figure 7.2) actually coined the term *genetics* (in 1906), and contributed to our knowledge of many ramifications of Mendelian inheritance.

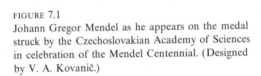

FIGURE 7.1
Johann Gregor Mendel as he appears on the medal struck by the Czechoslovakian Academy of Sciences in celebration of the Mendel Centennial. (Designed by V. A. Kovanič.)

BOX 7.B NOBEL LAUREATES IN GENETICS

The annual Nobel Prizes were established by Alfred Nobel, the inventor of dynamite. There is no prize specifically for genetics, or for biology for that matter. However, a number of geneticists have received the Nobel Prize in Physiology and Medicine for contributions to fundamental understanding of genetics, with most of these contributions developing the topics discussed in Chapter 6.

Year of award	Nobel Laureate	Nationality	In recognition of
1933	T. H. Morgan (1866–1945)	U.S.	Theory of the gene
1946	H. J. Muller (1890–1967)	U.S.	Radiation and mutagenesis
1958	G. W. Beadle (1903–)	U.S.	The function and
	E. L. Tatum (1909–1975)	U.S.	nature of the gene
	J. Lederberg (1925–)	U.S.	
1959	S. Ochoa (1905–)	U.S. (Spanish-born)	Extracellular
	A. Kornberg (1918–)	U.S.	synthesis of DNA
1962	F. H. C. Crick (1916–)	British	
	M. H. F. Wilkins (1916–)	British	Structure of DNA
	J. D. Watson (1928–)	U.S.	
1965	A. M. Lwoff (1902–)	French	Theory of gene
	J. L. Monod (1910–)	French	regulation
	F. Jacob (1920–)	French	
1968	R. W. Holley (1922–)	U.S.	
	F. G. Khorana (1922–)	U.S. (Indian-born)	The genetic code
	M. W. Nirenberg (1927–)	U.S.	
1969	M. Delbrück (1906–)	U.S. (German-born)	Foundations of molecular
	A. D. Hershey (1908–)	U.S.	genetics
	S. E. Luria (1912–)	U.S. (Italian-born)	
1970	N. E. Borlaug (1914–)	U.S.	See below
1973	K. von Frisch (1887–)	German (Austrian-born)	Behavior studies
	K. Lorenz (1904–)	German (Austrian-born)	
	N. Tinbergen (1907–)	English (Dutch-born)	

While Borlaug did not make a fundamental discovery in genetics, he did apply principles of evolution and genetics to oversee the "green revolution," for which he received the Nobel Peace Prize (see section 15.2). Other Nobel Laureates have been actively engaged in genetic investigations. For instance, Linus Pauling, one of two persons to have received two Nobel Prizes (he was awarded one for Chemistry, one for Peace), contributed to an understanding of sickle-cell anemia when he and his students separated abnormal hemoglobin from normal by electrophoresis (see Figure 7.4).

126

FIGURE 7.2
In 1956, John Kimber, a successful and public-spirited California
poultry breeder, in appreciation of the significance of genetics,
established the award of a medal by the U.S. National Academy of
Sciences to geneticists for meritorious work. The medal portrays
Charles Darwin, Gregor Mendel, William Bateson, and Thomas Hunt
Morgan. Among the contributors to genetics mentioned in this book,
the following have been recipients of the medal: W. E. Castle,
H. J. Muller, S. Wright, Th. Dobzhansky, T. M. Sonneborn, G. W.
Beadle, J. B. S. Haldane, C. Stern, N. W. Timofeev-Resovsky, and
B. McClintock. (Courtesy of John Kimber.)

We may illustrate the operation of Mendelian principles by considering the
transmission of sex-determining instructions in humans. Figure 7.3 is a photomicrograph of human chromosomes. Twenty-two pairs of the chromosomes are
common to both sexes: these are called **autosomes.** Normally, the two chromosomes of a pair have the same genes (although they may have different alleles),
and are called *homologous* pairs to indicate this gene-by-gene similarity. The
twenty-third pair comprises the sex chromosomes and is different in the male and
the female. The female has a homologous pair of **X chromosomes** (subscripted 1
and 2, on page 127, to indicate that they probably have different alleles for
some of their genes). The male has only one X. It pairs with the **Y chromosome,**

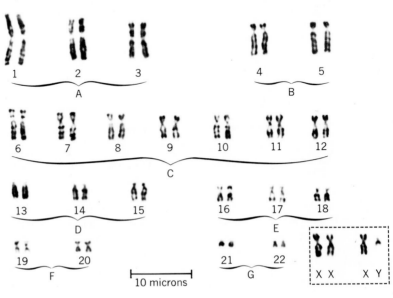

FIGURE 7.3
The 22 pairs of human
autosomes, plus the XX and
XY pairs that determine
femaleness or maleness, respectively. The chromosomes
have been stained to intensify
their different banding patterns,
which, along with size and
location of the centromere,
aid in the identification of
homologous pairs. (Courtesy
J. Tischfield.)

which is different in size and shape. It is this Y chromosome that carries the in-structional information that a zygote containing it should develop into a male rather than into a female.

In the course of human reproduction, the sex chromosomes segregate and recombine as shown here:

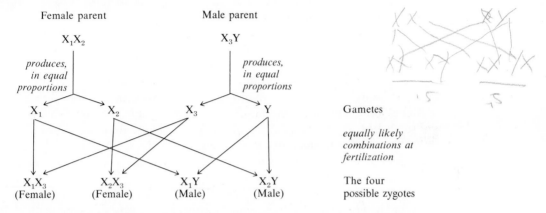

Female parent	Male parent
X_1X_2	X_3Y

produces, in equal proportions *produces, in equal proportions*

X_1 X_2 X_3 Y Gametes

equally likely combinations at fertilization

X_1X_3 (Female) X_2X_3 (Female) X_1Y (Male) X_2Y (Male) The four possible zygotes

The diagram should not be taken to imply that if two people mate and produce four offspring, they will necessarily have two male and two female offspring. If a large enough group of offspring (produced by many couples) is looked at, however, an approximate **1:1 ratio** of females to males is likely. This is the simplest Mendelian ratio, corresponding to that found in a **backcross** or the testcross (the mating of an individual having two like alleles or chromosomes with an individual having two unlike alleles or chromosomes).

The discreteness of the inheritance, or its particulate nature, is demonstrated by the fact that the genetic message indicating maleness remains intact as it passes from grandfather, through father, to son. It is not altered by the presence, in the zygote or adult, of the X chromosome from the male's mother. Genetic messages carried by individual genes on the autosomes and on the X chromosomes also remain intact throughout inheritance (with occasional but important exceptions).

Determination of human maleness by presence of the Y chromosome is similar to another Mendelian property, **dominance,** which will be discussed briefly here, and more fully later in this section and in Section 7.3. When an organism has two different alleles of the same gene, it is said to be **heterozygous** for that gene. For some genes, only one of the two unlike alleles expresses itself in the phenotype of the heterozygote. When this happens, the allele that is expressed in the heterozygote's phenotype is termed dominant; the allele not expressed, **recessive.**

The physical or conceptual location of the gene (or of the alternative alleles of the gene) within the structure of the chromosome is called the **locus** (plural, loci). The old idea of genes and chromosomes used the analogy of beads on a string. Modern ideas are more complex, but the beads-on-a-string analogy emphasizes the important features that chromosomes are linear, the genes are discrete, and adjacent genes do not overlap. Since there are in humans many thousands of genes and only 23 pairs of chromosomes, it follows that each chromosome includes a great many genes. Genes in the same chromosome are said to be **linked.** They tend to be transmitted through meiosis to the gametes in blocks, except that they are subject to crossing-over, the exchange of sections between homologous chromosome pairs (see Box 9.B).

It is not definitely known what function other than sex determination the Y chromosome performs in human males (see Sections 8.3 and 8.4). Very few genes have been identified as being located in the human Y chromosome, and most of these are disputed (see Box 8.A). No crossing-over between the X and the Y normally occurs, and therefore genes from the X are not transferred to the Y. Genes located in the X chromosome, and the possibly very few genes in the Y, are **sex-linked.** In the female, sex-linked genes, like autosomal genes in either sex, may be in the **homozygous** state (the genes at the same locus on the two X's are identical) or the heterozygous state (the individual has two different alleles of the gene). A recessive sex-linked allele is masked by its dominant allele in a heterozygous female. Since a male has only one X chromosome and the Y carries few or no homologous alleles, sex-linked genes are expressed in a male no matter whether they are dominant or recessive in the female. In the **heterogametic** sex (the male in humans), these unpaired genes are in the **hemizygous** state. We may note, by mentally extending the diagram on page 127 another generation, that the X chromosome of the grandfather (X_3 in our example) may be passed through his daughter to a grandson, but can not be passed directly to a son. Box 7.C shows an example of sex-linked inheritance. Note in particular the difference in outcome of mating 2 and the **reciprocal mating** 3.

We turn next to autosomal inheritance, which is the type that Mendel himself studied. If we designate two alleles A_1 and A_2 and observe the offspring of a cross between two homozygotes A_1A_1 and A_2A_2 (the **P** or parental generation), we see that all of their offspring (the first filial generation: F_1) are heterozygotes of genotype A_1A_2. Here, contrary to what we observed for sex-linked traits (Box 7.C), reciprocal crosses do not produce different results. Two of these F_1 heterozygotes mated with each other will, as Mendel discovered, produce an array of three different genotypes in the next generation (F_2):

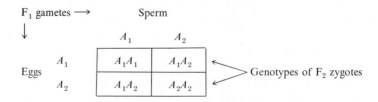

In a large F_2 population, the three genotypes will be very close to a ratio of **1:2:1.** Unless there is dominance, this will also be the F_2 phenotypic ratio. In Shorthorn

BOX 7.C SEX-LINKED INHERITANCE IN HUMANS

Over 120 phenotypic characteristics that are believed to depend on sex-linked genes have been described in humans, and more than half of them have been clearly proved to show such inheritance. One is the inability to distinguish green from red. We designate the allele for normal vision as B, and the one for red-green color blindness as b. The six possible mating combinations between people of different genotypes, and their expected offspring, are given in the table below.

Because this defect is sex-linked, we expect that its incidence would be much higher in males, an expectation borne out by observation.

In mating 5, we may observe that there are two genotypes among the female offspring, but they are phenotypically indistinguishable. One way to tell these genotypes apart is to analyze their progeny. The homozygotes "breed true," but the offspring of the heterozygotes show phenotypic segregation. Thus $X^B X^B$ females who mate exclusively with color-blind males produce only normal daughters; about half of the female offspring of $X^B X^b$ females who mate exclusively with color-blind males are color blind. But no matter what the genotype of the mate, $X^B X^B$ females have all normal sons (see crosses 1 and 2) while about half of the sons of $X^B X^b$ are normal and about half are color blind (see crosses 5 and 6).

Sex	Matings					
	1	*2*	*3*	*4*	*5*	*6*
♀	Homo-zygote normal ×	Homo-zygote normal ×	Homo-zygote color blind ×	Homo-zygote color blind ×	Hetero-zygote normal ×	Hetero-zygote normal ×
♂	Hemi-zygote normal	Hemi-zygote color blind	Hemi-zygote normal	Hemi-zygote color blind	Hemi-zygote normal	Hemi-zygote color blind
	Genotypes of parents					
♀	$X^B X^B$	$X^B X^B$	$X^b X^b$	$X^b X^b$	$X^B X^b$	$X^B X^b$
♂	$X^B Y$	$X^b Y$	$X^B Y$	$X^b Y$	$X^B Y$	$X^b Y$
	Gametes from					
♀	X^B	X^B	X^b	X^b	X^B, X^b	X^B, X^b
♂	X^B, Y	X^b, Y	X^B, Y	X^b, Y	X^B, Y	X^b, Y
	Genotypes of offspring					
♀	$X^B X^B$	$X^B X^b$	$X^B X^b$	$X^b X^b$	$X^B X^B, X^B X^b$	$X^B X^b, X^b X^b$
♂	$X^B Y$	$X^B Y$	$X^b Y$	$X^b Y$	$X^B Y, X^b Y$	$X^B Y, X^b Y$
	Phenotypic ratios of offspring					
♀	All normal	All normal	All normal	All color blind	All normal	1:1
♂	All normal	All normal	All color blind	All color blind	1:1	1:1

cattle, for instance, red coat color characterizes one homozygote, the heterozygote is roan, and the other homozygote white. If in the P generation crosses are made between red animals and white animals, all of the F_1 will be roan. In the F_2, the expected phenotypic ratio is 1 red:2 roan:1 white. The genotypes (and phenotypes) of the three classes of animal are r_1r_1 (red), r_1r_0 (roan), and r_0r_0 (white). Apparently a single dose of allele r_1 cannot manufacture enough of the enzyme needed to produce full redness, and allele r_0 does not code for an enzyme that produces red pigment. Therefore, the heterozygote is a dilute red, or roan, color.

If there is dominance, one of the homozygotes and the heterozygote are phenotypically undistinguishable, and the common phenotypic F_2 ratio of **3:1** is obtained. Dominance can be biochemically explained as the capability of a single dose of an allele to manufacture enough of the primary gene product to permit full expression of the dominant phenotype.

Mendel reported results from crosses of garden peas. He investigated seven independent sets of alternative characters, four of which are shown below. (It is customary to designate dominant alleles by capital letters, and recessives by the lower-case form of the same letter.)

Alternative characters of parents	F_1 *genotype*	F_1 *phenotype*	F_2 *phenotypes*	F_2 *genotypes*	*Observed phenotypic percentages in F_2*
Smooth *vs* angular wrinkled seed	*Ss*	smooth	5,474 smooth 1,850 angular	*SS, Ss* *ss*	74.74 25.26
Yellow *vs* green albumen	*Yy*	yellow	6,022 yellow 2,011 green	*YY, Yy* *yy*	75.06 24.94
Long *vs* short stem length	*Ll*	long	787 long 277 short	*LL, Ll* *ll*	73.96 26.04
Grey-brown *vs* white seed color	*Gg*	grey-brown	705 grey-brown 224 white	*GG, Gg* *gg*	75.90 24.10

Notice that the first two crosses included many more F_2 offspring than the last two, and that the F_2 percentages of the first two crosses come closer to the 3:1 ratio theoretically expected. Statistical laws indicate that the larger the number of observations, the closer the outcome will be to the expectation, provided, of course, the prediction was based on correct premises.

The discussion so far has been based on examples in which it is assumed that for each locus there are only two different alleles. Normally, one diploid organism can have only one or two alleles at each of its loci. But in a population, there may be numerous alleles for a single locus. A certain blood-group locus in cattle appears to have some 250 alleles. In sweet clover, over 200 alleles provide a genetic system to make the plants self-sterile, but for most combinations, cross-fertile. (For example, if a flower is of genotype $s_{37}s_{116}$, a pollen grain carrying either allele s_{37} or s_{116} will fail to fertilize the egg cells in that flower, and thus self-fertilization is not possible. Most pollen from other sweet clover plants will have other alleles, and thus the flower will be successfully cross-fertilized. Similarly, when the pollen from the plant, which carries either s_{37} or s_{116}, lands on flowers of other plants which do not have that allele, it will be able to cross-fertilize those flowers.) In *Drosophila melanogaster,* a group of three neighboring loci have numerous eye-color alleles with such flashy names as apricot, blood, buff, cherry, coral, ecru, eosin, honey, ivory, peach, and satsuma, and for nomenclatural contrast, white.

For some characters, the phenotype of the heterozygote is not the same as that of one of the homozygotes. It may be exactly intermediate between those of the two homozygotes, or it may be between them but closer to one than to the other. Phenotypic expression of the latter sort is called **partial dominance,** and if it is exactly intermediate, there is said to be no dominance. In the cattle example described earlier in this section, roan coat color is an example of such intermediate phenotypic expression in the heterozygote. A phenotype in which both alleles are expressed, as in the human ABO blood-group system, is also possible. Here the A and B alleles show **codominance,** as both A and B antigens are clearly and distinctly expressed in blood of AB individuals.

The distinction between these different kinds of effects may be complicated. For example, the allele for hemoglobin S can be viewed as dominant if resistance to malaria is considered: individuals whose genotype is Hb^AHb^S or Hb^SHb^S are resistant to malaria, while Hb^AHb^A individuals are highly susceptible. If anemia and the sickling phenomenon of the blood cells under low oxygen pressure are considered, however, heterozygotes show no dominance or partial dominance, depending on how the mild anemia and reduced sickling of the heterozygotes is quantified compared to the severe anemia and extensive sickling of Hb^SHb^S genotypes and to the absence of anemia and sickling in Hb^AHb^A genotypes. If hemoglobin production is considered, the alleles are codominant, since the heterozygotes produce both hemoglobin A and hemoglobin S (as indicated in the tabulation below) but in somewhat different amounts (Figure 7.4).

Phenotype	Genotype	Hemoglobin electrophoretic pattern Origin +4 +2 0 ↓ ↓ ↓ ↓	Hemoglobin types present
Sickle-cell trait	Hb^SHb^A		S and A
Sickle-cell anemia	Hb^SHb^S		S
Sickle-C heterozygote	Hb^CHb^S		C and S
Hemoglobin-C trait	Hb^CHb^A		C and A
Hemoglobin-C disease	Hb^CHb^C		C
Normal	Hb^AHb^A		A

(From D. L. Rucknagel and R. K. Laros, Jr. "Hemoglobinopathies: Genetics and Implications for Studies of Human Reproduction." *Clinical Obstetrics and Gynecology* 12:49–75, 1969.)

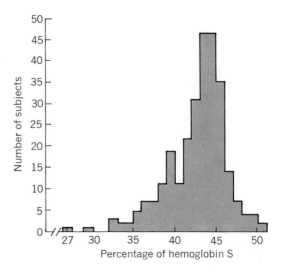

FIGURE 7.4
Distribution of the percentage of hemoglobin S in 272
subjects with sickle-cell trait. Each observation is the
average of duplicate measurements. (From W. E. Nance
and J. Grove, "Genetic Determination of Phenotypic
Variation in Sickle Cell Trait." *Science* 177:717.
Copyright © 1972 by American Association for the
Advancement of Science.)

7.3 GENOTYPE AND PHENOTYPE

Because of dominance, we would not be able to determine all genotypes unambiguously by observing the phenotypes among the females in Box 7.C or among the peas that Mendel investigated. This phenomenon of dominance is basically an interaction between alleles at the same locus. Another form of such allelic interaction that can have important evolutionary implications is **overdominance,** in which the expression of the character in the heterozygote is outside the range of average expression of the phenotypes of the two homozygotes. Suppose that in some organism there is a locus controlling height. Various forms of allelic interaction that might govern the expression of this trait are shown in the following diagram:

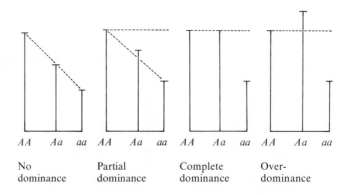

Dominance sometimes is a matter of the scale on which the measurement is made. For instance, if the expression of A_1A_1, A_1A_2, and A_2A_2 is measured as

121 cm, 36 cm, and 1 cm, A_2 is partially dominant. Should we choose to transform the centimeter measurements to logarithms, the figures change to 2.08, 1.56, and 0, and we have a reversal: now A_1 is the partially dominant allele. If we use square roots instead, the numbers become 11, 6, and 1, and no dominance is observed. However, no simple scale transformation can eliminate overdominance, nor change partial dominance to complete dominance.

Another source of variation between a genotype and its phenotypic expression is environmental. If a gene does not have a suitable environment in which to produce a specific substance, its phenotype may not appear as expected. Thus, the gene that makes a man sensitive to ragweed pollen would cause him to exhibit the character in the Midwest, but not in Antarctica where he would not be exposed to the irritant. Fuller implications of such genotype-environment interactions will be explored in Chapters 10 and 11.

Some ambiguity between phenotype and genotype accrues from the phenomenon of incomplete **penetrance.** Some genotypes do not, for a variety of reasons, have a phenotypic manifestation in all individuals possessing them. A number of such incompletely penetrant genes are known in humans. For instance, a gene that causes a bent and stiff little finger belongs to this category. Some heterozygous people have the bent stiff finger, others do not.

Finally, another form of gene interaction can also be responsible for modification of the predicted expression of a gene in the phenotype. This is interaction between genes at different loci, or epistasis, a subject that is discussed in Section 9.1.

7.4 DEFECTIVE HEMOGLOBINS REVISITED

At this point, we turn to the consideration of an important instance of fitness overdominance in humans. It was noted in Section 7.2 that the Hb^S allele may be considered to be dominant, partially dominant, or codominant, depending on which phenotypic property is examined. We shall now see that when fitness, or reproductive capacity, is considered, this locus exhibits overdominance in certain environments.

The group of diseases caused by defective hemoglobins includes, among others, (1) sickle-cell anemia, in which persons homozygous for the recessive allele show a severe and often lethal effect, (2) **hemoglobin C disease,** a milder form of anemia, and (3) **thalassemia,** for which many genes are imperfectly known, including one for β-thalassemia, an anemia caused by defective hemoglobin and by a partial suppression of the formation of the β-chain of normal hemoglobin. (Still another anemia-producing genetic defect is controlled by a number of sex-linked alleles that specify deficient production of the enzyme **glucose-6-phosphate dehydrogenase,** or **G6PD.** Under ordinary conditions, carriers of one of the G6PD deficiency alleles are normal, but if an antimalarial drug, primaquine or sulfanilamide, is administered to them, they will develop anemia. G6PD disease can also be generated in carriers by naphthalene or by inhaling pollen of the broad bean, *Vicia faba*. Indeed, the disease, which is called favism in Italy, was originally associated with prolonged diets of raw broad beans.

These various diseases are found in many populations of Africa, Asia, and the Mediterranean Basin, but not among northern Europeans, Japanese, American indians, and certain other populations. Computations based on current frequencies of the sickle-cell trait among West Africans suggest that at least 22 percent of the Negroes who were brought to the American continent two or three centuries ago had it. The present frequency of the trait among American blacks is about 10 percent. This reduction in frequency can be accounted for by independent estimates of the amount of gene interchange between American blacks and other racial groups (see Section 18.4), plus the changed selection against the Hb^S gene in the North American environment.

The remarkable thing about these diseases is their high frequency in many populations, despite the fact that they can be exceedingly severe and that natural selection would be expected to operate against the persons homozygous for the recessive alleles. Some examples of current frequencies, compiled from various sources, may be given. The frequency of the sickle-cell trait in many West African populations is about 20 percent, and especially along the rivers, nearly 40 percent of some populations are heterozygotes for Hb^S. A population in Ghana includes 27 percent Hb^C heterozygotes. On mainland Italy the frequency of heterozygotes for β-thalassemia reaches 20 percent and in some Sardinian populations it is 38 percent. In Israel, there are great differences in G6PD frequency; it is less than a half percent among male Israelis of European origin, and 60 percent or higher among Kurdish jews (Section 13.1).

These frequencies of hemoglobin and enzyme phenotypes reflect the different allele frequencies of different communities, demes, or gene pools. Since we are dealing with diploid organisms, the number of genes of a given locus in a population is twice the number of individuals. Allele frequency, then, refers to the proportion that a given allele constitutes of the total number of all alleles of a given locus in a population. For sex-linked genes, such as those specifying production of G6PD, the allele frequencies in males are easily determined since a phenotypic census gives a direct allele-frequency estimate. The concept of allele frequency as applied to autosomal genes is further developed in Chapter 13. Meanwhile, we can make use of estimated allele frequencies in connection with studies of the geographical distribution of the alleles.

When frequencies of the anemia-producing alleles are examined, it becomes apparent that they are high in areas of high incidence of malaria and low, or even zero, where malaria is not indigenous. For example (Box 7.D) in Sardinia, the frequencies of the alleles for β-thalassemia and G6PD deficiency are high in coastal areas where malaria is common, and are low in mountainous regions, which are free of malaria.

One form of malaria is caused by a protist parasite, *Plasmodium falciparum,* that is carried by infected mosquitoes and that enters the human bloodstream while the mosquito is biting. After a person has been bitten, red blood cells become infected and, if the infection becomes severe, the malarial symptoms of

The diagram illustrates the correlation observed in Sardinian villages between altitude (and the concomitant incidence of malaria) and the frequencies of the β-thalassemia and G6PD-deficiency alleles. Each unshaded circle gives the average allele frequency for the β-thalassemia allele in a group of villages at the altitude indicated; the shaded circles show similar figures for the G6PD-deficiency allele.

Of particular interest are Carloforte and Usini, shown by smaller circles. In these two villages the low frequencies of the β-thalassemia and G6PD-deficiency alleles are related to the geographical origins of the inhabitants. Carloforte is a coastal village close to the malarial plains, and would therefore be expected to have a high incidence of the defective alleles. The actual low incidence can be explained by these facts: the village was only established in 1700 and its founders were Genovese fishermen from nonmalarial areas, and, until recently, it was reproductively isolated. Indeed, according to genealogical records, the few carriers of β-thalassemia and G6PD-deficiency alleles there are of Sardinian ancestry.

Usini is another village with a relatively lower frequency of the defective alleles than expected, and for the same general reason. Its inhabitants are also of non-Sardinian origin, deriving from ancestors who emigrated from Genoa and Spain.

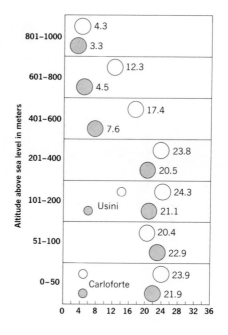

In fact, the local dialect shows definite influences of the Catalan language. More intermixture with Sardinians has taken place here than in Carloforte, as is reflected by the allele frequencies shown.

THE DIAGRAM IS ADAPTED FROM M. SINISCALCO ET AL.

chills, fever, and anemia follow. Carriers of defective hemoglobins, however, usually do not show these symptoms. It seems that their red blood cells do not provide as suitable a host for the parasite. One likely explanation is that the average life-span of the heterozygote red blood cell is reduced, compared to $Hb^A Hb^A$ cells, making it less likely that enough of the parasites can develop to produce malarial symptoms. The British biologist A. C. Allison infected volunteers with malaria, and found that 14 out of 15 normal persons (i.e., of genotype $Hb^A Hb^A$) developed significant symptoms of the disease, but only two out of 15 carriers ($Hb^A Hb^S$) of the sickle-cell allele did.

The picture with respect to fitness in a malarial environment is then as follows. Homozygotes for the hemoglobin variant alleles are at a selective disadvantage because they have severe anemia. Homozygotes for normal hemoglobin are at a disadvantage because they are subject to malaria. The heterozygotes, however, rarely develop worse than mild anemia and are much less susceptible to malaria. Thus, there is fitness overdominance. That is, in malarial environments, $Hb^S Hb^A$ heterozygotes leave more offspring than either severely anemic $Hb^S Hb^S$ homozygotes or malaria-prone $Hb^A Hb^A$ homozygotes. Both Hb^S and Hb^A alleles are passed on to the children of the heterozygotes, thus maintaining a high frequency of Hb^S in such populations.

This situation provides an example of selection in humans continuing to operate in our time. When people migrate to nonmalarial regions, or when the mosquito vectors are removed from a malarial environment by application of DDT or other methods, the heterozygotes lose their advantage and Hb^S frequencies begin to drop.

Comparison of the distributions of the Hb^S and Hb^C alleles indicates that where they coexist, the latter tends gradually to replace the former: hemoglobin C also affords antimalarial protection, but does not have such drastic effects when it is in the homozygous state as does hemoglobin S (most $Hb^S Hb^S$ homozygotes die long before reaching reproductive age).

In general, these diseases illustrate the interaction between human biological and cultural evolution. S. L. Wiesenfeld has presented interesting evidence in

BOX 7.E INFLUENCE OF ABNORMAL HEMOGLOBINS ON REPRODUCTION

Maternal hemoglobin	Fertility	Abortion rate	Maternal morbidity[1]	Maternal mortality	Perinatal mortality	Clinical severity (nonpregnant)[2]
S	Decreased	Increased	Increased A, M, S, C, P, Ph, Pn	10–20%	20–30%	Severe
A and S	Normal	Unchanged	Slight increase Pn	Unchanged	Unchanged	Negligible
C and S	Normal	Unchanged	Increased A, M, S, C, P, Pn, Hp	2–10%	Slight increase	Negligible to moderate
C	Normal	Unknown	Slight increase A, M	Unchanged	Unchanged	Mild

[1]A: anemia; M: megaloblastic crisis; S: acute sequestration; C: bone, abdominal, and cerebral crisis; P: pneumonia; Ph: antepartum hemorrhage; Pn: pyelonephritis; Hp: postpartum hemorrhage.
[2]See Figure 6.5 for some of the clinical manifestations.

FROM D. L. RUCKNAGEL AND R. K. LAROS, JR., "HEMOGLOBINOPATHIES: GENETICS AND IMPLICATIONS FOR STUDIES OF HUMAN REPRODUCTION." CLINICAL OBSTETRICS AND GYNECOLOGY 12:49–75. 1969.

support of the hypothesis that there is mutual feedback between the frequency of sickle-cell anemia and the dependence on agriculture among different African populations. Anthropological information suggests that it was the introduction of agricultural practices and crops from Malaysia that permitted Negroes in Africa to penetrate tropical rain forests and thus exploit new niches. This development led to increases in the population of people that could be supported. Clearing of the forests improved the environment for the *Anopheles* species of mosquito, which carry *Plasmodium falciparum,* and the mosquito populations also increased. Thus, the probability of malarial infection was increased, giving a selective advantage to the person carrying an allele for sickle-cell anemia or alleles of other genes conferring protection from malaria. The enlarged proportion of persons protected from malaria by their heterozygosity, in turn, permitted further expansion of the adopted agricultural practices. Mathematical models of the spread of the sickle-cell allele have been investigated, and there has been enough time to account for the evolutionary changes that have occurred in the 60–80 generations—that is, during the 1,500–2,000 years—since the introduction of Malaysian agriculture into Africa.

For perhaps the last twenty or more generations, the Hb^S frequencies have remained reasonably constant. But at present the situation has become dynamic once more, partly because of the control of the mosquito population, and partly because of medical advances in treating malaria.

Population genetics, ecological genetics, and molecular genetics of sickle-cell anemia are among the best understood of any gene or gene system available today. Sickle-cell anemia is not exclusively, but is largely, a disease of black people. In the 1950's and 1960's there were very few black geneticists. There was a disturbing tendency in those decades for geneticists to view sickle-cell anemia as a fascinating and informative case study, yet be little concerned with the people afflicted by it. It is estimated that sickle-cell anemia is a contributing cause to the deaths of over 80,000 African children per year, and at present 40,000 American blacks have sickle-cell anemia and, as a result, an expected average life-span of less than 20 years. Most $Hb^S Hb^S$ homozygotes die as a result of being weakened and anemic, which makes them easy prey to a number of infections and diseases such as pneumonia (see Figure 6.5 and Box 7.E). Even in relatively healthy $Hb^S Hb^S$ homozygotes, the abnormal hemoglobin sometimes causes the red blood cells to block blood vessels, causing excruciatingly painful, and for some, terrifying crises. The medical research establishment and the medical profession during the 1950's and 1960's exhibited at best insensitivity to the extent and severity of sickle-cell anemia, apportioning much greater efforts and funds to less severe or less frequent diseases that afflict the white population. This situation is changing in the 1970's. As screening and genetic counseling become more effective, fewer Hb^S homozygotes will be born. And as medical treatments are developed that permit not only survival of $Hb^S Hb^S$ homozygotes, but equal reproductive success of all three genotypes, then the ongoing natural selection at the Hb locus will respond to these medical advances and to other relevant changes in human culture and environment.

8

SEX

In Section 7.2 we used the inheritance of sex as a model of Mendelian genetics. Having considered autosomal inheritance, we now return to a fuller consideration of sex determination in humans and other organisms.

8.1 WHAT IS SEX FOR?

Common answers to this question include: "ego fulfillment," "reproduction of the individual," "reproduction of the species," "recombination of genes," "attaining orgasm," and "selling things on TV." While sex is frequently used for all of these purposes, all but the fourth can be done very well without sex. As was noted in Chapter 7, genetic recombination is efficiently accomplished during meiosis, both by the independent assortment of chromosomes to the gametes, and by exchange of segments between homologous pairs of chromosomes prior to their separation.

Sexual reproduction is both a progressive and a conservative evolutionary force. The replacement of fission by sexual reproduction in the course of evolution tremendously increased the potential genetic variability available. Each generation, a great many genotypes are produced and tested by natural selection in a sexually reproducing diploid species. By contrast, an organism reproducing itself by fission can multiply very rapidly and propagate well-adapted genotypes at the expense of others. Sexually reproducing organisms, on the other hand, very rarely and only in special circumstances leave offspring genetically identical to themselves; an organism whose genotype is highly successful leaves descendents that, on the average, are less well adapted than itself. Nevertheless, judging from the prevalence of sexual reproduction among living beings, this mode of propagation provides more advantages than disadvantages.

Meiosis, followed by sex cell (gamete) formation, followed by gametic fusion and zygote formation, provides a sufficient mechanism for genetic recombination, and it is the production of gametes that fundamentally defines sexual reproduction. Indeed, many sexually reproducing organisms normally or occasionally engage in self-fertilization. But recombination is more intense if the gametes originate in different individuals (Section 4.3: Process 2. Selection of mates), and many species have enforced this more intense recombination by sometime in their history evolving different mating types, or sexes.

8.2 SEX DETERMINATION SYSTEMS

The mechanisms of **sex determination** are highly varied, with different groups of organisms coding for maleness and femaleness in different ways. For some groups, evolutionary trees of sex-determining mechanisms can be reconstructed on much the same basis as the evolutionary relationship of morphological or biochemical traits. Only a few of the systems of sex determination found in nature will be reviewed here.

The system wherein it is the presence of a Y chromosome that determines development into males is shared by humans and other mammals with a plant of the pink family known as *Melandrium*. Populations of these plants, most of which are diploids, may include individuals with three sets of chromosomes (**triploids**) or four (**tetraploids**). The general term for multiplicity of sets is **polyploidy**. (As an aside, polyploidy makes it necessary for us to note one important exception to the general definition of evolution that was given in Section 4.3: Significant evolution, even the creation of a new species, can occur instantaneously through an increase in ploid number without an immediate accompanying change in allele frequency.) In diploids and triploids of *Melandrium,* a single Y produces males. But if a plant has tetraploid autosomes, a single Y does not suffice for this, and a **hermaphrodite,** possessing both male and female sex organs, is produced. Apparently, although the power of the Y to direct development is great, it is not complete, and a certain balance between genetic male determinants on the Y chromosome and the autosomes is involved.

In asparagus, sex seems to be determined by alleles at a single locus rather than by a whole chromosome. In the animal whose genetics is known best, the fruit fly *Drosophila melanogaster* ($2N = 8$ chromosomes), a system of balance between the number of autosomes and X chromosomes determines sex. The following table illustrates this mechanism.

Sex chromosomes	Number of sets of autosomes	Ratio of number of X chromosomes to sets of autosomes	Sex
XXX	2	1.50	super or meta female (sterile)
XX	2	1.00	female
XXY	2	1.00	female
XX	3	0.67	intersex
XY	2	0.50	male
XY	3	0.33	super or meta male (sterile)

There are, however, also single genes that govern transformation from a female genotype into that of a sterile male. In other *Drosophila* species, strains are known in which some females produce no males at all; preservation of such a line requires that every generation use males of other strains for reproductive purposes. This deficient production of male offspring can be cured by treatment, and can be transmitted to females of other strains by injection. Apparently, the abnormality is caused by infection of the females with a spirochaete, a parasite related to the organism that causes syphilis in humans. The infection is lethal to male zygotes; thus we are dealing here with a congenital disease selectively killing one sex.

Grasshoppers have no Y chromosomes: the males have a single X (X-null, or XO chromosome complement), the females two (XX). In birds and in moths the system is much like that in humans, but reversed: the males are XX and the females XY or XO, although no proof that the Y carries coding for femaleness as it does for males in humans (see Section 8.4) is available. Since the individual of XY chromosome complement produces two kinds of gametes, in birds and moths it is the female that is the **heterogametic** sex. The males, producing only one kind, are **homogametic.** Sometimes, the notation used for this type of sex determination is ZW and ZZ, to distinguish it from the kind found in humans.

In certain tropical fish, the guppies and the platyfish, both systems are known to exist within the same species. In domestic strains the chromosome complements for females are ZW and for males ZZ; in wild strains females and males are XX and XY, respectively. The two kinds of fish can be crossed with each other, and departures from the ordinary **sex ratio** of 1:1 are then observed. The known male chromosome complements are ZZ, ZX, ZY, XY, YY; the female,

ZW, YW, XW, and XX. The cross of XX by ZZ yields only male (ZX) offspring; the cross of XW by ZX produces a sex ratio of 1:3.

In some bees and related insects there are no sex chromosomes. Sex is determined by the males being haploid, that is, having only one set of autosomes. They develop from unfertilized eggs and present a curious paradox: they have no fathers, but do have grandfathers on the maternal side. Fertilized eggs develop either into queens, which are large fertile females, or into sterile female workers. In some species, the differentiation between the two types of females depends on nutrition. In others, it is genetically determined by two or three pairs of genes, all of which have to be heterozygous to give rise to a queen. For instance, in the stingless bee *Melipona,* a potential queen is genetically created when two 'caste' loci are heterozygous. They become queens in other respects, including full development of their reproductive systems, when given a favorable environment, including a rich supply of food. Otherwise, the larval forms of such double-heterozygotes grow into workers, but with the smaller number of abdominal ganglia characteristic of the queens.

An interesting point, perhaps having an important bearing on the evolution of social insects such as bees, has to do with the haploid-diploid method of sex determination common in bees and ants. Since the male that mates with the queen is haploid, he produces only one type of gamete, and all such gametes are genetically identical. The queen's genes are recombined in her gametes, as is normal. This results in sisters sharing about three-fourths of their genes, whereas mothers and daughters share only about one-half of their genes. Consequently, it is more advantageous (in the Darwinian sense of the passing on of one's own genes) for such a female to care for her sisters than to try to rear her daughters.

A parasitic wasp, *Habrobracon,* has a similar mechanism, with females being diploid and males haploid. However, there can be diploid males, and these are always homozygous for a locus at which multiple alleles exist. Femaleness, therefore, is determined by heterozygosity at one particular locus.

In the marine worm, *Bonellia,* sex determination is nearly entirely environmental. Larvae that are free-swimming throughout their entire larval stage become females. Some larvae attach themselves to the bodies of mature females and are turned into males by a masculinizing hormone from the females. A certain economy of not having any surplus males in search of mates is thereby achieved.

A similar type of economy is effected by the coral-reef fish, *Labroides dimidiatus.* Each group of these fish consists of a male and three to six or more females, which in turn have a dominance hierarchy. When the male dies or is removed, the most dominant female first attempts to repel males that approach her group, and if successful, then begins to display male aggressive behavior within a few hours, and courtship and spawning behavior in 2–4 days. Fertile sperm are released 14–18 days after the female begins sex reversal following loss of the group male.

Many other forms of sexual or parasexual (producing recombination) reproduction exist. Most plants, and a few animals, are hermaphroditic, i.e., have both sexes in a single individual. Most of the plants can reproduce by self-fertilization, others have to be cross-fertilized. Some hermaphrodites have a number of different mating types, with rules as to which can mate with which (Section 7.2). Enough, however, has been said to show the inventiveness of nature in maintaining genetic variability by sexual reproduction.

8.3 SEXUAL DIFFERENTIATION

Sex determination is the process that decides the gender of the zygote; **sexual differentiation** is the developmental pathway by which genetic instructions are carried out. In normal embryology of higher organisms, sexual differentiation must operate under the direction of powerful switch genes to avoid reproductively useless intersexes. Development must be **canalized** in such a way that either one or the other kind of gonads appears in the adult individual, rather than something in between. In normal hermaphrodites, such as pine trees, it is important that the right organs develop in the right places, i.e., that female structures develop in cones and male structures in the pollen-bearing strobili. Perhaps an analogy of a train arriving at a point on the railroad line where it can turn in one or another direction may be suggested. A switch must be used to send it to one of the alternative tracks, rather than have it make its own choice of tracks or even meander aimlessly in the field between tracks. Just as a train does occasionally jump the track, errors of development do occur, and result in intersexes, hermaphrodites (in this case, individuals possessing both male and female sex organs that belong to a species in which most individuals are either male or female), and **gynandromorphs** (see Figure 8.1), which are sex **mosaics,** with different parts of the body exhibiting properties of one or the other sex.

True hermaphroditism is rare in humans. About 200 authenticated cases have been reported, and of those cytologically investigated, most have the normal female complement of 44 autosomes and two X chromosomes. Human hermaphrodites have various forms, the most common being a combined ovotestis on one side and an ovary or testis on the other. Sometimes there is both a testis and ovary or ovotestis on each side, usually with fallopian tubes and a uterus. Sometimes there is a testis on one side and an ovary on the other, and sometimes the different kinds of gonadal tissue appear in unconforming places. Surgical procedures can partially, or sometimes fully, correct such developmental anomalies, with a choice of sex being possible. The cause of these anomalies, however, is not clear at present. In those instances in which the cause appears to be genetic, rather than being a disease or the effect of abnormal hormonal intervention during development, there are at least three possible explanations: One of the X chromosomes has acquired, by DNA exchange, a bit of the Y chromosome coding for testicular development; a mutant gene, presumably on the X, takes on functions that are similar to those of a Y; or the person is a sex-chromosome mosaic, possibly result-

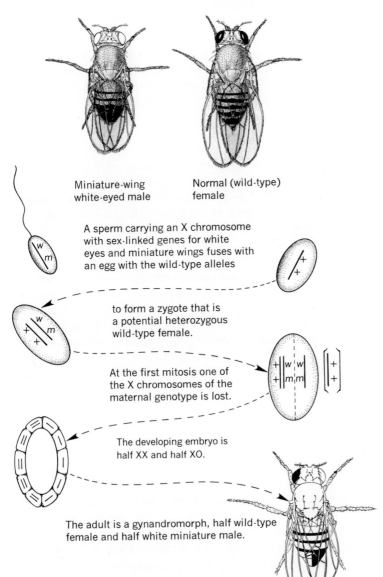

Miniature-wing
white-eyed male

Normal (wild-type)
female

A sperm carrying an X chromosome
with sex-linked genes for white
eyes and miniature wings fuses with
an egg with the wild-type alleles

to form a zygote that is
a potential heterozygous
wild-type female.

At the first mitosis one of
the X chromosomes of the
maternal genotype is lost.

The developing embryo is
half XX and half XO.

The adult is a gynandromorph, half wild-type
female and half white miniature male.

FIGURE 8.1
Normal and gynandromorphic *Drosophila
melanogaster.* (Adapted from A. M. Srb,
R. D. Owen, and R. S. Edgar, *General
Genetics,* 2nd Ed. W. H. Freeman and
Company. Copyright © 1965.)

ing from very early fusion of independently conceived fraternal twins, or separate fertilization of the egg and a polar nucleus followed by their fusion.

In normal human development, embryos develop without sexual differentiation for about six weeks. As the sex organs begin development, human embryos
become hermaphrodites. Then the relative developmental rates of the outer
(cortex) and inner (medulla) parts of the gonads signal that the switch has been

thrown, and the presence or absence of the Y chromosome appears somehow to affect these rates. If it is the medullas that grow, testes develop and begin to secrete androgen, which then influences continued development as a male. If the cortex is favored, ovaries develop, and in the absence of androgen, female development ensues. Later, both male and female hormones are produced (although in different amounts) by both sexes. Occasionally, even after normal adulthood is reached, **sex reversal** occurs, as in the coral-reef fish. In mammals, however, complete functional reversal does not seem to occur spontaneously.

Particularly instructive is sex reversal among the domestic fowl, where it is relatively common. In the avian gonad, the male component, the medulla, begins to develop first in both sexes. In due course of time, the female component, the cortex, is stimulated to faster growth and begins to produce feminizing substances in XY fowl, which inhibit medulla growth. But only the left ovary develops. The right one remains rudimentary. An adult hen, unlike a mammal, has only one working ovary. Now, if the left ovary is destroyed (spontaneously by disease, or artificially by surgery), compensatory growth begins in the right gonad. Depending on the proportions of medullary and cortical elements, this replacement gonad will become an ovary, an ovotestis, or a testis. If it becomes a testis, full sex reversal and production of sperm is possible. The sperm, however, are not functional. Sex transformation of this type in chickens has long been known. It can also be reversible, if a partially destroyed left ovary regenerates. Indeed, there is a record of a cock that was publicly burned in Basel in 1474 "for the unnatural crime of laying an egg." The one reported instance of a hen, which had laid fertilized eggs (thus being a mother), later turning into a cock and siring offspring, is very likely not reliable.

A somewhat similar situation obtains in cattle twins. As fetuses, they usually share a common placental circulation, and thus, if of opposite sex, may exchange either sex-influencing substances, or even cells with chromosomal constitutions of the opposite sex. In opposite-sex fraternal twins, such exchange causes the female co-twin to become a sterile **freemartin.**

In general, sex differentiation in higher organisms, although programmed by the genes, is mediated by hormones. In creatures such as the *Bonellia* marine worm described in Section 8.2, hormones seem to determine the sex. In the vertebrates, hormones alone, without additional surgical intervention, seem much less likely to reverse the sex indicated by the organism's genetic constitution. Nevertheless, experiments in which fertilized frog and rabbit eggs were dipped in solutions of female sex hormones resulted in all of the frogs and 90 percent of the rabbits produced being female.

Because of the normal control of sexual differentiation by hormones, gynandromorphs are not as a rule found in higher vertebrates, since all parts of the body share common hormone circulation in the blood. But where the sex differences are based more on chromosome constitution and less on hormones, or are controlled in some other ways, sex mosaics are possible. Some abnormalities of human sexual development produced by errors of chromosome distribution into germ cells are discussed in Section 8.5.

8.4 HUMAN SEX CHROMOSOMES

In 1949, M. L. Barr and his associates found some dark-staining material in the nuclei of cells from the brain of a female cat. Such material, now named a **Barr body**, is an inactivated X chromosome.

Since we first began to understand gene action, we've been puzzled about why male cells can get along with one X chromosome, while female cells have two. In *Drosophila*, "dosage-compensation" genes have been invoked to explain the phenomenon, and indeed, no Barr bodies have been seen in these flies. Several years ago, it was suggested that in mammals one of the two X's becomes inactive in each cell in the course of development. This proposal is now referred to as the Mary Lyon hypothesis, after the British geneticist who elaborated it. In some cell lines the inactivated X chromosome may be of paternal origin, and in other cell lines of the same individual, of maternal origin.

At the time of inactivation, just which chromosome is inactivated in a given cell appears to be determined at random. This hypothesis is now generally accepted, although some voices dissenting about its generality are still heard.

There is much evidence for the Mary Lyon hypothesis. Normal human females have a single Barr body; normal human males have none. In a series of abnormal chromosomal complements with more than two X's (see Section 8.5), the number of Barr bodies is always one less than the number of chromosomes (Figure 8.2).

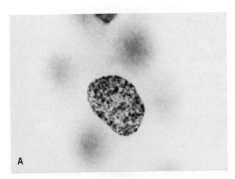
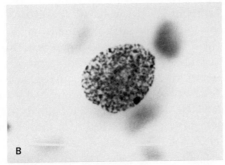
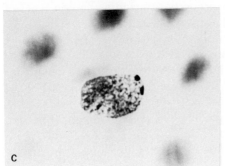
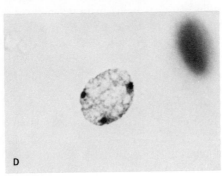

FIGURE 8.2 Photomicrographs of cells with various numbers of Barr bodies. **A:** none (one X chromosome). **B:** one (two X's). **C:** two (three X's). **D:** three (four X's). (By Arthur Robinson.)

Freemartins, despite their masculinization, have a Barr body in accordance with their chromosomal complement for femaleness. Samples of red blood cells from women heterozygous for G6PD deficiency (which is sex-linked) are mixtures of cells, some capable and others incapable of producing the enzyme.

Given that some human hermaphrodites are a result of an X with a bit of Y, the inactivation of the X^Y in one gonad or the normal X in the other during early development could explain how persons have one testis and one ovary, or a later differential inactivation within the same gonad could explain an ovotestis (Section 8.3). Cells from female mules are particularly useful for investigating this subject, as the horse-origin and donkey-origin X chromosomes can be cytologically distinguished, and they have X-linked G6PD genes that code for electrophoretically distinguishable enzymes. When the majority of Barr bodies are donkey X chromosomes, the majority of G6PD is of the horse type, and vice versa. Interestingly, similar work with kangaroo hybrids indicates that the X chromosome of paternal origin is preferentially switched off—a finding not in keeping with general finding of random inactivation. (Perhaps it is kangaroos, or kangaroo hybrids, that are peculiar.) The mottled coat of tortoiseshell cats—which, with very rare exceptions, are females—is thus explained. The fur-color genes are X-linked, and in a heterozygote about half the skin patches express one gene, and the rest express the other gene.

The question of whether the entire X chromosome is inactivated has been investigated using hair follicles of human females heterozygous at two widely separated enzyme loci on the X, *G6PDa : G6PDb*, and *HGPRT⁻ : HGPRT⁺* (the latter locus codes for the enzyme hypoxanthine-guanine phosphoribosyl transferase—see Box 12.A). It was observed that enzyme activity in individual hair follicles of such a female fell into three genotypic classes: *G6PDa* and *HGPRT⁻* (the maternal-origin X); *G6PDb* and *HGPRT⁺* (the paternal-origin X); and *G6PD a* and *b* plus intermediate *HGPRT* activity (follicles made up of a mixture of cells, some with the maternal X, some with the paternal X, active). Thus, an entire X chromosome, not parts of one or the other, appears to be inactivated, and normal XX mammals are mosaics, having some of their development generally in accord with that described for the abnormal fly illustrated in Figure 8.1.

Finally, some indication that female identical twins are more variable than male identical twins in many measurable properties (including certain types of behavior), as would be expected if females are mosaics, has been reported.

Because of the discovery of Barr bodies, it is now possible to identify the sex of an unborn child at as early an embryonic age as three weeks. This is done by obtaining fluids from fetal membranes, culturing cells from them, and looking for Barr bodies. As a complementary technique, intense fluorescence of part of the Y chromosome following the staining of such cells with fluorescent agents allows identification of male embryos early in pregnancy. The possibilities go far beyond the early satisfaction of parents' curiosity: There are implications both for increasing the sensitivity of genetic counseling and for establishing a practice of aborting embryos of the unwanted sex.

Let us briefly examine two other topics pertaining to human sex chromosomes. First, it may be noted (Box 7.C) that by now some 70 X-borne genes in humans have been definitely recognized as such, and another 50 or so postulated on basis of reasonable evidence. They code for a great variety of traits of physiological, biochemical, or anatomical expression. Besides the G6PD deficiency and the red-green color-blindness genes already mentioned, the list of X-linked genes includes those for atrophy of the optic nerve, toothlessness, night blindness, one form of muscular dystrophy, brown teeth, a form of diabetes, total color blindness, sensitivity to the smell of hydrogen cyanide, and others (Section 19.2).

Second, we may give brief consideration to the question of genes other than sex-determinants that may be carried on the Y chromosome. The property of such genes, of course, would be their transmittal only from father to son and never to a daughter. Several pedigrees of this type of inheritance have been published, including one for a defect described as porcupine skin and one for webbed toes. Fuller examination of the material has not lent credence to them but the possibility that one other trait, hairy ears (Box 8.A), is Y-borne still exists. Meanwhile, it has been found from the study of chromosomal abnormalities that the Y chromosome may have other than sex-determining effects.

8.5 X AND Y ABNORMALITIES IN HUMANS

The original proof that the Y chromosome determines sex in humans came from cytological investigation of abnormal genotypes. Because of errors in the distribution of chromosomes in the course of meiosis, aberrant individuals were produced. One such error, **nondisjunction,** causes unequal numbers of chromosomes to enter gametes. Figure 8.3 shows a diagrammatic representation of this process.

If during a particular meiosis sex chromosomes are nondisjunctional, a gamete containing only a set of autosomes, or a set of autosomes plus two X's, or plus two Y's, or plus both an X and a Y, could be produced. If these abnormal types of gametes unite with normal gametes, the zygotes would have complements for

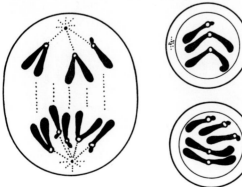

FIGURE 8.3
The mechanism of nondisjunction schematically illustrated. Instead of each daughter cell receiving four chromosomes, one pair does not segregate and enters the same cell. As a result, gametes containing three and five chromosomes are produced. (From Curt Stern, *Principles of Human Genetics,* 2nd Ed. W. H. Freeman and Company. Copyright © 1960.)

BOX 8.A INHERITANCE OF HAIRY EARS

The photograph shows the left ear of a 54-year-old man from south India and illustrates the trait "hairy ear," which some investigators believe is caused by a gene on the Y chromosome. Because of the age at which this trait develops, and difficulties of classification of phenotypes, others do not feel compelled to accept the hypothesis that the gene is sex-linked on the Y chromosome.

A seven-generation pedigree of a family in which the hairy ear characteristic was studied is reproduced below, in one of the conventional forms of representing pedigrees. Marriages are indicated by short horizontal dashes from which longer vertical lines lead to offspring. Double lines are used to indicate consanguineous marriages (matings between near relatives). Persons included

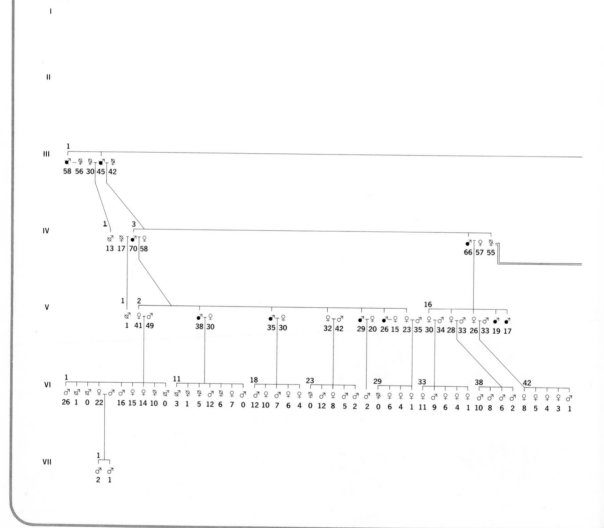

[148]

without indication of parentage are mates drawn from the population at large, unrelated to this immediate family. A black circle indicates possession of the trait being investigated, in this example, hairy ears. When a sign is crossed out, the person in question was dead at the time the study was made. A number below a crossed-out symbol tells the age of the person at the time of death. Numbers given for all other persons are their ages at the time the study was made. Close examination of the pedigree will show that all the male descendents of the original patriarch who lived beyond a certain age appear to have inherited the gene, while no females have exhibited it.

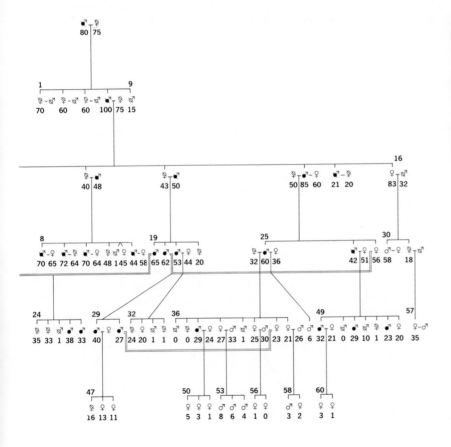

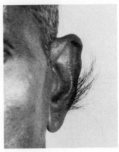

♂ Male
♀ Female
♂ Affected male
♂ ♀ Dead
= Consanguineous marriage
? Not observed

THE PHOTOGRAPH AND PEDIGREE ARE REPRODUCED COURTESY OF THE AUTHOR OF THE INVESTIGATION, KRISHNA DRONAMRAJU. THE PEDIGREE IS REPRINTED FROM JOURNAL OF GENETICS, 57:230, 1960.

the sex chromosomes of YO or XO, XXY or XXX, XYY, and XXY. Indeed, all of these types except the first, and others with higher numbers of sex chromosomes, have been observed (Box 8.B). YO apparently cannot survive, probably because of a complete absence of the genetic information normally coded for by the X chromosome. If a Y chromosome is present, a male phenotype develops, and if it is absent the phenotype is female. The fact that, unlike the X, the human Y chromosome can be dispensed with may suggest that no essential genes are carried on it. In *Drosophila*, however, Y-less males are sterile (recall from Section 8.2 that maleness in *Drosophila* is determined by a balance between autosomes and X chromosomes rather than by presence of a Y). A number of fertility genes or regions on the *Drosophila* Y chromosome have been identified.

Some abnormal sex chromosome complements produce no obvious phenotypic effects; others do. In particular, XO females display a series of morphological features such as short stature. The intelligence of most of them is impaired, and almost all are sterile, owing to the failure of the ovaries to develop. (An interesting side note is that female XO mice are fertile. At least one case of an XO human giving birth to a son is recorded.) The totality of properties associated with the XO chromosome complement is designated as **Turner's syndrome** (Box 8.B).

BOX 8.B ABNORMALITIES OF THE SEX CHROMOSOMES IN HUMANS

The incidence of various abnormalities of sex chromosomes in humans is relatively high. One estimate places the average number of people who are afflicted with Turner's syndrome at one in 2,500, and an estimate for Klinefelter's syndrome, made from a study of a white Philadelphia population, is one in 500. All X and Y abnormalities together are estimated to occur at the rate of one in 250 births. Since many such abnormalities cause abortion, it is probable that abnormal complements of sex chromosomes constitute as many as one percent of all conceptions.

Some abnormalities are caused by deletions or additions of parts of the X chromosome, instead of whole missing or extra X or Y chromosomes. Two true hermaphrodites were found to be mosaics, one for loss of a section of the X, and the other for duplication of a section of the X.

The following table lists general descriptions of the bearers of various known complements of whole chromosomes.

		Number of X chromosomes				
		0	*1*	*2*	*3*	*4*	*5*
	0	Not found	Turner	Normal female	Normal or sterile	Female	Female
Number of Y chromosomes	*1*	Not found	Normal male	Klinefelter	Klinefelter	Klinefelter	Klinefelter
	2	Not found	Normal or abnormal male	Klinefelter	Klinefelter	Klinefelter	Klinefelter

Males with supernumerary X chromosomes show lengthened limbs and other associated features. As their testes fail to develop, they too are sterile. The abnormality is known as **Klinefelter's syndrome.** (The rare tortoiseshell tomcats probably are sterile because they are Klinefelters.)

There are a number of other pathological manifestations of some of the abnormal complements of sex chromosomes. Of particular interest is the fact that the effect of the abnormalities may extend not only to such physical traits as stature, but also to some behavioral characteristics. Thus, in the general British population about 0.2 percent of males and 0.08 percent of females show abnormal numbers of Barr bodies, but the percentage in mentally subnormal institutionalized persons is 1.0 and 0.4 percent, respectively.

The significance of XYY chromosome complement among human males is still the subject of investigation and controversy. Publicity given early reports led to the stereotype of XYY individuals as physically aggressive and violent (an impression reinforced in the United States by false reports that Richard Speck, the Chicago mass murderer tatooed "born to raise hell," was an XYY). In most well-designed studies of the chromosomes of socially deviant males, higher proportions of XYY chromosome complements were found than occur in the general population, but instances of apparently well-adjusted XYY individuals are also documented. In mental-penal and penal settings, men with XYY chromosome complements do not appear to be concentrated among the most dangerous, violent, and physically aggressive inmates. In general, their offenses are similar to or less serious than those of XY's there.

About one in 1,000 newborn male infants is XYY. By contrast, in the wing for mentally retarded men in the Carstairs (Scotland) maximum security hospital, 7 of 197, or 36 times the frequency in newborns, were XYY. Similar studies have also yielded rates much higher than reasonable random expectation of XYY men in mental-penal institutions. It is interesting to note that the frequency of detected XYY's is higher in whites than blacks in the United States, and to date, there is no extensive documentation associating the XYY chromosome complement with deviant behavior in any nonwhite group.

Three general hypotheses may be proposed to explain the high frequency of XYY whites in mental-penal institutions. (1) Men with XYY chromosome complements tend to join groups, or are attracted to social settings, where deviant behavior is frequent. (2) XYY's tend to be much taller than XY males, and are more likely to have severe acne. Some external correlate of manifestations of the XYY chromosome complement, such as greater height, either makes social adaptation less likely, or institutionalization more likely or likely to be of greater duration. (3) The XYY chromosome complement brings about, directly or indirectly, some neural aberration that tends to produce deviant behavior. Of these, the first two seem less likely, but they are by no means mutually exclusive.

Discovery of the XYY chromosome complement in an infant hardly predicts antisocial behavior with the confidence, for instance, that discovery of trisomy 21 (the chromosomal abnormality causing Down's syndrome) predicts mental retardation. Nevertheless, some officials at one time suggested screening for and immediate institutionalization of XYY infants, for their own and society's good. That this will be done now seems unlikely, but there remains a dilemma for the physician or genetic counselor about whether and how to advise the parents of an XYY child, or whether the XYY individual himself should be informed as he matures. Clearly, we need more information on XYY's in normal surroundings. The fluorescence technique for detecting Y chromosomes will enable more efficient screening of both newborn and older XYY individuals, so that such information may be acquired.

Another interesting variant class is males with XX chromosomal constitution. These people have male psychosexual identification, testes without evidence of ovarian tissue, and absence of developed female genital organs. XX males appear to be less than one-tenth as frequent at birth (roughly one in 25,000) as either XXY or XYY. Also unlike the XO, XXY, and XYY, the frequency of the XX male among the mentally retarded is not unusually high. While both XXY and XYY are taller on the average than XY males, the XX males tend to be somewhat shorter than XY's, but they are taller on the average than XX females. As with hermaphrodites, there are theories explaining XX males based on gene mutation and based on a DNA exchange resulting in an X with a bit of Y. The theory that is perhaps best supported suggests that an XX male may be a mosaic, in which a cell line containing a Y was present in early embryogenesis. As in the freemartin, the Y triggered male differentiation before it was eliminated from the embryo, leaving a slightly short, phenotypically normal male with a female chromosomal constitution.

The X and Y chromosomes themselves can vary in length. Especially for the Y, it has been found that there are considerable differences between individuals. Because the measurable length of a given chromosome may be affected by how cells are prepared (usually white blood cells in culture), in order to standardize the measurement, Y length is usually expressed as a ratio; that is, it is divided by the length of some other chromosome or chromosomes in the culture. This technique becomes more sensitive when coupled with measurement of the fluorescent portion of the Y, which seems to account for much of the total variability in its length. Significant differences in such ratios have been reported between different ethnic groups, with Indians, for example, showing relatively short Y's, and Japanese, long Y's.

Variability in Y length has been suggested as a possible diagnostic tool in cases of disputed paternity, since the single Y is transmitted from father to son. Two limitations of this method are (1) it can only be used for boys, and (2) it cannot be used to determine which of two brothers is the father; their Y's, both originating from the Y of the boy's grandfather, would be identical.

As a final word of sex-chromosome abnormalities we may note that they can lead to the production of identical twins of opposite sex. At least two such cases have been reported. The origin of identical twins is discussed in Section 11.4. Here it need only be said that they are produced when a single zygote divides and separates into two organisms. Should the Y chromosome be lost at this stage, as was the *Drosophila* X chromosome in Figure 8.1, an XY zygote may produce a normal XY male and a twin XO Turner female.

Sex ratios are sometimes given as percentages of males in the population. The alternative convention, to which we shall adhere, is to calculate the number of males per 100 females: thus, a sex ratio of 106 indicates that there are 106 males for every 100 females.

The **primary sex ratio** is that at the time of fertilization. Since many zygotes fail to be implanted and fetuses may abort before their sex has been determined, the actual primary ratio for humans is not known. The use of the Barr body technique for early identification of the sex of embryos will eventually bring us closer to the knowledge of the primary sex ratio. For the present it may be inferred to be higher than the **secondary sex ratio,** calculated at the time of birth, largely because the number of male embryos aborted or stillborn is somewhat higher than that of females. Up to four months of embryonic age, sex ratios of the stillborn tend to run (these figures are not very accurate) in the neighborhood of 107.

The secondary sex ratios tend to vary: in the United States the ratio is about 106 for whites and 103 for blacks, in Greece 113, in Cuba 101. The **tertiary sex ratio** refers to any specified time after birth. The comparison of tertiary ratios at different ages shows a decline with age. Some approximate ratios for U.S. whites are as follows:

birth	106
18 years of age	100
50 years of age	95
57 years of age	90
67 years of age	70
87 years of age	50
100 or more years of age	21

Figure 8.4 shows a similar sex ratio decline with age. Thus, in terms of survival in the human species, the male is weaker than the female. Deaths of infants younger than one year (1961, U.S.) are one-third more numerous among boys than girls, and the higher mortality persists throughout life. Of the 64 specific causes of death listed by the U.S. census, 57 show a lower rate among females at all ages. Of the remaining seven, diabetes and pernicious anemia cause more deaths among females than among males. The others include breast and uterine cancer, and childbearing, which do not directly affect males.

The reasons for the differences in viability between the two sexes are obscure. Deleterious or lethal sex-linked recessive traits can account for some of it, because they would be expressed in hemizygous males but not in heterozygous females. This is unlikely, however, to be the full explanation. Differences between common male and female behaviors, leading to lethal accidents being more common among males (including war-related casualties), are one contributing cause. It used to be that men smoked more cigarettes than women. Social

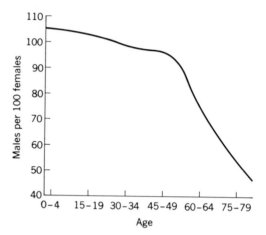

FIGURE 8.4
1960 data of human sex ratios by age
in England and Wales. (Redrawn
from A. S. Parkes.)

customs have changed, with cigarette smoking becoming increasingly a female activity. It will therefore be interesting to follow statistics in the coming years on deaths among men and women attributed to smoking-related causes.

The sex ratio is known in part from theoretical considerations, and in part from experimental evidence, to be under genetic control. In general, a 1:1 sex ratio appears to be the best evolutionary strategy for a species, though, perhaps, not under all circumstances. Its advantages include the facts that it provides for maximum genetic variability and maximizes the probability of male-female encounter. Among the experimental evidence indicating genetic control is the observation of differences in sex ratio between selected strains of mice of the magnitude of 40 percent (111.8 *vs* 71.8). In crosses between such strains, incidentally, it is the father's genotype that apparently determines the sex ratio of the progeny.

There are many possible ways in which a high primary sex ratio can be obtained. Among them are: (1) more Y-bearing than X-bearing sperm are produced; (2) more Y-bearing than X-bearing sperm reach the eggs; and (3) Y-bearing sperm have a competitive advantage over X-bearing sperm in effecting fertilization.

All of these are supported by observations on mice. The first, in particular, is highly probable, since the phenomenon of **segregation distortion** (Section 4.3), unequal representation of the two sex chromosomes in gametes produced by heterozygotes, has been well established in nonhuman material.

Some other general features of the human secondary sex ratio may be listed. More males than females appear among firstborn than among subsequent children. The ratio in one study dropped from 106.6 for firstborns to 104.5 for children born to couples who had already produced ten or more children. The sex ratio in single births is higher than in twins, which in turn is higher than among triplets. Proportionately more males are found among legitimate than illegitimate children. Formerly, proportionately more males were born to parents in rural than in urban communities. There are differences in sex ratio between socioeconomic classes: in England and Wales, the upper- and middle-classes show a secondary sex ratio of 106.1, skilled workers, 105.7, and unskilled laborers, 103.4. All of these facts are consistent with the general idea of the relative weakness of the male sex. The higher sex ratios in higher socioeconomic classes may be associated with better nutritional and physiological state of the mothers, who can then provide a better uterine environment, in which more males would survive.

But the picture has many other more puzzling aspects. The age of the father has a bearing on the sex ratio. Even the father's occupation has been claimed to have an effect. There are seasonal differences, with a higher sex ratio in the births between June and August than at other times of the year. Similarly puzzling are the sex ratio differences between different countries and between regions of the same country. Also, the slight but significant tendency for the sex of successive children in a family to be the same has yet received no adequate explanation.

A rise observed in the secondary sex ratio during or shortly after wars (Figure 8.5) has been commonly misinterpreted as the way in which Nature restores the proportion between the sexes to compensate for the soldiers killed. There is no conceivable mechanism—at least no reasonable one—by which Nature could know about the need for adjustment, nor by which it could mediate the rise, and much more sensible explanations can be found. Many of the factors listed that accompany a higher secondary sex ratio are present during wars. The less than one percent rise observed after World War II could be easily accounted for either by these factors or by the amplitude of normal fluctuations in the secondary sex ratio from year to year.

The obvious question concerning human sex determination is whether we could ever come to dictate the gender of babies to be produced. In some lower forms, such as silkworms, methods for producing only males or females have been developed (see Section 17.3). In higher forms many attacks on the problem have been made. For instance, changing the acidity or alkalinity of the vagina, by introducing chemicals such as dilute vinegar or baking soda prior to intercourse, has been claimed to change the relative rates of travel of X- and Y-bearing sperm. Depth of penetration at the time of ejaculation, and the presence or absence of female orgasm may also influence the probability of which type of sperm first reaches the egg, thus affecting the proportion of female and male conceptions. This is not at present reliable technology.

FIGURE 8.5
Secondary sex ratios in England and Wales. Note rises at the times of the two world wars. (Redrawn from A. S. Parkes.)

More promising are methods in which X-bearing and Y-bearing sperm are separated outside the body and then used in artificial insemination. Some positive results have been reported in rabbits and in cattle, for which it would be a very useful technique from the economic point of view. With the ability to determine the sex of sperm by using Y fluorescence, research on separation of X- and Y-bearing sperm can now progress rapidly and on a massive scale, compared to earlier attempts which could be tested only by the expensive and time-consuming process of producing offspring to verify altered sex ratios.

Various forms of separation of the two kinds of human sperm have been tried. They include (1) ordinary centrifugation, based on differences in the weight of X-bearing and Y-bearing sperm; (2) electrophoresis, based on the assumption that the two kinds of sperm have different electric charges; (3) high-speed centrifugation, based on density differences; and (4) differential sedimentation methods. A more visionary possibility, permitting natural matings, might lie in immunizing wives against the X or Y chromosomes of their husbands. Particular antibodies interfering with implantation of zygotes of one or the other sex might thus be produced. Still another distant prospect is to fertilize an egg outside the mother's body, culture the zygote, identify its sex from the number of Barr bodies, and implant it within the mother only if it is of the desired sex. Although external fertilization of eggs of rabbits and pigs has been accomplished, the actuality of reported successes with human eggs is not clear. However, an alternative technique is available by which eggs fertilized naturally may be removed from the mother's body before they are implanted, and grown in tissue cultures to the stage at which sex determination is possible. (The practicability of either of these procedures may, however, be questioned by considering another alternative: with the same moral accommodation and simpler technological and medical manipulations, the sex of a normally conceived and implanted zygote can be determined and it could be aborted if it were of the undesired sex.)

Should any of these or other methods be found efficient—and there is little reason to doubt that working techniques will be developed reasonably soon—new problems of personal and social responsibility will face us. Some social critics think that even today determination of the number of children a given couple may have is the responsibility not only of the parents but also of society. The making of choices to predetermine sex of offspring would have similar implications. If left to the parents, what would govern the decision? It might be the prediction of happiness of a girl, living in a world overpopulated by boys. It might be the Freudian proclivities of the parents. It might be, in the view of some, even the welfare of the commonwealth. For instance, with population size a general problem, family size might be more effectively limited if people could achieve the desired mixture of male and female offspring. As it is now, there is a tendency for couples with two like-sexed children to have a third, in an attempt to get a child of the opposite sex. But if the option were to result in a sexual imbalance—even 60 percent boys and 40 percent girls (not bad at first glance, but actually a 50 percent surplus of males)—it might have far-reaching effects on our views of such social conventions as monogamous marriage, and the practice of homosexuality and of prostitution.

Development of an egg without fertilization by a sperm, or **parthenogenesis,** is a normal method of reproduction in some organisms. It can also occur among organisms that normally reproduce sexually, such as some species of *Drosophila,* in which its occurrence may be genetically determined. This has been demonstrated by experiments in which its frequency has been increased by selection. Such experiments have had as subjects both *Drosophila* and the domestic turkey. In various strains of turkeys, several generations of selection led to increases in parthenogenesis from 16.7 to 40 percent, from 4.0 to 21.1 percent, and from 1.1 to 18.6 percent. The interesting revelation was that parthenogenesis occurs in high frequency in the presence of viruses in eggs that have been produced by vaccinated hens or by hens whose dams had been vaccinated. This suggests that viruses may interfere with the normal genetic message specifying development. All of the parthenogenetic turkeys are male, and some of them have proved to be fertile and heterozygous for some genes determining plumage color. This could be the result of a number of different processes. Among them are failure of reduction of the egg, fertilization of the egg by the polar body, and the doubling of the chromosome number after reduction, with the last process always leading to the production of a homozygote. These same processes are relevant to other diploids among which parthenogenesis occurs.

One laboratory reported that it had experimentally produced parthenogenetic rabbits. Extensive repetition of this experiment, which stimulated parthenogenetic development by heat shocks or treatment of eggs with sodium chloride, has failed to create additional surviving parthenogens, although many treated eggs have developed to an early stage.

In humans, claims of parthenogenetic births have been often advanced, especially by unmarried women. If the offspring are boys, such a claim can be rejected out of hand, unless miraculous intervention forms part of it. Since human females have no Y chromosome, they can give rise parthenogenetically only to girls. Biological tests for parthenogenetic origin of daughters are also available. One involves blood groups: no blood group present in a parthenogenetic daughter would be absent from her mother. A very sensitive test is provided by grafting of tissues. Since all the proteins of a parthenogenetic daughter would be manufactured under the direction of genes that the mother supplied, none of them would act as a foreign antigen *vis-à-vis* the mother. Grafts from such a daughter to the mother would therefore always be accepted, barring rare cases of somatic mutation. Grafts from a nonparthenogenetic daughter would be rejected, because they would contain antigens derived from the paternal genes. There are one or two documented cases in humans that meet the tests of reasonable certainty of virginity, and compatibility of the mother with the daughter's blood groups and tissue grafts. Perhaps the best explanation is that the last polar body (Box 7.A) failed to be excluded from the egg cytoplasm, but instead rejoined the egg nucleus following meiosis, effecting a special sort of self-fertilization. Being very closely inbred, most such fetuses would probably abort early in pregnancy. But some may come to term, and the children born may be fairly, or even fully, normal. All would be girls, since both egg and polar body would have X chromosomes.

Further fanciful speculation on immortal humans, *H. continuus,* derives from this topic. Should human parthenogenesis be possible, women could produce by this process daughters, who in turn could be used to produce similar grand-daughters. These could be kept in storage and used to provide the mother with an infinite series of acceptable tissues and organs to replace those worn out as she ages. Males, of course, would be biologically completely superfluous under such a development, and might be kept only for amusement. Some men, however, find the prospect of being maintained as a pet in Fremlin's world (Section 15.1), a world populated by sixty quadrillion women who have numerous daughters in cold storage, a somewhat depressing one.

8.8 SEXISM

It is clear that, in most cases, the sex of a human is genetically determined. Furthermore, as a result of the chromosomal basis of human sex determination, there are imposed consistent genetic differences between males and females. These are due to the presence or absence of the Y, the uniform hemizygous state of the X in males as contrasted to a mixed population of active and inactive X's in females, and the functional masking of certain kinds of X-linked recessive genes in heterozygous females that are expressed in males. The secondary sexual characteristics that develop under the influence of the different hormones trig-gered by these alternative sex-chromosome complements are probably even more significant.

Sexism, like racism, builds on real differences. In addition to these real dif-ferences, sexism creates another layer of different expectations, which may in significant degree be self-fulfilling. The present battle to counter sexism utilizes a number of tactics, including demands for equal rights, equal legal responsibilities, and equal pay for equivalent work. It will be waged in the context of some dif-ferences that are clearly biologically real, and other differences that are artificially constructed. There is an increasing body of evidence, based on both observation and experiment, that indicates many of the conventional patterns of masculine and feminine behavior can be altered, perhaps even interchangeably so. These behavioral patterns that are psychologically labile and sex associated may arise and be perpetuated through ignorance, or through the desire of members of one sex to exploit or, in some cases, to protect the other.

The accommodations that may be worked out, in order to be stable, must re-spond to biological reality, the diversity of psychological needs and desires within both sexes, and the needs of our culture and society. This process, largely political, will demand both extensive scientific knowledge and considerable sensitive skill, if indeed these adjusted relationships are to be stable and significantly to improve the total quality of human life.

9

GENE INTERACTION

So far we have considered Mendelian inheritance of single genes or of such units as the sex chromosomes. In all organisms, however, Mendelian phenomena occur simultaneously and with interacting effects at many loci. We now turn to the more complex considerations involving more than one gene.

9.1 POLYHYBRID RATIOS

The cross used in Mendelian investigations in which differences in only a single trait are considered leads to an F_2 ratio of 1:2:1 if there is no dominance, or to an F_2 ratio of 3:1 if there is dominance. These are ratios of **monohybrid** crosses. Mendel, however, did not confine his observations to them. He also studied crosses in which the parental types differed in two characters (**dihybrid** crosses) and in three (**trihybrid**). The general term for crosses in which more than one contrasting trait is considered is **polyhybrid.**

In his experiments with peas, Mendel studied a total of seven different traits, some singly and some in combination. It so happened that in the multiple trait experiments, he studied characters that were determined by genes located on different chromosomes. (The significance of this will become apparent in Section 9.2.) From these experiments, he established that in the transmission from the F_1 to the F_2 generation, genes may **assort independently** of each other.

Thus, if we designate the allele that codes for round-shaped seed as R, and its recessive variant, which codes for angular seeds, as r; the dominant allele for yellow albumen (the nutritive part of the seed) as Y, and its recessive green-producing allele as y, then a dihybrid cross may be represented as follows:

P generation

phenotype	round, yellow	×	angular, green
genotype	$RRYY$		$rryy$
gametes	RY		ry

F_1 generation

phenotype   round, yellow

genotype $RrYy$

gametes RY, Ry, rY, ry

The expected proportions of the four kinds of gametes produced by the F_1 is $1:1:1:1$ (or, in terms of frequencies $\frac{1}{4}:\frac{1}{4}:\frac{1}{4}:\frac{1}{4}$). If they unite randomly to form the F_2 zygotes, we expect these zygotes to have the genotypes presented in the chart below. The F_2 genotypes in all compartments of the chart (called a Punnett square, after an early British geneticist, one of the first to study the inheritance of genes in populations) are, in this example, expected in approximately equal frequencies. The genotypic frequencies are expected to be the products of the frequencies of the gametes contributing to the genotypes; in this case, all are $\frac{1}{4} \times \frac{1}{4} = \frac{1}{16}$. In other words, expected genotypic frequencies may be calculated from the frequencies at which the different kinds of gametes are expected to come together to form zygotes.

Sperm

		RY	Ry	rY	ry	Gamete genotype
		$\frac{1}{4}$	$\frac{1}{4}$	$\frac{1}{4}$	$\frac{1}{4}$	Expected frequency
RY	$\frac{1}{4}$	$RRYY$	$RRYy$	$RrYY$	$RrYy$	
Ry	$\frac{1}{4}$	$RRYy$	$RRyy$	$RrYy$	$Rryy$	
rY	$\frac{1}{4}$	$RrYY$	$RrYy$	$rrYY$	$rrYy$	
ry	$\frac{1}{4}$	$RrYy$	$Rryy$	$rrYy$	$rryy$	

Eggs

For each locus separately, a 3:1 phenotypic ratio prevails. In combination, the phenotypic ratio becomes (3:1) (3:1) = 9 (round, yellow) : 3 (round, green) : 3 (angular, yellow) : 1 (angular, green). The genotypic ratios can be readily observed from the Punnett square on page 160.

One of Mendel's experiments with the dihybrid cross is presented below. The numbers he obtained in the F_2, indeed, conformed with the expectations that we may determine from the Punnett square. In Mendel's experiment, below, the center (double heterozygote) compartment expected value (recall that this is the product of the gamete frequencies) is $\frac{2}{4} \times \frac{2}{4} = \frac{4}{16} = .25$ (observed = 138/529 = .261); the homozygous recessive (*rryy*) compartment expected value is $\frac{1}{4} \times \frac{1}{4} = \frac{1}{16} = .0625$ (observed = 30/529 = .0567); and the homozygous angular, heterozygous yellow (*rrYy*) compartment expected value is $\frac{1}{4} \times \frac{2}{4} = \frac{2}{16} = .125$ (observed = 68/529 = .129).

F_2 genotype for shape

	Expected frequencies ⟶	RR $\frac{1}{4}$	Rr $\frac{2}{4}$	rr $\frac{1}{4}$	Total	Approximate ratio
YY	$\frac{1}{4}$	38	60	28	126 ⎫	
Yy	$\frac{2}{4}$	65	138	68	271 ⎬ 397 yellow	3
yy	$\frac{1}{4}$	35	67	30	132 green	1
		138	265	126	529 Total plants	

(F_2 genotype for color)

Total 403 round / angular

Approximate ratio 3 : 1

The combined phenotypic ratio is 301 (38 + 60 + 65 + 138 round, yellow) : 102 (35 + 67 round, green) : 96 (28 + 68 angular, yellow) : 30 (angular, green), or approximately 9:3:3:1.

In the same way, the trihybrid F_2 ratios expected are (3:1) (3:1) (3:1) = 27:9:9:9:3:3:3:1. Such expressions can be extended to any number of genes, but study of the higher polyhybrid ratios becomes experimentally impractical, because of the increasing numbers of organisms required.

So far we have considered genes that are located on different chromosomes and that affect independent biochemical pathways. But in many cases, polypeptides, enzymes, and their various altered substrates interact with each other and modify phenotypic expressions of the various genotypes. Interaction between products of genes at different loci is known as **epistasis.** Originally, Bateson suggested this term for one form of interaction only, but today the tendency is to use it for any kind of interaction between nonallelic genes.

Box 9.A shows a variety of dihybrid ratios, some of which show epistasis. The various combinatorial schemes, many of them worked out by Bateson and his

BOX 9.A DIHYBRID MENDELIAN RATIOS

The following table presents various results that have been reported for independent assortment of genes (each having two alleles) at two loci. The first row shows an ordinary Mendelian F_2 ratio, combining two independent characters as in the classical Mendelian experiments.

The second row also shows a 9:3:3:1 ratio, but in this example, and those in rows 3–9, genes at two independent loci are interacting to affect a single character. In the particular pedigree from which row 2 was taken, the grandparents (P generation) were rose-combed hens (*AAbb*) and pea-combed roosters (*aaBB*). The F_1 were all *AaBb*, and had a comb shape called walnut, a phenotype not shown by any of the P generation. The walnut-shaped comb is caused by an epistatic interaction between one or more dominant genes at each of the two loci. Another new epistatic phenotype, which did not appear in either the P or F_1 generations, is the double-recessive single comb, which was shown by $\frac{1}{16}$ of the F_2.

Examples 2–9 present various combinations of epistatic expressions and the different F_2 phenotypic ratios that result. In the example in row 10, there are no interactions: both *A* and *B* have the same effect on kernel color, which is determined merely by the dosage of dominant alleles, ranging from four for the deepest color to none for white.

| | | | Genotype and number | | | | | | | | | |
Organism	Character	Ratio	AABB 1	AABb 2	AaBB 2	AaBb 4	AAbb 1	Aabb 2	aaBB 1	aaBb 2	aabb 1	Type of interaction
1. Peas	Seed shape and color	9:3:3:1	Gray-brown, round				Gray-brown, angular		White, round		White, angular	None: regular F_2 ratio
2. Chickens	Comb shape	9:3:3:1	Walnut				Rose		Pea		Single	Regular with interaction
3. Mice	Color	9:3:4	Agouti[1]				Black		White		Albino	Complementary (for agouti)
4. Squash	Color	12:3:1	White						Yellow		Green	Inhibitor (*A*) of color
5. Swine	Color	9:6:1	Red				Sandy				White	Complementary (for red)
6. Sweet peas	Color	9:7	Purple				White					Complementary (for purple)
7. Chickens	Leg feathers	15:1	Feathered								Feather-less	Duplicate genes
8. Chickens	Color	13:3	White						Colored		White	Inhibitor of color
9. Mice	Color	10:3:3	White spotted				White		Colored		White spotted	Complex interaction
10. Wheat	Kernel color	1:4:6:4:1	Dark red	Medium dark red		Medium red		Light red	Medium red	Light red	White	None: multiple genes

[1] Black hair with a yellow-reddish band, giving gray appearance.

associates, produce the different F_2 phenotypic ratios. Explanation of two of the examples will give some idea of how the different ratios come about. Apparently, two interacting enzymes are needed to produce color in sweet peas (example 6). Absence of either (or both) of these enzymes will lead to colorlessness. Hence, the 9:7 ratio. In chickens, it seems that two different dominant genes control the appearance of feathers on the legs. Presence of one dose of either produces feathering, and only the double recessive has featherless legs; the ratio, therefore, is 15:1.

9.2 CHROMOSOME ORGANIZATION

As you may recall, each species is normally characterized by a constant chromosome number. These numbers range from a single pair in the roundworm *Ascaris*—although it should be mentioned that in the somatic cells of this organism they break up into a number of smaller chromosomes—to over 800 pairs in the marine animal *Aulacantha.* In plants the number of chromosome pairs may vary from two to more than 500, with some ferns having the highest number.

We may ask what happens when two genes entering a dihybrid cross are located on the same chromosome. In such a situation independent assortment fails, and the alleles entering the gametes of the F_1 tend to retain the association they had in the P generation. For example, in *Drosophila,* an F_1 male from a cross between a female with scarlet eye color and curled wings (both recessive genes) and a male normal for both eye color and wing shape will be heterozygous at the two loci. If this F_1 male is backcrossed to a female of the same genotype as his mother, the expected ratio among the progeny (should the traits assort independently) is 1 scarlet eyes, curled wings : 1 normal eyes, curled wings : 1 scarlet eyes, normal wings : 1 normal eyes, normal wings. However, because these two genes are located on the same chromosome, the F_1 male will produce only two kinds of gametes: one containing the scarlet and curled recessive genes and the other the dominant normal alleles. As a result, only two kinds of individuals will appear among the backcross offspring, one like the male and the other like the P-generation female.

Complete linkage, however, is not the rule. In the process of meiosis there may be an exchange of sections between members of homologous pairs of chromosomes, a process known as **crossing-over.** In *Drosophila* males, crossing-over is usually suppressed; hence, the results described. But different species differ in their behavior. Thus, in the silkworm, it is also only the homogametic sex (but this time the male) that shows crossing-over. In some birds, the homogametic male sex shows less crossing-over than the female. In mice, the frequency of crossing-over is higher in females. In the pea, it is equal in the male and female.

Crossing-over is a very complex process, and much is still to be learned of its physics, chemistry, and biology. For our purposes, we will treat it in a simplified way (see Boxes 9.B and 9.C), ignoring details and present uncertainties.

BOX 9.B CROSSING-OVER

The process of crossing-over may be schematically and numerically represented as follows. The data deal with the color of poppies, and were reported originally by J. Philp. Only two loci are shown here, but the principle applies to situations with many genes, and multiple crossovers also occur. The genes in question are the dominant for white petal edge, *W,* and the recessive *d,* which dilutes normal color and produces some other effects. If one parent is a double recessive

$$\frac{dw}{dw},$$

any cross-overs will have no effect, since exchange between the upper and lower chromosomes would merely reconstitute a *dw* gamete (bottom part of the diagram). If the other parent has the genotype

$$\frac{dW}{Dw},$$

a crossover will result in the production of four kinds of gametes, as shown in the center part of the diagram. The particular form of representation is based on the fact that crossing-over occurs when the chromosomes are in a double-stranded stage.

In Philp's experiment the numbers of the different phenotypes (1) expected on the basis of independent assortment, (2) expected on the basis of complete linkage, and (3) actually recovered, were as follows:

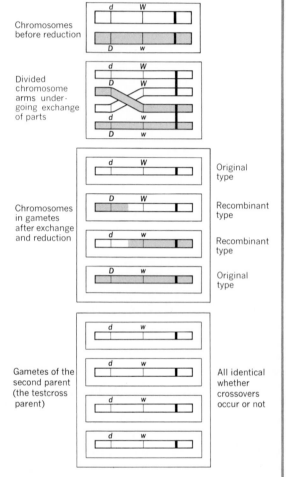

Phenotypes	Dilute with white petal edges (ddWw)	Normal with white petal edges (DdWw)	Dilute (ddww)	Normal (Ddww)
1. Expected (independent assortment)	199.5	199.5	199.5	199.5
2. Expected (complete linkage)	399	0	0	399
3. Observed	400	3	4	391

The heterozygous parent in the cross received a *dW* chromosome from one grandparent and a *Dw* chromosome from the other. In the offspring, out of a total of 798 plants, 791 were of the grandparental types. Seven out of 798, or 0.9 percent, were new *recombinant* types.

BOX 9.C CYTOLOGICAL PROOF OF CROSSING-OVER

The diagram illustrates the classical experiment reported in 1931 by Curt Stern, giving the cytological proof of crossing-over. In the experiment he used two abnormal X chromosomes of *Drosophila melanogaster*. Because of their abnormalities, these chromosomes and their cross-over recombinants were visually distinguishable when viewed through a microscope. The one represented first in the diagram was a broken X chromosome, a small portion of which had become attached to an autosome; thus a sector appeared to be missing from it. The recessive allele for carnation eye color and the dominant allele for bar-shaped eyes were present in the main portion of the X. In the other, a small piece of the Y chromosome, shown in outline, was attached to the X chromosome. By making the cross shown and comparing the phenotypes of the offspring with the cytological appearance of their chromosomes, Stern was able to demonstrate that recombination between the genes for eye color and eye shape was accompanied by a physical exchange of parts of the X chromosomes. A similar

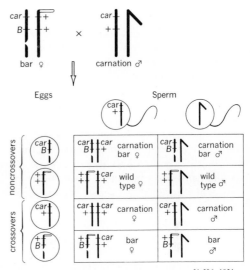

THE DIAGRAM IS ADAPTED FROM STERN, BIOL. ZENTR., 51:586, 1931.

experiment with corn was performed by the American geneticist, Barbara McClintock.

The understanding of linkage and crossing-over was greatly increased by the concept that genes are arranged linearly on the chromosome. This concept was first proposed and later brilliantly confirmed by the first geneticist to receive a Nobel prize, **Thomas Hunt Morgan** (1866–1945; Figure 7.2), and a group of his students. They assumed that the probability that crossing-over occurs between any two loci on the same chromosome is proportional to the physical distance separating them. Because their assumption proved to be correct, it is possible, by making crosses among individuals differing at three or more loci on the same chromosome, to construct a **linkage map.** Such a sequence of loci on a chromosome may be deduced much as the arrangement of stations along a railway may be deduced by the number of hours a train, traveling at a constant rate, takes to cover the distance between them. Thus, if it takes the train two hours to travel from A to B, three hours from A to C, and one hour from B to C, it is reasonable to infer that the order of the stations is A, B, C, with relative distances of two and one between them. If the train speeds vary, because of different topography or other factors, the relative map distances of two and one will probably not be very accurate, but the station order will be. The analogy holds for chromosomes, in that genetic map distances do not precisely measure physical distances between loci on a chromosome, but mapping determinations of the order of the

BOX 9.D LINKAGE MAPS

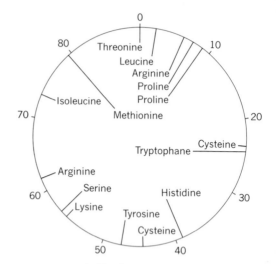

The bacterium *Escherichia coli* has one circular chromosome, although at certain stages of the life cycle there may be two to four replicates of this chromosome within a single bacterium. Between one-fifth and one-third of all the metabolic reactions in this organism are already known. Over one hundred loci on the chromosome have been mapped, some of which are shown in the illustration to the right. All the genes listed in this illustration are associated with requirements for some amino acid. Many loci are usually associated with requirements for a particular amino acid: for instance, at least eight are known for arginine (two of which are mapped in the illustration). Other known genes control such biochemical characters as ability to metabolize various energy sources, resistance to lethal effects of certain substances, and presence of antigens; others cause variation in morphological traits essentially through biochemical processes.

Of all plant linkage maps, the most complete is for corn, with about 100 loci known on its ten chromosomes. *Drosophila melanogaster,* the representative of the animal kingdom best known to geneticists, has about 500 loci mapped on its three autosomes and X chromosome. The diagram below illustrates the correspondence between the position of some *Drosophila* sex-linked genes as determined by frequency of crossing-over between them (mapped along the dark bar) and the location of nearby points on the chromosome (pictured at the top of the diagram) as established from cytological evidence.

About a hundred genes have been mapped in the mouse, and 30 in the chicken.

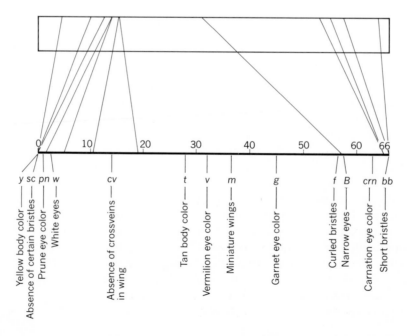

THE DROSOPHILA DIAGRAM IS REDRAWN FROM TH. DOBZHANSKY.

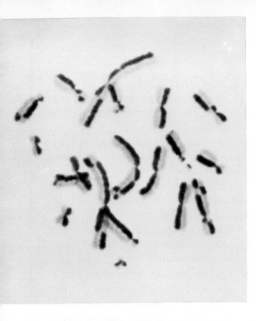

FIGURE 9.1
Chromosomes from a Chinese hamster cell culture treated and prepared for photography in a special way that causes the newly replicated strands to be light while the "old" strands are dark. Many of these somatic chromosomes clearly show the location of exchanges (crossovers) between sister strands (chromatids). (Courtesy of Sheldon Wolff.)

loci on the chromosome are accurate. There are many complexities in chromosome mapping that we shall ignore here. Box 9.D gives some examples and Box 9.E gives information on human autosomal linkage.

Some points about crossing-over need emphasis. The first is the evolutionary significance of this process. Without crossing-over, the number of different genotypes possible in a species would be limited by the number of chromosomes. With crossing-over, the number of potential genotypes is astronomical, since, theoretically, an exchange may occur between any pair of adjacent nucleotides. Given multiple allelism and crossing-over, the number of possible different human genotypes is fantastically large, exceeding the total number of particles in the universe. The contention that no two human beings that ever existed, or will ever exist, would have identical genetic constitutions is not unrealistic under these circumstances. Identical twins might be an exception, but even they would differ from each other if mutation took place in one (or more likely both) shortly after twinning. (In any case, because of the way cytoplasm is distributed, identical twins may still differ because of early differences in cytoplasmic material.) Another point about crossing-over is that it can occur in somatic cells as well as in germ cells (Figure 9.1). This process in a heterozygote is one possible cause of mosaicism (Box 9.F).

9.3 COADAPTATION

The epistatic gene interactions that were considered in Section 9.1 were analyzed on a formal Mendelian basis. When we do Mendelian analyses, we are operating as if the genes under consideration are functioning independently of the other genes in the organism. Directly or indirectly, however, all genes must be interacting to produce an adapted individual and an adapted population. The complicated biological processes that transform single-celled zygotes into a successful

BOX 9.E LINKAGE IN HUMANS

Classical genetic techniques requiring matings between individuals of known genotypes and enumeration of large numbers of offspring do not seem promising for applications to a long-lived, politically difficult, expensive organism with relatively few offspring per family. Some milestones in human chromosome mapping are:

1911—First assignment of a gene to a human chromosome (a gene for color blindness to the X)

1937—First linkage estimate (of color blindness and hemophilia on the X)

1951—First autosomal linkage estimate (between the Lutheran and Lewis blood-group genes)

1956—Verified correct count of human chromosomes (46, not 48 as previously thought)

1962—First determination of a group of three linked loci (with addition of location of G6PD gene on the X)

1968—First determination of a linkage group of three autosomal loci

1968—First assignment of a gene to a particular autosome (Duffy blood-group gene to chromosome 1)

1968—First use of cell culture to assign a gene to a particular group of chromosomes (thymidine kinase gene to a chromosome in Group E)

1970—First assignment of a gene to a particular chromosome arm (haptoglobin gene on chromosome 16)

1971—First determination of linkage from cell culture

Thus, until recently, the human genetic map was a virtual blank. But two new techniques are now available and being developed, and in laboratories around the world the task of mapping human genes is under way. One of these uses the very recent advances in staining techniques, which make the identification of individual chromosomes unambiguous and permits recognition even of parts of chromosomes.

Development of the second technique goes back to 1960, when it was discovered that mixtures of two different mouse cell cultures occasionally produce a third cell line, with characteristics of both "parental" cell lines. The cells fuse, and then their nuclei fuse, and cell division continues with an approximately tetraploid chromosome set (i.e., most of the chromosomes of both cell lines are retained). In 1967, human and mouse cells in cell culture were successfully fused. Now, one might ask, what is the purpose of that scary trick? It turns out that in this mouse-human cell hybrid, and similarly in the subsequently produced Chinese-hamster–human cell hybrid, the human chromosomes do not fare too well. Over a number of cell generations, most of the mouse (or hamster) chromosomes are retained, while most of the human chromosomes are expelled. Different clones can be established with different surviving human chromosomes among the chromosome complement of the mouse or hamster. Since many of the gene products of the human cell line are different from those of the hamster or mouse cell lines, the human enzymes or other gene products can be identified as being associated with the one or few particular human chromosomes remaining in the different cell clones, and with each other in linkage groups. While the fact of a human-rodent hybrid suggests frightening or bizarre monsters, the hybrid cell cultures (thus far) have produced no monsters but much useful and interesting information in this important line of research.

It is even possible to use techniques of molecular biology, selecting cell lines that *need* a particular human gene product that is missing from the hamster or mouse lines. After such selection, the new cytological staining techniques are employed to see precisely which human chromosomes have been selectively kept in the cell hybrid line. Mapping genes within chromosomes makes use of chromosomal translocations, in which a bit of a human chromosome is attached to a different chromosome and then maintained in the line, producing its unique gene products. Here the new staining techniques are particularly important for identifying such bits of chromosome.

Knowing the locations of many genes within the human chromosome complement will be important for various types of genetic engineering (Section 22.2). It may also prove a valuable tool for genetic counseling, as genes linked to known defects in the parents can be checked in fetal tissue. A defective gene may not express itself in early embryos, but if the appropriate linked genes can be identified in cultures of the fetal cells, a decision affecting possible treatment or abortion can be made with greater certainty.

population of mature organisms must have extensive regulatory aspects. This is to say that alternative alleles at a locus are selected not only for what they themselves do, but also for their interactions with the totality of all alleles carried by an individual, and, beyond that, with the total contents of the gene pool.

The evolutionary process of selection for a balanced combination of genes in an individual, and of individuals in a population, is known as **coadaptation.**

BOX 9.F PRODUCTION OF MOSAICS

Mosaics may result from somatic crossing-over (Figure 9.1), as well as from losses of chromosomes as illustrated in Figure 8.1. Suppose that in a heterozygote for eye color, there was crossing-over during the first division after the zygote was formed. (*BB* and *Bb* are genotypes expressed as brown eyes; *bb* as blue.) The course of development outlined in the diagram might occur.

The photograph shows a human mosaic. Genes affecting both eye color and hair color must have been included in the chromosomal event (or, less likely, events) that produced this mosaic.

Not all mosaics are caused by chromosomal abnormalities. Thus, Siamese cats and Himalayan rabbits have a system of regulation in which genes producing melanizing enzymes are turned on only at certain temperatures. Differences in the temperature of the different parts of the body during development may lead to a mosaic or variegated phenotype. In plants, variegation in leaf color depends on differences between green and colorless plastids in the cytoplasm (see Box 2.C and Section 9.4).

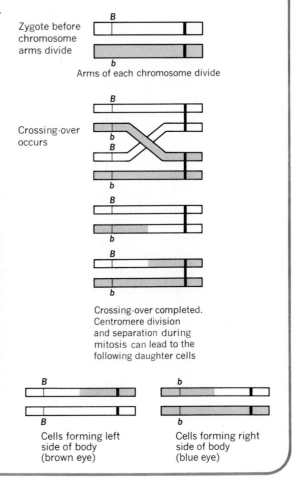

Zygote before chromosome arms divide

Arms of each chromosome divide

Crossing-over occurs

Crossing-over completed. Centromere division and separation during mitosis can lead to the following daughter cells

Cells forming left side of body (brown eye)

Cells forming right side of body (blue eye)

THE PHOTOGRAPH IS REPRODUCED COURTESY OF Y.R. AHUJA AND PERMISSION OF L. GEDDA FROM ACTA GENET. MED. GEMELL.

Darwin used this term to describe correlated modifications of different structures in the course of the evolution of an organism to produce a harmonious living being. Its meaning, however, can now be extended to the genic level on the one hand, and to the species level on the other.

One example of genic coadaptation is provided by *synthetic lethals.* As already noted, many genes have detrimental effects, some of which may be fatal or sterilizing. But, in *Drosophila* at least, some genes are harmful or lethal only when present in some epistatic combination with certain others. In analyzing the effects of whole chromosomes on viability, it is sometimes found that a chromosome, which had no detrimental effects originally, becomes lethal after crossing-over. However, another crossover can remove the lethal effects. Clearly, an interaction between genes on different parts of such a chromosome must exist. Another and even simpler example of interaction, or balance on the genic level, is provided by overdominance. Still other examples can be found in the biochemically determined *position effect,* that is, the modification of genic expression by a change in the location of a given gene on the chromosome, as may happen when an inversion (Figure 3.6) occurs.

On the population level, coadaptation can take the form of *frequency-dependent selection.* That is to say, the relative fitness of a given allele may depend on its frequency in a gene pool. When a certain allele is rare, it may confer a reproductive advantage on the individual carrying it; if it becomes common it may prove to be disadvantageous. A nonbiological analogy in human society would be in the supply and demand situation controlling, let us say, the number of watchmakers in a community. When few watchmakers are available, they prosper, but as their numbers increase, their prosperity may vanish. The exact biological nature of frequency-dependent selection is not known, but it can logically be related to such things as exploitation of particular niches or substrates; to the likelihood that predators will search for common forms, while being less alert for rare ones; or to mating patterns in which unusual individuals are preferred as mates.

Another consideration dealing with interactive adaptation relates to the phenomenon of **pleiotropy,** the multiple phenotypic effects that a particular gene may have. A gene is, in general, responsible for a single specific product, a polypeptide (generally a protein). Many proteins are enzymes, and the enzyme that a single gene codes for may participate in a variety of reactions, each one of which may have ever-spreading consequences in the functioning of an organism. Figure 9.2 shows two examples of the pleiotropic action of single genes. The first part presents a series of developmental events resulting from the presence of a lethal gene in the rat. The second shows the effects of the frizzle gene in the chicken, stemming from the production of defective feathers.

Pleiotropic effects mimic and may be confused with very close linkage. If crossing-over between two nearby genes is very rare, in many instances they may be considered to lie at a single locus. Only in the lower organisms, which can be grown by the millions in the laboratory, or those in which biochemical techniques

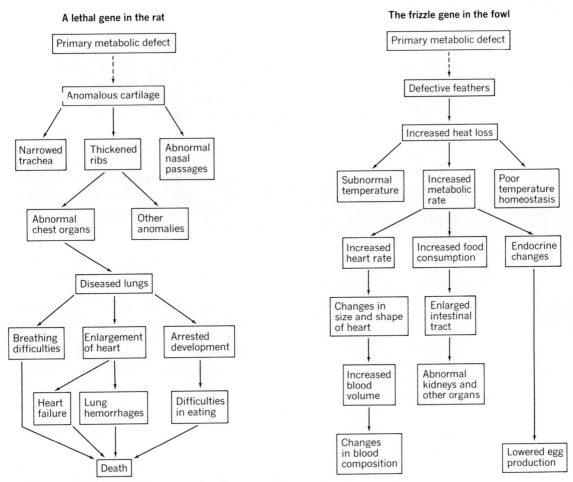

FIGURE 9.2
Pleiotropic effects of two genes. (*Left:* after H. Grüneberg; *Right:* after W. Landauer.)

permit study of the fine structure of genes, is it possible to distinguish unequiv-ocally tight linkage from pleiotropy.

In many organisms, a function lost due to a deleterious mutation can be re-covered through a **suppressor mutation.** That is, a second mutational event at a different locus may introduce there an allele that corrects the detrimental effects of the first mutation. There are several ways in which a suppressor mutation may lead to the recovery of function: by changing the frame of reading of the DNA (Figure 6.3), by altering one of the tRNAs, by changing the substrate to something the new enzyme (i.e., the enzyme coded for by the deleterious mutant allele) can usefully act upon, or by creating a use for the altered product.

9.4 NONNUCLEAR GENETIC EFFECTS

So far we have largely confined ourselves to the action and interaction of genes carried on chromosomes. But the products of genes also interact with extrachromosomal determinants of the phenotype. We may briefly mention here two kinds of such interaction, **cytoplasmic inheritance** and **maternal effects.** In a sense, the former might be included in the latter, because the greatest part by far of the cytoplasm of the zygote is contributed by the mother (i.e., is the cytoplasm that is originally in the egg).

Cytoplasmic inheritance may take many forms. In plants and some protists, self-reproducing plastids, some of which function in photosynthesis, are commonly transmitted through the maternal cytoplasm; for example, those referred to in Box 2.C. In some instances, they, as well as mitochondria, have been demonstrated to contain DNA and thus could be described as nonnuclear genes. In other protists, such cytoplasmic inclusions as antigens are known to be under the control of nuclear genes, or to show various kinds of interactions with nuclear gene products. In general, nonnuclear genes are difficult to study because the cytoplasm is not as regularly distributed to daughter cells as the nuclear material. Furthermore, although the bulk of the zygote's cytoplasm comes from the egg, the sperm also makes some cytoplasmic contribution.

Two generalizations concerning cytoplasmic genetic systems, long anticipated but only recently established, are: (1) organelles (particularly chloroplasts and mitochondria) contain double-stranded DNA of high molecular weight, as well as other molecular structures required for protein synthesis; and (2) organelle biogenesis is controlled by interactions between chromosomal and cytoplasmic genes. Experimental approaches being used to study these phenomena include not only genetic analyses, but physical studies of the implicated DNAs, studies of RNA and protein synthesis in isolated cell fractions, hybridization of cytoplasmic DNA and RNA, and structural studies of organelle proteins.

Thus far these studies have yielded the following conclusions and speculations: Cytoplasmic genes appear to be less frequently mutated than nuclear genes. Chloroplasts and mitochondria contain ribosomes that differ from the cytoplasmic ribosomes of nuclear origin, both in their physical properties and their responses to physiological regulation. The structural proteins of mitochondrial and other cytoplasmic membranes are insoluble, possibly requiring that transcription and translation occur at the site of membrane formation. The cytoplasmic DNAs are replicated at different times than nuclear DNA. This may provide some regulatory flexibility and responsiveness to environmental signals. It is fair to say at this time that we recognize two kinds of genes: NUCLEAR, and UNCLEAR.

In one peculiar type of maternal inheritance, the direction of coiling in certain snails, the phenotype of the offspring is determined not by its own genotype, but by that at a single locus, *Rr,* in its mother. Thus, in reciprocal crosses of homozygous "right-coiled" and "left-coiled" snails, the direction of coiling expressed in the two genetically identical F_1 families (both Rr) is different, being left-coiled if the mother was genetically rr and the father RR, but right-coiled if the mother was RR and the father rr. An F_2 can be produced by crossing right- or left-coiled Rr mothers with right- or left-coiled Rr fathers. Genetically, the F_2 will be $\frac{1}{4}$ RR, $\frac{1}{2}$ Rr, and $\frac{1}{4}$ rr, but they will all be right-coiled, in response to the dominant R gene in their heterozygous mothers.

Cytoplasmic factors in some animals can have practical significance for humans. For instance, in different strains of mosquitoes, cytoplasmic differences lead to incompatibilities between the chromosomes of one strain and the cytoplasm of another. Crosses between two such strains result in eggs that fail to develop. Massive numbers of males of one such strain may be released at appropriate times in a locality in which an incompatible strain may be a vector of malaria or other diseases. A high proportion of the native female mosquitoes, the ones inseminated by the visitors, would then fail to reproduce, because in mosquitoes all zygotes derive from sperm introduced at first copulation. Complete eradication of mosquitoes is possible in some places by this method.

In mammals, maternal effects mediated through differences in prenatal intrauterine environment or in postnatal nutrition provided by the mother are more common than those based on cytoplasmic differences. A classic example is illustrated in Figure 9.3. Some further details of this phenomenon are discussed in Section 11.4.

While we have indicated the similarity between "maternal effects" and "cytoplasmic inheritance," there is an important difference. Maternal effects are due to some phenotypic or genetic property of the mother that (when not shared by the offspring) produces an effect in the offspring for only one or at most a few generations. Cytoplasmic inheritance means inheritance. It is due to nonnuclear genes, and will last for as many generations as the offspring receive those genes from their parents.

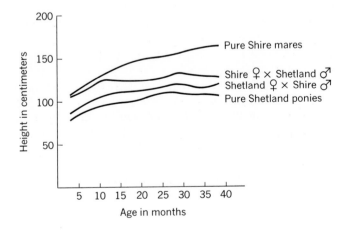

FIGURE 9.3
Demonstration, from matings between breeds of horses, of the effects of maternal environment. The larger mothers have the larger hybrid offspring. All data are on females; hence, sex linkage cannot account for the differences between reciprocal crosses, and they must be attributed to maternal effects. (Redrawn from A. Walton and J. Hammond.)

10

POLYGENIC INHERITANCE

So far, our emphasis has been on phenotypic expressions of single genes, and interactions within and between a relatively few loci. There are, however, many characters that depend on a larger number of genes. Their inheritance presents special problems that we shall consider next.

10.1 METRIC AND MERISTIC TRAITS

This chapter, and a good share of the next five, will be devoted to a discussion of the genetics of **polygenic** traits. The important points can be summarized as follows.

1. Most **metric** and **meristic** traits are affected by genes at many loci. (Meristic traits are countable, for example, the number of kernels in an ear of corn; metric traits such as height are measurable on a continuous scale. But not all metric and meristic traits in all organisms are necessarily polygenic. For instance, human

dwarfism can be caused by a single gene, and stem height in peas—one of Mendel's characters—segregates into clearly defined "tall" and "short" classes in response to two alleles at a single locus.)

2. The effects of alternative alleles at each of the segregating loci are usually relatively small and, in a sense, interchangeable. That is, apparently identical phenotypes may result from many different genotypes (Box 10.A).

3. The many loci affecting a particular character may be scattered throughout the genome. (The genome is the physical totality of the organism's genes—in the DNA of the chromosomes and cytoplasmic organelles.) For instance, in *Drosophila*, genetic analysis based on many gene markers and on cytological observation has been brought to a high degree of precision, and it is possible to estimate the effects of individual chromosomes or even parts of chromosomes on polygenic characters. Figure 10.1 provides an example of one such analysis.

4. The phenotypic expression of most polygenic characters is subject to considerable influence by environmental factors during development.

5. Most populations maintain great genetic variability for polygenic traits. The particular biological mechanisms that maintain such variability against loss due to selection or drift are not fully understood at present, and are under active investigation in population genetics.

6. Nothing biochemically exceptional has been discovered about genes controlling polygenic inheritance; in the past, however, it was proposed that genes

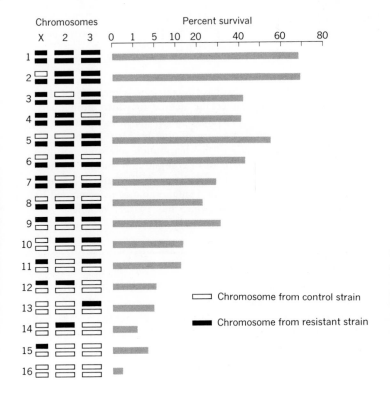

FIGURE 10.1

James F. Crow's analysis of the effects of the three large chromosomes of *Drosophila melanogaster* on resistance to DDT. The sixteen different combinations of chromosomes from a resistant strain and a control strain indicate that each of the three chromosomes carries some genes for resistance. (From Crow, *Annual Review of Entomology* 2:228, 1957.)

BOX 10.A THE NORMAL DISTRIBUTION

Phenotypic manifestations of genes with major effects tend to be discontinuously distributed. The top diagram shows the distribution expected in the F$_2$ of a monohybrid cross, such as one involving stem length in peas. But in reality, such distribution would prevail only if the two phenotypes were classified simply as long or short. Since there is some environmental modification of stem length, the phenotypes would probably cluster around some **mean** value. (Other measures of central tendency are the **mode,** the value of the most numerous group, and the **median,** the value of the middle item in the whole array.) If the relative frequencies are plotted against the phenotypic values, curves such as those shown in the second diagram are often obtained. The distribution within each phenotypic class becomes continuous. For the total population, however, a discontinuity between *AA* and *Aa,* on the one hand, and *aa,* on the other, still exists.

If we now remove the condition of dominance, the expected distribution of phenotypes can be computed from the coefficients of the expanded binomial $(a + b)^N$, shown in the third part of the illustration, which is known as Pascal's triangle. When the relative frequencies of phenotypic classes are plotted, a *normal* continuous distribution, characterized by a bell-shaped curve is approached, as shown on the bottom diagram. The larger the number of classes and the more individuals in the sample, the closer will be the approximation.

A characteristic property of a normal distribution is that the mean, mode, and median coincide. In addition to the measures of central tendency,

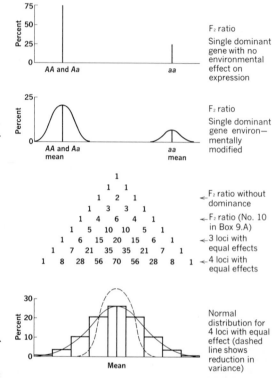

F$_2$ ratio
Single dominant gene with no environmental effect on expression

F$_2$ ratio
Single dominant gene environmentally modified

← F$_2$ ratio without dominance
← F$_2$ ratio (No. 10 in Box 9.A)
← 3 loci with equal effects
← 4 loci with equal effects

Normal distribution for 4 loci with equal effect (dashed line shows reduction in variance)

we shall also have recourse to a measure of dispersal, that is, the degree to which individuals tend on the average to deviate from the mean. A statistical measure of this property is known as the *variance* of the distribution. We shall not discuss the methods of computing it, but the dashed line in the lowest diagram illustrates a normal distribution with a small variance as compared to the solid curve, which has a larger variance.

controlling major effects were qualitatively different from those controlling minor effects. Both "kinds" of genes control production of enzymes or other proteins, and it is reasonable to think there is only one kind of structural (as contrasted to regulatory) gene. In fact, some genes that have a "major" effect on one character may have "polygenic" pleiotropic effects on others. For example, a gene for white feather color in chickens also moderately inhibits growth rate.

7. Single-family progenies are not particularly useful for study. The unit of study must be larger groups of many progenies. The data studied are not frequencies of offspring in discrete classes (which may be expressed as Mendelian ratios), but instead, are continuously varying measurements (see Box 10.A) of individuals that may best be described by averages and dispersion about those averages.

10.2 CORRELATION AND CAUSATION

The statistical techniques that have been developed for the study of continuous variation form the foundation of biometrical genetics. They stem from **biometry,** the statistical study of biological observations, a science developed by Francis Galton (Box 10.B).

In particular, great use in biometry is made of analyses of variance, and of covariance of quantitative characters, that is, the manner in which one trait changes with changes in another. Also very useful are the **correlation coefficient** (usually indicated by r), and regression analyses, which measure the degree of association between two or more variables.

Suppose it is found in comparing the speed at which an automobile travels and the distance it takes to brake to a stop, that for every increase of 10 miles of speed per hour, 20 feet of stopping distance is added. If this exact relation holds in every test, the correlation coefficient is positive and has a value of 1.0. If it were observed that increasing the number of people on a construction project reduces the time for its completion, the correlation would be negative. In reality, few perfect correlations are found: there is almost always some variation in the variables compared—one automobile may take 40 feet to stop, and another, traveling at the same speed, only 35. The numerical value of the coefficient is then reduced. If the additional distance required to stop is not a constant but is dependent on speed—or if so many workers are added to the construction project that they get in each other's way and are unable to perform their tasks efficiently—the relationship will not be linear and may even change sign. Correlations range from -1.0 to $+1.0$, with negative or positive sign indicating a negative or positive relationship between two variables. An r of 0.0 indicates no correlation, i.e., complete independence, between the two variables.

A very important consideration must be kept in mind when speaking of correlation. In the examples given, the speed of the automobile and the number of people employed may reasonably be assumed to be *causes,* while the distance traveled and the time taken to complete the job are *effects.* But generally speaking, we cannot infer merely from the presence of a high correlation which of two correlated items is the cause and which is an effect. Or, just as important, whether either one is a cause: they both may be the effects of a common cause.

For example, if a comparison were made between the number of storks found in the various districts of a country and the number of babies born in them, a positive correlation might well be established. It would, however, be rash to conclude that storks bring babies. A more plausible explanation is that storks abound

BOX 10.B SIR FRANCIS GALTON (1822–1911)

Francis Galton, one of the most eminent Victorians, was a man of insatiable curiosity and remarkable industry and ingenuity. His experiments on the genetic consequences of blood transfusion have already been cited in Section 3.6. He is credited with being the founder of biometry, although this honor might more properly be given the Belgian astronomer L. A. J. Quetelet (1796–1874). Galton was also the founder of mental testing, and a pioneer student of the individuality of fingerprints. (Fingerprint classifications actually were independently devised by an Englishman, Henry Faulds, in 1880, and then adopted by the British police in 1901). Galton conceived the idea of using twins in research on heredity. He coined the term **eugenics** for programs of improvement of the human species by selective breeding (although Plato had earlier suggested this possibility, basing it on the success of dog breeders.) In his book on hereditary genius (1883) he examined pedigrees of judges, statesmen, military commanders, literary men, scientists, painters, divines, oarsmen, and wrestlers, coming to the conclusion that eminence in these fields is largely governed by biological inheritance. Galton's views on eugenics were elitist, and if they were not the product of his class, they were certainly not in spite of his social origins.

There seem to be very few subjects that did not interest Galton. He even published an article in a scientific journal discussing the optimum ways of cutting a cake to preserve its freshness. In 1872 he created a furor amid the Victorian establishment by publishing an investigation on the efficacy of

prayer. Reasoning that the most-prayed-for people are royalty, and acknowledging that those who daily or weekly prayed for the health and long life of the Royal Families were doing so in good faith, he failed to find evidence that members of these Royal Families outlived people having similar access to good housing, nutrition, and medicine but lacking multitudes of subjects who offered prayers for their health. He even extended his study from those prayed for to those who pray, that is, the clergy. Here, once more, the net conclusion was not conducive to the belief that praying is beneficial. Galton's influence in genetics, psychometrics, anthropology, and many other fields, has been exceeded by few.

in the countryside and are rare in cities, and that birthrates are lower in urban than in rural communities. The causal factor here is the degree of urbanization of the districts, and the two variables originally compared are both effects. Hence, shooting half the storks in a district should not be expected to lower its birthrate.

Interpretation of correlations can thus be very tricky. Galton, for instance, assumed that the correlation between parents and children for eminence in a profession was produced by community of inheritance. But, as parents and children also tend to have a similar environment, at least two causes contributed to the correlation. We shall see later that there are biometrical techniques that can help to untangle nature from nurture, or genotype from environment. Such techniques are highly effective in experiments with plants and animals, in which much of the environment can be randomized and made independent for parents and offspring, but less so in the study of humans. Meanwhile, we may profit by viewing the relationship between genotype (G), environment (E), and phenotype (P) as shown in the diagram, which uses single-headed arrows to point from cause to effect and double-headed arrows to indicate correlation due to a shared cause:

This may also be written as $G + E + r_{GE} = P$. The diagram may be extended to two generations, with subscripts S for sire, D for dam, and O_1 and O_2 for two offspring:

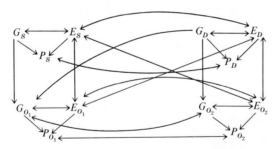

Some relationships pictured in this messy-looking, but useful, diagram are:

The genotype of the sire (G_S) is correlated (r_{GE}) with the environment of the sire (E_S). Likewise for the genotypes and environments of the dam ($G_D : E_D$) and both offspring ($G_{O_1} : E_{O_1}$, $G_{O_2} : E_{O_2}$). An example of a genotype-environment correlation might be that high school students genetically endowed to run fast tend to go out for track, and thus to receive training and conditioning that help them run even faster; students not thus endowed may engage in more sedentary pursuits, to the detriment of their running performance.

The genotype of offspring one (G_{O_1}) is jointly caused by gametes from the sire and dam. Thus, this offspring may share some genetic sources of ability to run fast with either or both parents. The genotype of offspring two shows a causal relationship from the dam, but the true father is neither known nor shown. Thus, the genetic correlation between these half-sibs ($G_{O_1} : G_{O_2}$) can be related to a common inheritance from the mother.

The two children share elements of their environments ($E_{O_1} : E_{O_2}$). They also share elements of the husband's environment ($E_S : E_{O_1}$, $E_S : E_{O_2}$), which may be subtle, such as a pressure to go out for track exerted by the trophies in the husband's den. There is a strong possibility these two genetically—and phenotypically—different children will not respond in the same way to this parental environment.

If the parents were relatives, there would also be a double-headed arrow between G_S and G_D. Or, even if they are not related, if the mother's choice of this male to father her first child were influenced by any correlation between his and her abilities to run fast, to the degree that ability is genetic, it would generate a genetic correlation between G_S and G_D due to this nonrandom element in the choice of mate.

Other possible causes, effects, and correlations can be observed in the diagram. To disentangle the various components represented in this diagram is no simple matter, and for human data it is possible only with a low degree of accuracy—with some exceptions, which we shall examine.

10.3 HERITABILITY

Three of the architects of biometrical genetics and its closely related fields were the British statistician **R. A. Fisher** (1890–1962; Figure 10.2) the American geneticist **Sewall Wright,** and J. B. S. Haldane (Box 2.G). Much of what appears in these chapters is based on their work and that of the American geneticist Jay L.

FIGURE 10.2
Sir Ronald Fisher, one of the foremost statisticians of the twentieth century. He formulated several principles important in genetics and in the study of evolution, and developed a number of powerful tools of biometry. (Courtesy of Godfrey Argent.)

Lush and the British geneticist and biometrician Kenneth Mather. Among the important concepts developed by Lush is that of **heritability.**

The concept, or term, "heritability" can be found being used in three different ways:

1. The degree of genetic control of the development of a character.
2. The ratio of genetically caused variability to total variability of a character in a population.
3. A measure of genetic response to selection.

Of these, only the second is really correct. But knowing such a ratio is rarely an end in itself. Either meaning (1) or (3) is usually applied to the ratio, and it is here that errors of interpretation may result. Before we discuss these further, we will pursue the construction of (2), the heritability ratio.

If a quantitative character is measured in a population, it will exhibit some phenotypic variability. Part of this variation arises from the fact that the different individuals constituting the population have different genotypes, and part, that they developed in different environments. Ignoring for the moment the interaction between genes and environment, we might say that if it were possible to equalize all the environments, all environmentally caused variation in the character would disappear, but genetically caused variation would still be present. It is this fraction of the total phenotypic variation that is called heritability. Usually designated by the symbol h^2, heritability may be determined by a variety of statistical methods. If we designate phenotypic variation and its genetically and environmentally caused components by P, G, and E, we have the relation

$$G + E = P.$$

Dividing by P, we obtain

$$\frac{G}{P} + \frac{E}{P} = \frac{P}{P} = 1,$$

or

$$h^2 + e^2 = 1,$$

where e^2 is the complementary term to heritability, indicating the fraction of the phenotypic variation that would be left were all genetically caused variation removed.

It is clear that h^2 values can range from 1, when no environmentally caused variation is present (as in blood groups), to zero, when all variation is environmental.

Heritability ratios are most cleanly constructed by the statistical techniques of analyses of variance: The environmental and genetic components contributing to total variation can be estimated, and the ratio constructed directly. An example of this will be presented in Section 11.2.

BOX 10.C DETERMINATION OF HERITABILITY FROM SELECTION EXPERIMENTS

Let us assume that there is a normal distribution of a metric character in a population. (This assumption is generally reasonably satisfied.) We may first consider two extreme (and rarely encountered) conditions: When all variation in the character is determined by environmental differences, h^2 is 0, or it is 1 when variation observed among the phenotypes is entirely due to genetic differences. We may understand these two conditions better if we examine the effects they predict when applied to (3) on page 181, response to selection. For this example, we will restrict reproduction to individuals that exceed the average of their generation, i.e., those individuals whose measurement falls to the right of the mean of the bell-shaped curve. We will call these individuals that reproduce the "selected parents." If $h^2 = 0$, provided that the environments of the parents and the progeny are not correlated, the average expression of the trait in the offspring would be not at all dependent on the phenotypes of the parents. Whether the parental phenotypes were above or below the mean, the average phenotype

of the filial generation would still equal the average of the total parental generation and not that of the selected parents (left curves).

If $h^2 = 1$, most of the offspring (having inherited the genes of the selected parents) would exceed the mean of the parental generation (right curves). Note that some offspring are below the mean of the parental generation, but that the mean of the offspring equals the mean of the *selected* parents, and that many offspring exceed the most extreme individuals in the parental generation.

Intermediate heritability values predict intermediate results (center curves). It is such intermediate heritabilities, and means of offspring between those of the selected parents and the total parental generation, that are typically found in real situations.

Note that in all three sets of curves, the difference between the mean of the selected parents and the mean of the population they were selected from (the **selection differential**) is the same. In each case, the offspring are redistributed from the

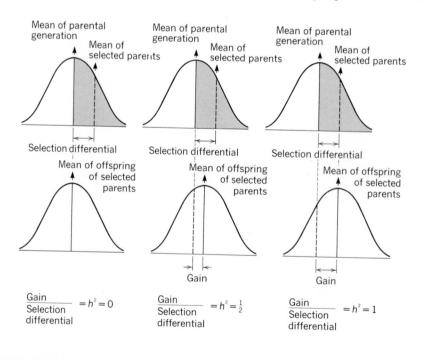

truncated distribution of the selected parents to form a normal distribution. The change in mean (**gain**) differs, and depends on the heritability, which may be estimated by dividing the gain by the selection differential, as shown in the diagram.

Actual examples of differences in the performance of offspring whose parents were selected for characters of moderate (0.25) and high (0.75) heritability are presented in the diagrams at the right (from experiments carried out at the University of California at Berkeley).

The top diagram shows the November egg production of a group of Leghorn pullets, a meristic character with heritability of about 0.25. The more prolific layers among them were selected to be parents, and the performance of their daughters is shown by the shaded distribution in the next diagram. The poorer layers from the original group also produced progeny. The egg production of their daughters is shown by the outlined distribution in the same diagram. Note, on the average, they laid fewer eggs, but there was considerable overlap with the egg production of daughters of the better layers.

The third and fourth diagrams show distributions from a similar experiment, in which the average weight of eggs produced (rather than their number) was the performance character selected for. The heritability of egg weight is very high (about 0.75). After one generation of selection, the means of the two offspring populations were well separated, and there was little overlap compared to that in the experiment with egg number.

Precise heritability figures for human characters cannot usually be obtained by such methods. Actually, using selection is not the problem, as we can simply compare offspring of groups of parents classified as high or low for expression of a specific trait, as in the following example. The difficulty lies in disentangling genetic correlations from environmental correlations. For instance, in the set of data at the right, a clear relationship between longevities of parents and their offspring is shown. But in these data, how much of the effect can be attributed to G, and how much to E, is impossible to tell.

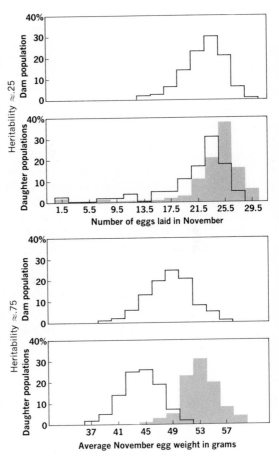

Country	Date of study	Life-span of offspring from short-lived parents, in years	Life-span of offspring from long-lived parents, in years
U.S.	1903	39.4	41.8
U.S.	1918	32.8	52.7
U.S.	1931	40.9	45.8
China	1932	31.5	39.8
Scandinavia	1951	36.2	48.1
U.S.	1956	49.5	57.9

DATA IN THE TABLE ARE FROM F. J. KALLMAN.

Essentially, h^2 is a measure of correlation between genotype and phenotype. As an example, if we represent the genetic and phenotypic correlations in the lower part of the diagram appearing in the previous section by r_G and r_P, the relationship $r_P/r_G = h^2$ may be used to estimate heritability. Theoretical genetic correlations can be computed. For instance, since each parent provides half of the genes of each offspring, the genetic correlation between parent and immediate progeny is 0.5. Because identical twins have identical genotypes, the genetic correlation between them is 1.0. In a similar way, it can be shown that the genetic correlation between **full-sibs** (sisters or brothers with parents in common) is 0.5. Between half-sibs (having one parent in common) it is 0.25, between first cousins 0.125, and so forth.

Phenotypic correlations can be computed from observed data. Thus incubator-hatched chickens can be reared and maintained in a random environment independent of the environment of their parents, except for the food material that the hen provided in the egg. If it is then found that the phenotypic correlation between full-sister hens for the average weight of eggs they lay is 0.375, the h^2 of this trait is 0.375/0.500 or 0.75. (The less closely related the individuals considered in computing heritabilities, the less useful is this method, because its reliability goes down with reduction in r_G.)

Another method, described in Box 10.C, is provided by experiments on artificial selection. Figure 10.3 shows the range of values of h^2 reported for chickens. Many other heritability studies have been carried out. In cattle, birth-weight heritability is 0.40–0.50, and milk production is about 0.30; in sheep, staple length of wool is about 0.25; in swine, body weight at different ages is around 0.30.

Now, to uses (1) and (3) of the heritability ratio: Since

$$h^2 = \frac{G}{P} = \frac{G}{G + E},$$

it follows that changes in any component of the fraction will cause changes in its value. Indeed, it is found that by inbreeding (see Chapter 21), genetic variation within groups is changed, and h^2 for a given character also changes. Similarly, if changes in environment cause increases in environmentally generated variation, h^2 will drop; if they cause decreases, h^2 will rise. Thus, it is important to realize that the heritability is a property not only of a character, but also of the population and of the environmental circumstances to which the population is subjected. Heritability ratios applied to other populations or under other circumstances will be at some risk of error.

The heritability of anatomical and physiological traits of plants and animals is of great concern to agriculturists, since the choice of breeding methods used in plant and animal improvement depends on it (Sections 14.3 and 15.2). In defense of heritability, most h^2 estimates (where reasonable control of the experiments is available) are not too bad. Furthermore, selection indexes (which depend upon heritability estimates to weight properly the information from individuals and their relatives) are reasonably robust. This means that moderate errors in either

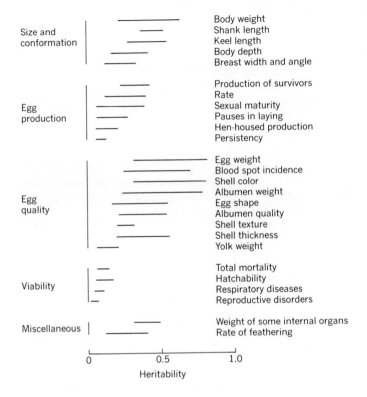

Size and conformation
- Body weight
- Shank length
- Keel length
- Body depth
- Breast width and angle

Egg production
- Production of survivors
- Rate
- Sexual maturity
- Pauses in laying
- Hen-housed production
- Persistency

Egg quality
- Egg weight
- Blood spot incidence
- Shell color
- Albumen weight
- Egg shape
- Albumen quality
- Shell texture
- Shell thickness
- Yolk weight

Viability
- Total mortality
- Hatchability
- Respiratory diseases
- Reproductive disorders

Miscellaneous
- Weight of some internal organs
- Rate of feathering

0   0.5   1.0
Heritability

FIGURE 10.3
Range of heritabilities reported for various economically important characters of the domestic fowl. (I. M. Lerner.)

the estimation or application of heritabilities result in only a small departure from optimum selection efficiency.

Another problem with heritability used as a measure of breeding value is that not all genetically caused variation in the parents can be reliably recovered in the offspring by means of selection. In particular, when genetic interactions exist, creating dominance or epistatic genetic variation (Boxes 9.A and 11.A), the heritability ratio tends to overestimate the gain possible by selection.

Another error frequently committed in the name of heritability is to imply that, if stature has an h^2 of 0.75, the height of any given individual is determined three-fourths by heredity and one-fourth by environment. This misunderstands both the concepts of heritability and of developmental genetics. Heritability is a population concept, and $h^2 = 0.75$ means that three-fourths of the *variation* in stature within the population is associated with genetic differences, and the remainder with environmental differences among its members. And development proceeds as an *interaction* of an individual's genotype and environment, not as fragmented contributions of each.

Although variable characters of low heritability can be more easily modified by manipulation of the environment, high heritability must not be confused with the impossibility of environmental influences affecting a phenotype. A person is born with a fixed genetic endowment, but its phenotypic expression may be extensively affected by either a highly favorable environment or an unfavorable one. The degree of modification possible increases as our knowledge of genetics and development progresses. Many environments are or may be totally outside the range of experience of the population in which heritability was estimated. For example, the syndrome of characters associated with phenylketonuria (Section 6.4, Box 6.F) has very high heritability. But because we learned the nature of the metabolic defect, the syndrome can now be extensively changed by a diet low in phenylalanine.

In humans, the h^2 of stature or of skin pigmentation is mostly a matter of intellectual curiosity. The heritability of such human characteristics as intelligence, temperament, and behavior, however, is of social significance today, and will perhaps be increasingly so in the future (Chapter 12).

11

NATURE AND NURTURE

Many not very fruitful debates have been conducted on the theme of heredity *versus* environment. An important theme of this chapter, and for that matter of much of this book, is that organisms are the products of the combined action of their genes *and* their environments, and that neither heredity nor environment can influence an organism's development in isolation from the other. Yet, some characteristics are influenced relatively more by hereditary variation than by environmental variation, and for other characteristics the reverse is true. It is possible to analyze variation, and to assign relative weights to the various kinds of environmental and genetic variation that influence variation of characteristics in populations. In humans, twins of one-egg and two-egg origin provide a useful but imperfect opportunity for such analyses.

11.1 CONTROVERSY

Few disagreements or controversies among informed and well-meaning persons have raged for so long and with such bitterness as that over the nature-nurture interaction. Extreme views of what are the important determinants of differences in human mentality and behavior have been held by proponents of various theories since the time of the Greek natural philosophers. The most disparate views may be represented by the ideas of the English philosopher John Locke (1632–1704), who believed that minds are blank at birth (*tabula rasa*) and that all later differences are derived from individual experiences, and those of the French diplomat and writer J. A. Gobineau (1816–1882), who, in his essay on the inequality of human races, declared that differences in abilities are entirely innate.

Locke's position was forcefully restated by the French philosopher C. A. Helvetius (1715–1771), who believed that "the inequality of minds is the effect of a known cause, and this cause is the difference of education," Gobineau's viewpoint was later amplified by the English-German political theorist Houston Stewart Chamberlain (1885–1927), and acted upon by racists in various places, and in a particularly drastic form, by Adolf Hitler and his followers in Europe. Regrettably, not only demagogues but also some scientists of reputation, even a few geneticists and psychologists, have upheld near-extremist positions clearly unsupportable by scientific facts; both extremes of the controversy have had such supporters.

The stances taken by people on this subject are all too frequently correlated with their social and political views, and beclouded by prejudices generated by upbringing, social milieu, and psychological factors. Hence, it would be surprising if the contents of this book were detached from and free of influences in the writers' personal histories. What is to follow should, therefore, be understood as a mixture of reported data and authorial interpretation. Although we are trying to be impartial and enlightened, the warning *caveat lector* must, in fairness to the reader, be spelled out.

11.2 PARTITIONING OF VARIATION

In Section 10.3, we discussed the division of the total phenotypic variation among the members of a population into two component parts: that due to genetic differences and that due to environmental ones. Not only can each of the components be partitioned further, but some of the phenotypic variation may be the result of interaction between genotype and environment.

As examples of genotype-environment interaction, consider a Texas range cow and a Guernsey milk cow on a Wisconsin dairy farm. Both probably thrive and produce milk. But the Guernsey, being specifically bred for milk production in this sort of an environment, will likely far surpass the Texas range cow in both quantity and quality of milk. If two other cows, ideally the twin sisters of the two on the Wisconsin farm, are required to forage on a Texas range, the result may be very different. The Texas range cow will probably not be as well fed on the Texas range as her twin on the Wisconsin farm. But she is adapted to the Texas range, and she will probably survive and produce enough milk to feed her calves adequately. The Guernsey, if she survives at all, will probably not be able to

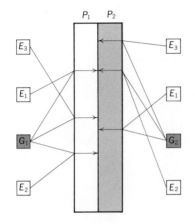

FIGURE 11.1

Some examples of genotype-environment interaction diagrammed on a vertical scale of phenotypic expression. From comparison of the sets of phenotypes produced by genotypes G_1 and G_2 in each of three environments, it may be seen that E_1 is best of the three for G_1 but worst of the three for G_2. Moving G_2 to E_2 equalizes the phenotypes if G_1 stays in E_1, while E_3 can make G_2 superior to G_1 in any of the three tested environments. The genotypes dealt with here could be for intelligence; the environments, schooling or parental socio-economic class.

produce enough milk to feed calves adequately on the Texas range. Thus, these two very different bovine genotypes will not only produce different amounts of milk in each of the two environments, but will probably reverse rank order. Figure 11.1 schematically shows other genotype-environment interactions.

In general, natural populations are adapted to their habitats by selection. If an organism finds itself in a significantly different environment, or if the environment changes, the phenotype that develops may be very different from that which might have developed in the environment where it evolved. The principle is demonstrated in experiments with a strawberry-like plant, *Potentilla glandulosa*, in which strains from sea level, an intermediate elevation, and the high mountains were grown in all three environments, with the results shown in Figure 11.2. Note in particular the relative reproductive potential of these three *Potentilla* genotypes in each of the environments, particularly as evidenced by the numbers of flowers produced.

It is possible to estimate the interactions between genotype and environment in plants and animals experimentally, but there are few data on humans that are suitable for this purpose. Hence in interpreting the heritability of human traits, it is important to keep in mind that the h^2 estimates are subject to error because of lack of information on the interaction between genotype and environment.

The genetic fraction of the phenotypic variation can be further subdivided into fractions attributable to dominance, epistasis, and straightforward additive effects of genes. Although this kind of refined analysis has not been done for many organisms, partitioning of the genetic variation in a variety of different ways is feasible even for humans, as can be seen in the diagram in Box 11.A.

The phenotypic variation due to environment is also subject to subdivision. Thus, in many experiments, e^2 can be divided into the effects produced by tangible influences and those produced by intangible ones. Among the former may be the effects of maternal influences, which will be common to all offspring of a given dam; effects resulting from the nutritional state of health of the mother during a particular pregnancy; effects common to a group of individuals raised

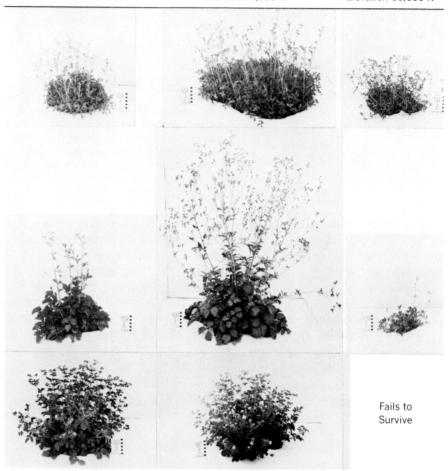

| At Stanford
Elevation 100 ft | At Mather
Elevation 4,600 ft | At Timberline
Elevation 10,000 ft |

Fails to
Survive

FIGURE 11.2

The top row shows the appearance of a genotype from a subalpine population of *Potentilla glandulosa* grown in three environments; the second row that of a genotype from a mid-elevation meadow; and the third row that of a genotype from near the coast. In each row, the plants shown came from cuttings from a single individual. (Courtesy of J. Clausen and the Carnegie Institution of Washington.)

in a specific environment; and so forth. The environmental variation not thus accounted for is considered to be due to intangible and uncontrolled differences in the environments of the members of the population studied.

Box 11.A provides two illustrations of the partitioning of variation in mammals, that on human birth weight being the most completely analyzed example of partitioning in humans as yet available.

Partitioning of the variation in metric traits is one way of approaching the more general problem of nature and nurture. It is only recently that techniques which provide reasonably meaningful estimates of the roles of these two forces have been perfected.

There are few good data available on heritability in humans. Yet it appears that in a human population that is homogeneous for income, geographical location, and genetic origin, the genetic fraction of phenotypic variation of such morphological metric traits as weight or stature is generally similar to that found in other animals. But with such complex traits as intelligence or personality, we are still facing many unsolved issues not only of interpretation, but also of methodology.

11.3 PARTITIONING VARIATION USING CLONES

A particularly useful way to partition variation is to gain experimental control of one or more of the major variables. Even with the best of equipment, such as controlled growth chambers for research with plants, environmental variability and gradients cannot be fully eliminated. But it is possible to achieve very uniform genetic material in a number of ways.

One of the classic methods is to inbreed (Section 21.3), ultimately deriving nearly **pure lines.** Using such pure lines, with their highly reproducible results, was one of the keys to Mendel's success with peas. But even prolonged inbreeding does not rid a line of all its genetic variation. Furthermore, in a species in which inbreeding is not the normal mode of reproduction, inbred lines tend to be sickly and difficult to maintain. The results can be criticized as not being representative of normal members of the species.

An attractive alternative, which can at present be used with a few higher plants, entails inducing a post-meiotic haploid cell to form an embryo and then double its chromosome number. This leads to instant homozygosity at every locus in the genome. Such an organism is said to be **isogenic.** Such plants are also distinctly abnormal in species that are normally outcrossing. But there is a way out of this problem.

Either different inbred pure lines, or different haploid-origin isogenic lines, can be crossed. The offspring of such a cross will be heterozygous at all loci that differ between the lines, and these plants are thus similar to the original outcrossed parents. The advantage is that all the F_1 offspring of a given cross are genetically identical (except where there has been a new mutation). Thus the desired experimental control can be obtained. (The experimenter has many F_1 families, each genetically very uniform but genetically different from each other, and with a level of heterozygosity and homozygosity essentially like that of normal members of the species.)

There are even more direct ways in which genotypes drawn from a genetically heterogeneous, outcrossing population can be expanded into genetically uniform lines. Nuclei from the cells of a frog can be transplanted into unfertilized frog

The tabie shows the partitioning of variation in body weight of mice of different ages. In this experiment, weight at birth had an h^2 of zero. The negative value for the postnatal maternal influence is a statistical artifact and hence may be viewed as zero. As the animals grew older, the influence of the genetic component increased, and that of the prenatal maternal effects decreased. The postnatal influence of the mother at first rose, but by the time of weaning (3 weeks) it had begun to drop. (The symbol m_u^2 stands for intrauterine effects, m_n^2 for nutritional ones.)

| | | Source of Variation | | |
| | | Maternal | | |
Age in weeks	Genetic (h^2)	Prenatal (m_u^2)	Postnatal (m_n^2)	Environmental (e^2)
0	0.00	0.52	−0.14	0.62
1	0.11	0.27	0.30	0.31
2	0.23	0.25	0.38	0.14
3	0.30	0.12	0.30	0.29
6	0.26	0.17	0.23	0.34

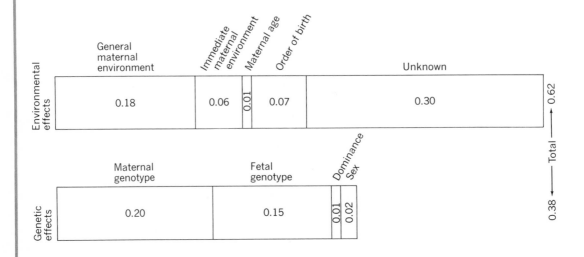

eggs, whose own nuclei have been destroyed by irradiation. Such transplants give rise to genetically identical individuals. This process may be repeated to produce any number of frogs, which would (provided the same frog supplied the replacement nuclei) belong to the same **clone,** a term originally applied to plants of identical genotype reproduced in a vegetative fashion—like *Potentilla*, which is pictured in Figure 11.2. (The possible immortality of humans, should such methods become available, could thus be extended to males as well as females.)

The use of clones for both experimentation and the normal practice of horticulture and agriculture is well established for many (but not most) higher plant

The diagram illustrates the partitioning of sources of variation in human birth weight. It may be seen that the greatest of the recognized causes of variation is the maternal genotype for ability to provide an optimum environment for the fetus. The influence of the paternal genotype is included in the 15 percent that the fetal genotype contributes to the variation in birth weight. The other factors are self-explanatory.

The partitioning of genetic causes of variation into their additive, dominance, and epistatic components is complicated, and beyond the scope of this book. The analyses depend on the fact that different kinds of relatives share different propor- tions of the various kinds of genetic variation. Data from these different sorts of relatives can be compared to determine the presence and relative importance of the different kinds of genetic components of variation. The table below presents the proportion of variation shared for additive (A), dominance (D), and first- and second-order epistatic ($A \times A$ = additive by additive; $A \times D$ = additive by dominance, etc.) genetic components. (Note that the proportion of epistatic variation shared can be calculated by multiplying the appropriate additive and dominance variation shared in any given relationship.)

| | | | Proportion of variation shared | | | | | | |
| | | | Two-locus epistasis | | | Three-locus epistasis | | | |
Relationship	A	D	$A \times A$	$A \times D$	$D \times D$	$A \times A \times A$	$A \times A \times D$	$A \times D \times D$	$D \times D \times D$
Identical twins	1	1	1	1	1	1	1	1	1
Fraternal twins	$\frac{1}{2}$	$\frac{1}{4}$	$\frac{1}{4}$	$\frac{1}{8}$	$\frac{1}{16}$	$\frac{1}{8}$	$\frac{1}{16}$	$\frac{1}{32}$	$\frac{1}{64}$
Full-sibs	$\frac{1}{2}$	$\frac{1}{4}$	$\frac{1}{4}$	$\frac{1}{8}$	$\frac{1}{16}$	$\frac{1}{8}$	$\frac{1}{16}$	$\frac{1}{32}$	$\frac{1}{64}$
Mother-daughter	$\frac{1}{2}$	0	$\frac{1}{4}$	0	0	$\frac{1}{8}$	0	0	0
Half-sibs	$\frac{1}{4}$	0	$\frac{1}{16}$	0	0	$\frac{1}{64}$	0	0	0

THE TABLE ON BODY WEIGHT IN MICE IS FROM EL OKSH ET AL. THE DATA USED IN THE DIAGRAM ARE FROM L. S. PENROSE AND M. KARN.

species. Techniques for producing plant clones include planting "eyes" (potatoes), dividing bulbs (iris), grafting (navel oranges), and rooting cuttings (bamboo). Recently, the ability to recover plantlets from tissue culture has extended this option to many additional species (orchids), and by controlling aging in the donor plants, the rooted-cutting technique is being extended to others (pines).

The classical experiments of Clausen, Keck, and Hiesey on genetic and physio- logical adaptations to environment made use of this technique. In Figure 11.2, the three plants in a row are members of the same clone, and thus have identical genotypes. The variability expressed within each row is therefore entirely the result of environmental differences. The variability in each column is the combined expression of genetic differences between each clone, and local environmental differences at each planting site unique to each plant. In the actual experiment, each clone was replicated more than once within each of the three major environments, in order to sort out this local environmental variation from the genetic differences. An overview of the rows and columns in Figure 11.2 demonstrates clone (genotype) by environment interaction of the different-origin clones in environments similar to each of the three origins.

In order to understand better what is meant by environmentally caused and genetically caused components of variation, and how these may be partitioned, consider the following experiment. A group of individuals is randomly sampled from a population, so that it is representative of the larger population. Each individual is cloned twice, and (very early in development) each newly produced member of these clones is randomly placed in some environment of interest. The effect of this is to eliminate correlations between genotype and environment (Figure 11.3).

The variation in a character or characters (measured after development is complete) within each clone is all nongenetic, and can be assigned to environmental heterogeneity. Thus, the average of within-clone differences = the average environmentally caused variation = E.

If we measure differences within random pairs of individuals in the sample that are *not* members of the same clone, this is the same as measuring variation

FIGURE 11.3
Clones a–f, each
with two members
located randomly
in a varying
environment.

between random individuals in a population that has not been cloned. Thus, the average of unrelated-pair differences = the average phenotypic variation = P.

Such between-phenotype differences are due to the same environmentally caused differences found within clones, plus the genetically caused differences that do *not* exist within clones. Thus, as indicated in Section 10.3, $P = E + G$.

We can now subtract the average environmentally caused variation observed within clones from the average phenotypic variation observed within random unrelated pairs, and thus estimate the average variation caused by differences in genotypes: $G = P - E$. If we observe as much variation within clones as between unrelated individuals ($P = E$), this is strong evidence that most or all of the population's variation in this character is environmentally caused. However, if P is larger than E, there must be some additional factor adding to variation in P, and experiments with clones (if properly done) cleanly identify the cause as genetic differences between the clones. These genetic and environmental components of phenotypic variation can then be used to construct h^2 and e^2 ratios (Section 10.3), and for other purposes in studying the population architecture of species.

Similar experiments can be done using families within which individual relationships are known (full-sibs, half-sibs, etc.), in which the interpretation is based on a theoretical knowledge of proportions of genetic variation shared by different kinds of relatives (Box 11.A). A thorough study of a species' genetic architecture will employ information from a hierarchy of related groups. But clonal studies are particularly appealing, as the statistics are simple and the interpretation is generally clear.

The most useful material available to students of the genetics of human behavior comes from studies of twins. Identical twins are the most common form of human clones. Before proceeding with the discussion of the genetics of intelligence and personality, therefore, we will consider the subject of these studies.

11.4 TWINS

There are two kinds of twins: **monozygotic,** also known as **identical** or one-egg twins, and **dizygotic, fraternal,** or two-egg twins. Monozygotic twins originate from a single zygote that splits into two at an early division and develops into two genetically identical individuals (see exceptions in Sections 8.5 and 9.2). Such identical twins have been produced experimentally in animals, as, for example, is shown in Figure 11.4. Monozygotic twins sometimes fail to separate, with the result that conjoined twins are produced (Box 11.B).

Dizygotic twins arise when, through superovulation, two eggs are present in the oviduct and both are fertilized. The genetic relationship between these fraternal twins is exactly the same as that between ordinary full-sibs, as in each instance the zygotes form from different eggs of the same mother and different sperm of the same father. There is one case of half-sib twin boys, born and raised in Austria. It is substantiated by testimony of the husband, the wife, and her

BOX 11.B CONJOINED TWINS

THE DRAWING OF CHANG AND ENG IS REPRODUCED COURTESY OF THE NATIONAL FOUNDATION.

Normal monozygotic twins become separate very early in their development, but late duplication or failure of separation sometimes also occurs, with a frequency estimated at 1 in 600 twin births. The doubling in such cases involves anything from a finger to complete functional individual. Most conjoined twins are born dead, but many have survived for a long time. On occasion surgical separation is possible, as in the case of two San Francisco girls shown in the photographs below.

The most publicized conjoined twins and source of the eponymic designation "Siamese" were Chang and Eng. They were sons of a Chinese father and a Chinese-Siamese mother. In later life they settled in North Carolina and took the family name of Bunker.

As shown in the drawing, the twins were united by a band of tissue. It was very short at first but gradually expanded as Chang and Eng began walking. Despite the fact that conjoined twins are thought to be monozygotic, the Bunker brothers displayed many differences. Mark Twain made the most of this in his famous short story on the Siamese twins. Chang, at maturity, was about an inch shorter, was somewhat deaf, and had an irritable temper. Eng was always healthier and stronger, and was known for his amiable disposition.

After making a considerable fortune with the help of P. T. Barnum by exhibiting themselves, the two brothers retired to raise tobacco. At 42 they married two sisters, and produced a total of 22 children. Eventually, their wives quarreled and the two families set up separate residences, with the twins spending three days in one and then three days in the other. On the whole they got on remarkably well with each other, with only one fight leading to an exchange of blows between them on record. The occasion was precipitated by Chang's heavy drinking, which did not stimulate Eng, who was not an imbiber. At the age of 63, Chang contracted pneumonia (he had had a stroke earlier) and died within four days. Eng immediately became progressively weaker and survived his brother by only two hours, his death being diagnosed as due to fright.

There were other famous conjoined twin pairs. One such twin, Rosa Blazek, gave birth to a normal son.

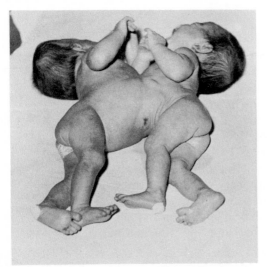

THE PHOTOGRAPHS ARE BY ROBERT DAVID WONG, MEDICAL ILLUSTRATOR/PHOTOGRAPHER, SAN FRANCISCO. THEY ARE REPRODUCED BY PERMISSION OF PETER A. DE VRIES AND THE NATIONAL FOUNDATION,

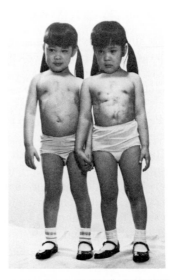

AND FIRST APPEARED IN "CONJOINED TWINS," A PUBLICATION OF THE NATIONAL FOUNDATION (APRIL 1967).

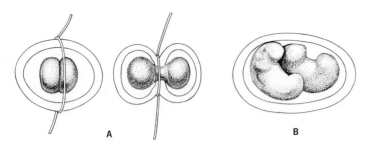

FIGURE 11.4
The classical experiment of the German embryologist and Nobel Prize winner Hans Spemann, producing monozygotic twins in a salamander. A shows the constriction of a zygote with a fine hair loop along the plane of the first cleavage; and B, the resulting twin embryos. (After Spemann, *Embryonic Development and Induction.* Yale University Press, 1938.)

lover as to timing of coitus, and by the blood groups of the two boys, each being consistent with that of one of the putative fathers and inconsistent with the other.

Unlike normal monozygotic twins, dizygotics can be of either the same or of opposite sex. One way of estimating the incidence of monozygotic twinning in the population uses the fact that approximately equal numbers of males and females are produced throughout a group of dizygotic twins. Suppose that in a group of twins (of both kinds) the distribution of sexes were 120 pairs ♀ ♀ : 140 pairs ♀ ♂ : 120 pairs ♂ ♂. With rare exceptions (Section 8.5), opposite-sex twins are dizygotic, and thus all 140 ♀ ♂ pairs are reasonably identified as fraternal twins. Hence, on a 1:2:1 basis, the expectation of sex distribution among the dizygotics is 70 ♀ ♀ : 140 ♂ ♀ : 70 ♂ ♂. The excess of twins of the same sex in the above example is 100 pairs. Among the 380 twin pairs observed, 100 (26 percent) are thus inferred to be monozygotic.

Determination of whether particular pairs of twins of the same sex are monozygotic or dizygotic is not easy. Because dizygotic twins share the same intrauterine environment, and because after birth most twins receive more similar treatment than siblings born at different times do, additional resemblances between them may be generated. However, discrimination between the two kinds of twins is made much easier by information now available about blood-group and serum-protein differences. The best test is provided by grafting of tissues: identical twins, being of the same genotype, will accept grafts from each other; fraternal twins, each having some antigens foreign to the other, will reject them.

This is not entirely true in cattle. It may be recalled from the discussion of the freemartin in Section 8.3 that cattle twins share fetal circulation. Apparently, early exposure to foreign antigens induces a degree of tolerance that makes grafts from one fraternal twin to the other in cattle possible. This discovery is paving the way for the development of methods of overcoming immunogenetic incompatibilities in humans (Section 20.3).

Multiple births include triplets (unusual), quadruplets (rare), and larger groups (until recently, births of quintuplets were exceedingly rare events, and human

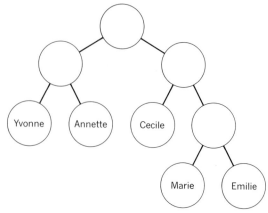

FIGURE 11.5
Probable origin of the Dionne quintuplets
from a single zygote.

sextuplets or higher multiples were not known). Recently, there has been a rise
in the number of human multiple births owing to the use of ovulation-inducing
drugs or hormones to combat sterility. Children produced in this way are of poly-
zygotic origin. However, the Dionne quintuplets, the first to be successfully raised
in modern times, were monozygotic. Figure 11.5 diagrams their probable origin
by successive divisions and separations.

The frequency of births of monozygotic twins does not show much variation
between different populations, but that of dizygotics does. For instance, the frequencies reported by N. Morton per 1,000 births are:

Race of mother	Monozygotic	Dizygotic
Johannesburg Negro	4.9	22.3
Nigeria Negro	5.0	39.9
Antigua (West Indies) black	3.9	11.5
U.S. black (1922–54)	3.9	11.8
U.S. white (1922–36)	3.8	7.4
Caucasoid	3.7	7.0
Caucasoid-oriental	5.0	3.2
Oriental	4.5	2.4
Japan (1926–31)	4.2	2.7
Japan (1956)	4.1	2.3
"Part Hawaiian"	5.5	2.4

These differences may be interpreted as indicating that there is little or no
genetic causation of monozygotic twinning, or at least that there is little difference
between racial groups. Such twins are probably produced by developmental accidents. The tendency for dizygotic twinning apparently has a moderately large
genetic component, which is different among racial groups. The genetic component likely acts by means of the hormones controlling ovulation.

Twins provide a useful tool for human genetic research, although its use is beset by many difficulties. Since identical twins have all of their genes in common,
and fraternal twins have (on the average) only one-half, comparison of intrapair

differences might be thought to provide an accurate way of determining genetic architecture. Unfortunately for genetic research, the environments of monozygotic twins are not necessarily fully comparable to those of dizygotic twins (partly for genetic reasons), so that observed resemblances and differences may not be accurately attributable to genetic or environmental causation.

The estimation of h^2 is improved when identical twins reared apart can be found. Several studies have been done on such pairs, but the total number of such twins investigated is still small. Such separated pairs are becoming less useful as adoption agencies become more efficient in matching the environments of the adopting homes to each other and to those of the biological parents. A return to the old hit-or-miss, near random element in adoption would serve genetics (but possibly not the agencies' clients) much better.

Methods other than the estimation of heritabilities may also be used to evaluate the effects of nature and nurture. Comparisons of intrapair correlations of identical and fraternal twins is one, and evaluation of **concordance** *vs* **discordance** within the two types of twins is another. (When both twins show a given character, they are said to be concordant; if only one does, they are discordant.) Figure 11.6 shows an example of concordance among identical triplets for a number of features.

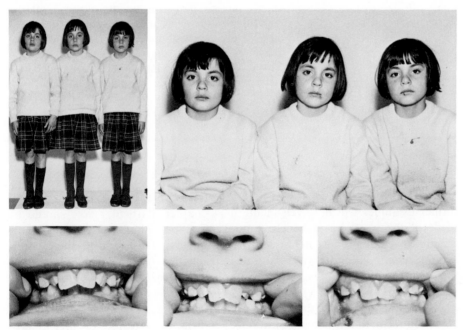

FIGURE 11.6
Triplets. Note concordance in tooth shape, size, facial expression, and dress, listed in the probable order of decreasing genetic and increasing environmental causation. (The lower photographs are from *Acta Genet. Med. Gemell.*, April 1966. Reproduced by courtesy of L. Gedda.)

BOX 11.C A TWIN MISCELLANY

The following is a sampling of data available on resemblances and differences among twins. First is a series of presence-absence observations, indicating concordance (if one twin has it, does the other twin also have it?). A higher concordance value means some greater common causal factor, with genetic factors being strongly implicated when monozygotic (MZ) twins have a much higher concordance than dizygotic (DZ).

	Percentage concordance	
Observed disease or behavior	*MZ twins*	*DZ twins*
Tuberculosis	54	16
Cancer at the same site	7	3
Clubfoot	32	3
Measles	95	87
Scarlet fever	64	47
Rickets	88	22
Arterial hypertension	25	7
Manic-depressive syndrome	67	5
Death from infection	8	9
Rheumatoid arthritis	34	7
Schizophrenia (1930's)	68	11
Criminality (1930's)	72	34
Feeble-mindedness (1930's)	94	50

Fingerprint ridge counts seem highly heritable, with correlations obtained very closely agreeing with those expected under the assumption that causation is entirely genetic.

	Correlation coefficients for fingerprint ridge counts	
Relationship	*Expected from genetic correlation with $h^2 = 1.0$*	*Observed*
MZ twins	1.00	0.95
Full-sibs	0.50	0.50
Sibs	0.50	0.50
Parent–offspring	0.50	0.48
Parent–parent	0	0.05

It appears that there are some genetic components in as imprecise an observation as the pooled causes of death among the elderly. Recalling the uniformity of the X chromosome in male identical twins, but the likely heterogeneity of inactivated X's in females (Section 8.4), the contrast between the sexes in difference in time of death is perhaps not surprising.

Box 11.C gives a miscellany of findings on resemblances between twins, and further examples will be given in the next chapter. Particular note should be made of the instances in which, contrary to expectation, twins reared apart resemble each other more than twins reared together. This could be due to interaction or competition between twins reared together—another kind of environmental influence on expression of genotype. In this case, identical twins who are in frequent association may make a conscious or subconscious effort to be distinct, and thus emphasize their differences in behavior. (As a German student of twins commented, it is indeed normal for identical twins reared together to have one assume the role of minister of foreign affairs and the other that of internal affairs.) Identical twins reared apart do not have the frequent opportunity to observe each other, and are thus liable to behave more similarly than if they were reared together.

Sex of twins	Differences in age at death (60–69-year-old group)	
	MZ twins	DZ twins
Male	4 years, 2 months	6 years, 3 months
Female	9 years, 6 months	10 years, 7½ months

If correlation coefficients of monozygotic twins reared together and apart are similar, and higher than for dizygotic twins reared together, the genetic component is inferred to be relatively large. If, however, the correlations of monozygotic and dizygotic twins reared together are similar and higher than for monozygotic twins reared apart, the environmental component is inferred to be relatively large.

	Intrapair correlation coefficients		
	MZ twins		DZ twins
Measure	Reared together	Reared apart	Reared together
Binet I.Q. test	0.97	0.67	0.64
Otis intelligence test	0.92	0.73	0.62
Stanford achievement test	0.96	0.51	0.88
Height	0.93	0.97	0.65
Weight	0.92	0.87	0.63
Introversion-extroversion	0.42	0.61	
Neurotic tendency	0.38	0.53	

In the last two categories, interaction between twins raised together apparently makes them more dissimilar than twins raised apart.

Finally, it has been shown that dogs trained in tracking (who unerringly can follow trails of individuals through fresh trails laid by other people, including close relatives) are frequently confused by trails of identical twins, and cannot distinguish the odor of one from the odor of the other on objects such as handkerchiefs. Francis Galton first suggested such a possibility in 1875, and it was brought to the attention of Professor Penrose of the Galton Laboratory in a 1952 letter from Professor John MacArthur containing the following passage:

"In a prospector's camp in Northern Ontario was a nearly blind Great Dane, Silva. She was not a friendly dog, but she had a passion for the prospector and fawned on him delightedly whenever he came to camp.

One day a stranger appeared for the first time and to the surprise of those present, Silva greeted him with great affection. Upon enquiry by the camp crew it was found that the stranger was the identical twin of the prospector, and that he had never been there nor seen the dog before."

12

BEHAVIOR

In this chapter, behavior is discussed in several dimensions, with evidence drawn from several kinds of animals, but with the emphasis on human behavior. We review mental retardation and mental disease, for which the evidence for a genetic basis is in several cases clearcut. Experimentation with animals frequently shows that forms of what may be called both normal and abnormal behavior can be genetically enhanced by selection. More indirect lines of evidence have shown that variation in human personality traits may also have significant genetic components. Some of the arguments for and against such genetic components affecting human intelligence are presented. The possibilities of significant changes in average intelligence, or in the distribution of intelligence, within and among human populations are examined. Finally, the question of instinct is briefly addressed. The information and opinions on many of these topics have been developed in a background of great general interest, and considerable controversy.

12.1 CONTROVERSY AND IDEOLOGY

If there is a genetic basis for behavior, should we attempt to discover it, and if so, how might a detailed understanding of it be used? With better understanding may come the ability to avoid or reduce mental disease, as well as various destructive or antihuman (but not inhuman) forms of behavior. But it may also result in an increased ability of unscrupulous leaders to manipulate and control their subjects.

In this book we decry governmental repression of open scientific inquiry (in Chapter 23 in a discussion of Lysenkoism in Russia). But perhaps some repressions are justified. Psychological and cultural damage is done to American blacks, or others, while scientists test uncertain hypotheses. Is it worth it? On the other hand, how much damage is done if scientists refrain, or are restrained, from poking about in socially painful areas, and how much intellectual integrity is lost?

The case of sickle-cell anemia (Section 7.4) is perhaps instructive. Scientists studied the disease for two decades while doing little for the diseased. The beginning of this decade witnessed widespread and justifiable outrage at the lack of effort devoted to treating those afflicted. Yet for centuries, tens of thousands of children per year had died of sickle-cell anemia, many of them following painful and unrelieved crises, before the scientists began to understand the disease. And now, perhaps a little belatedly, there is reason to expect that much or even most of the suffering and death caused by this disease will be stopped. In this case, scientific poking about seems to have been a good idea.

In Section 24.2 we support the Stanford scientists who broke off research with a bacterial system that might have proved dangerous to humans. Many (probably most) scientists agree with us on this point, and thus that the rights of scientists and the search for truth are not absolutes, untouched by any other value system.

For research in behavior genetics, there is no simple answer. The search for understanding and thus prevention or cure of the Lesch-Nyhan syndrome (Box 12.A) is clearly supportable. An understanding of human instinctive behavior (Section 12.6) should be the most controversial, for here lies a key to controlling warfare and thus possibly to survival itself, but here also lies the possibility of mass control and thus a form of slavery. The present controversy over intelligence (Section 12.5) is minor by comparison, but is current and very real. It is difficult at present to assess what the dangers and benefits of research into mental abilities may be, other than to recognize a considerable potential for each.

In the United States, the emphasis on environment as a significant factor in determining human behavior has been strongly influenced by the concept that all men are created equal. In their 1973 book, *Introduction to Behavioral Genetics,* G. McClearn and J. DeFries have collected several Greek definitions of equality. These seem more in keeping with biological reality than does ideological insistance on absolute equality, and we recommend them as guardians of the American ideal:

isonomia: equality before the law

isotimia: equality of honor

isopoliteia: equality of political rights

isopsephia: equality of the right to vote

isegoria: equality of the right to speak

Th. Dobzhansky, in his 1971 Tryon Lecture on Behavior Genetics, reminded the audience that only DNA is inherited. Characters such as intelligence, friendliness, and aggression are developed while DNA-directed growth and differentiation proceeds in real and unique environments, and even DNA can't be DNA in the absence of an appropriate environment. So let us now look at a few examples of the interaction of genes and environment in the development of behavioral characteristics.

12.2 MENTAL RETARDATION AND MENTAL DISEASE

There is a segment of the human population that, although appearing to be part of the continuum of phenotypes for intelligence, may be genotypically discontinuous from the rest of the population. Whether there is such a segment at the upper end of the distribution, that is, whether there exists a class of geniuses discontinuous from the distribution of the rest of us, is not known. But at the lower end, such a class of genotypes, which express themselves in the form of mental retardation, undoubtedly exists. It is a large problem, for society as well as for the individual families affected. The President's Committee on Mental Retardation estimated that 3 percent of the population under age 65—close to 6 million Americans—suffer some serious mental retardation.

A classical example is Down's syndrome, to be discussed further in Section 16.1. Down's syndrome, or trisomy 21, is caused by a chromosome imbalance (in particular, by presence of an extra chromosome 21). It is important to note that most additions or deletions of whole chromosomes, or large parts of chromosomes, are associated with an average reduction of mental ability. The *cri du chat* syndrome, caused by a deletion in chromosome 5, is another example. Severe mental retardation is a part of this syndrome, which received its name from the characteristic mewing cry of afflicted infants. The effects of some abnormal sex-chromosome constitutions were discussed in Section 8.5. Observations of the effects of such chromosome imbalances suggest that normal mental function is developed from a balanced set of instructions from many genes, located on many, perhaps most or even all, of the chromosomes of the human genome.

Additional early evidence, drawn from twin data of the 1930's (Box 11.C), supported the hypothesis that there is powerful genetic influence on the incidence of mental retardation—concordance among identical twins was found to be greater than 90 percent. There is a high frequency of moderate to severe mental retardation (25 percent versus none among nearly 100 outcrossed half-sibs) among the inbred children of matings between close relatives. This high frequency indicates that many genes carried in the population, when made homozygous, have a deleterious effect on normal mental development. As further evidence, a Minnesota investigation found that among first-degree relatives (parents and children; sibs) of mentally retarded persons, 28.0 percent were also mentally retarded; among second-degree relatives (such as cousins), 7.1 percent; and among individuals of still more distant kinship, only 3.1 percent.

Information on specific causes of mental retardation is now accumulating rapidly. More than thirty types of mental retardation associated with amino acid abnormalities have been described, and others are related to abnormal metabolism of hormones, carbohydrates, and lipids. With such specificity of facts comes the opportunity for greater understanding of the disease. We present one syndrome in some detail, as an example.

Galactosemia results from a defect in carbohydrate metabolism, namely, milk sugar (galactose) is not effectively converted to glucose. The failure derives from homozygosity for an abnormal allele, instead of presence of the allele coding for the enzyme galactose-1-phosphate uridyl transferase (G1PUT). As a result, galactose-1-phosphate, normally an intermediate metabolite in the conversion, accumulates. This leads to nausea and vomiting, inducing malnutrition in galactosemic infants. Their livers and spleens grow abnormally, and cataracts form on their eyes. They become seriously mentally retarded, and typically die in about one year. The frequency of children resulting from matings among near relatives is about 4 percent in the U.S. white population. But 31 percent of the galactosemic children in the U.S. white population are the result of such matings. So this is one of those mentally retarding genes that turn up where there is close inbreeding.

The disease can be produced in laboratory animals by feeding them a high-galactose diet. Conversely, replacing the milk in the diets of galactosemic children with galactose-free substitutes dramatically prevents development of the symptoms, and saves the lives of the children. Whether such children will live and reproduce as normal adults is not yet clear, as the first children successfully treated are only now entering young adulthood. But they have made it that far, and the prognosis is hopeful. Unfortunately, unless children bearing the defect are put on a galactose-free diet very early, the serious mental retardation ensues, and is not reversible by techniques now available.

The activity level of the enzyme G1PUT can be detected in red blood cells as well as in other tissues. Furthermore, a person who is heterozygous at the G1PUT locus can be distinguished from both the normal and the galactosemic homozygotes by characteristic enzyme activity levels. Thus, normal sibs of galactosemics can be informed whether they are carriers of the gene or not (two-thirds of them are). And suspected heterozygotes can be tested and advised on the likelihood of having galactosemic children (Chapter 22).

PKU (Section 6.4) is an example of defective metabolism of a particular amino acid due to homozygosity for an abnormal allele; in this disease it is phenylalanine metabolism that is disturbed. Like galactosemia, it is manipulable by dietary treatment: a low-phenylalanine diet prevents development of symptoms if begun early, but cannot reverse the mental retardation of advanced cases. As of 1972, 43 states required that infants be tested for PKU at birth. PKU differs from galactosemia in that not all PKU individuals become mentally retarded if they do not receive treatment, and unfortunately, heterozygous individuals cannot yet be confidently distinguished from normal individuals. Thus, genetic counselors cannot obtain certain knowledge of the genotypes of suspected carriers.

Not all forms of severe mental retardation are genetic in origin. In many cases,

BOX 12.A THE LESCH-NYHAN SYNDROME

The Lesch-Nyhan syndrome is caused by a genetic error that affects purine metabolism, with devastating effects on the central nervous system, resulting in distinct mental deficiency and a reproducible pattern of abnormal behavior. The disease was first studied by William Nyhan and Michael Lesch in 1965. All patients thus far studied have exhibited spastic movements, abnormal posture, and difficulty with speech, and many have difficulty swallowing. The patients have unusually pleasant personalities, with perhaps greater intelligence than standard tests detect. They are frequently favorites of hospital staff. Yet, if not restrained or protected, they display aggressive and self-destructive behavior. It is compulsive, and they do not like it at all. In one case, when restraints were removed, the child generally appeared terrified. His hand would go directly to his mouth and he would begin tearing the flesh, screaming all the time. Lip tissue is invariably bitten away in Lesch-Nyhan patients, unless the teeth are removed. Some children have learned to lacerate themselves with braces or catch themselves in the spokes of wheelchairs. They are frequently aggressive, and will bite others.

Other clinical symptoms include about a twenty-fold overproduction of uric acid; the bladder may become coated with uric acid crystals, and recurrent abdominal pain may accompany execretion of crystals in the urine. Gout and arthritis frequently develop in older patients as a result of the high concentration of uric acid in the blood. Pedigree studies suggested X-linked inheritance, as the trait was passed unexpressed from grandmothers through mothers to sons in whom it was expressed.

Autopsies, performed on boys with and without the syndrome who had died, surprisingly did not reveal any abnormal anatomic brain pathology in Lesch-Nyhan victims. High levels of the enzyme HGPRT (Section 8.4) were found in brain tissue of normal boys, particularly in the basal ganglia. The brains of the dead boys with the syndrome had no detectable HGPRT activity. It is significant that abnormalities of function of the basal ganglia dominate the clinical manifestations of abnormal behavior. Thus, the abnormality in the brain of a Lesch-Nyhan child is biochemical, rather than anatomical.

Sex linkage has been verified by a number of techniques, including: study of additional pedigrees of a mild form of HGPRT deficiency in males, none of whom had affected sons; use of mouse-human hybrid somatic cell lines to assign the gene to the X chromosome; investigation of clonal lines of cells from heterozygous females, which develop either as normal or HGPRT-deficient lines due to inactivation of one or the other X; and assays for HGPRT levels in hair follicles of heterozygous females, giving similar inactivation results (as predicted by the Mary Lyon hypothesis—Section 8.4). This latter technique provides a method for screening for heterozygotes among the female relatives of Lesch-Nyhan patients. Amniocentesis (Box 19.B) can be used to screen early embryos of those heterozygotes who choose to have children. Male embryos with low HGPRT levels could be aborted, or treatment to prevent development of the disease might be begun, if such becomes available.

A knowledge of these biochemical effects on brain development is one step towards a better understanding of behavior. It may also lead to effective treatment of this disease. Recently, treatment with the drug allopurinal has been effective in reducing uric acid buildup, thus reducing gout and related clinical consequences. Unfortunately, it does not prevent the neurological deterioration. It is possible to induce the syndrome in laboratory animals, and this offers a promising experimental approach for study of the disease and its genetic as well as clinical implications. The photograph shows the paw of such an animal, documenting the effects of the compulsive chewing that characterizes the syndrome.

THE PHOTOGRAPH IS REPRODUCED COURTESY OF R. PALMOUR AND P. GLICK.

mental retardation is caused by one of various kinds of trauma, particularly those that occur during birth. And evidence continues to accumulate that the failure to achieve full mental development by people classed as moderately or mildly retarded derives as much—or more—from socioeconomic causes as from genetic ones. Poor nutrition, poor health care, and the subtle mental malnutrition associated with the hopelessness of poverty are undoubtedly implicated. But the evidence isolating root causes, and the suggested cures such knowledge would bring, is difficult to obtain.

One behavior, or disease, or form of entertainment, manipulation, or escape, that early-on was considered a candidate for "hereditary control" is excessive consumption of alcohol. Alcoholism, which is not quite the same thing, has remained a social issue of great importance and is one of the principal public health problems in the United States (and several other countries) today. One estimate suggests that 10 million Americans are significantly addicted to alcohol (i.e., suffer from alcoholism). Whether there are indeed important genetic factors contributing to excessive use or addiction is still a good question, as little appropriate research has been done. The confounding of inheritance with the effects of having as a role-model a parent who is alcoholic, or with the effects of sharing a family environment conducive to alcohol use, are obstacles to the simple interpretation of the observation that drinking runs in families. Twin studies indicate there may be a significant heritable component affecting how often and how much a person drinks, but there is not convincing evidence for genetic control of "lack of control," i.e., addiction. One small but conceptually better study used half-sibs: where it was possible to evaluate children of one alcoholic parent, some of whom were raised in homes with an alcoholic parent-figure, others in homes where neither parent-figure was alcoholic; and conversely, some children of parents who were not alcoholic were raised in homes with an alcoholic parent-figure, while others were not. This study still did not control the general (nonfamily) environments of the children, but the comparisons strongly suggested that genetic factors are important, and the mere presence of an alcoholic in the home is relatively unimportant.

A recently discovered behavioral syndrome, which is more self-destructive than alcoholism, is the Lesch-Nyhan syndrome (Box 12.A). Once one recovers from the shock of its clinical manifestations, this disease is both elegant and fascinating as a study of abnormal behavior with genetic causation.

Schizophrenia is a mental illness that hospitalizes more than a quarter of a million persons in the United States alone, accounting for about half the patients in public mental hospitals. Two million Americans either have schizophrenia, or will fall victim to it in their lifetimes. It is a disease with a great variety of expression among different individuals, and many investigators are convinced that it includes several different diseases. Affected persons tend to display abnormal responses (or no responses at all) to environmental stimuli and particularly to other people. They are withdrawn from reality, and may have delusions or persecution complexes and manic-depressive fluctuations in mood. They show symptoms of what has been referred to as split personality: in fact, the name of the disease derives from the Greek words for "split mind." Schizophrenics are more likely to exhibit a variety of emotional disturbances than nonschizophrenics.

In spite of tremendous effort on the part of psychiatrists, no generally successful treatment has been discovered, although drug therapy appears to be promising. There is a wide range of opinions about the causes of the disease. One school of thought is that schizophrenia stems entirely from up-bringing and early social environments that produce disturbances in interpersonal relations. Environmental effects appear to be very important; so important that they have led many to minimize or even disbelieve a possible genetic causation. Evidence in favor of a genetic component comes from two main sources: the concordance of monozygotic twins, which varies greatly from study to study, but is practically always higher than the comparable dizygotic concordance (Box 12.B); and adoption studies of progeny of schizophrenic parents, and of a few monozygotic twins brought up apart. The evidence from these sources supports an important genetic contribution, and suggests that family environmental influences are relatively small.

During the nineteenth century, physicians at mental asylums noticed a tendency toward abnormal behavior among visitors whose relatives were committed and classified as schizophrenics. The physicians did not regard the visitors' abnormal behavior as typical schizophrenic behavior, and in 1863 applied the term "schizoid" to it. After more than a century of nomenclatural confusion, there is now some evidence that—at least genetically—schizoid and schizophrenic are the same thing. Schizoid personality and schizophrenia occur in about the same frequency among monozygotic twins of identified schizophrenics, and 88 percent of such co-twins exhibit either one or the other. Since the monozygotic twins are genetically identical, it appears that the same genotype is compatible with either, or sometimes neither, manifestation, probably depending on environmental variables and influenced by age of the twins (young twins may later develop one or the other manifestation). Similarly, about 45 percent of the sibs, parents, and children of a schizophrenic are either schizoid or schizophrenics, which suggests that there is a single dominant gene exhibiting about the same penetrance (90 percent) as that suggested by the data on monozygotic twins. About 66 percent of the children of two schizophrenics are either schizoid or schizophrenic, lending further support to the hypothesis of causation by a single dominant gene with incomplete penetrance. But the case for a genetic cause of schizophrenia is much shakier than that of, for instance, the Lesch-Nyhan syndrome. Other genetic or environmental explanations, with useful predictive properties leading to effective treatment, may soon replace this hypothesis, which is thus far based on circumstantial evidence.

Although schizophrenia has been the most thoroughly studied of the psychoses from a genetic point of view, other personality disorders with a genetic basis have also been described. Thus, significantly large heritability indexes have been reported for tendencies described by such terms as depression, psychopathic deviation, and anxiety, all of which perhaps form part of the total range of inherent attributes entering into the determination of temperament and personality.

BOX 12.B SCHIZOPHRENIA

If schizophrenia is a polygenically determined character and has a continuous phenotypic distribution, the concordance-discordance method of analysis is not necessarily the appropriate one. In the past, however, it was viewed as an all-or-none trait, and data on twins have been collected on that basis. The figures in the table at the bottom of the box represent an extract from material gathered by I. I. Gottesman and J. Shields from many sources. Several more recent studies support the general trend presented in the table.

Another way of expressing the heritability of a character is to compute *recurrence risks;* that is, the likelihood of a trait being manifested in relatives of a person known to have it. For instance, in the simplest situation, if a child shows a recessive phenotype, the probability that a full-sib will also show it is one out of four. For situations in which the exact mode of inheritance is not known, such risks are computed from empirical observation. F. J. Kallman has estimated that the rates (per 100) of occurrence of schizophrenia among relatives of twins, one of whom is known to be affected, are:

Parents	9.2
Spouse	2.1
Step-sib	1.8
Half-sib	7.0
Full-sib	14.3
DZ co-twin	14.7
MZ co-twin	85.8

One more figure may be added to this table from a study by L. L. Heston on psychiatric disorders in children separated from schizophrenic mothers when not more than two weeks old, and raised in foster homes. He found that 166 per 1,000 had such disorders (about 16 times the frequency in the population at large), thus indicating either a high heritability or the possibility that the mothers supplied their children with a toxic intrauterine environment. Among other attributes of these children, Heston found a greater than average incidence of mental deficiency, neurotic personality disorders, and criminal behavior.

Country	Year	Monozygotic twins		Dizygotic twins	
		Number of pairs studied	*Percent concordance*	*Number of pairs studied*	*Percent concordance*
Denmark	1965	7	29	31	6
Germany	1928	19	58	13	0
Great Britain	1953	26	65	35	11
Japan	1961	55	60	11	18
Norway	1964	8	25	12	17
United States	1946	174	69	296	11

12.3 ANIMAL BEHAVIOR

To animal breeders, behavior genetics by any name is not new. Dog breeders have developed a variety of breed behaviors, including those that typify pointers, retrievers, and sheep dogs, while sheep breeders have selected for gregariousness.

In chickens, the establishment of pecking order was noted to have a genetic component, and it was further observed that small groups of closely related hens in a larger flock behave essentially as single units in performing many behaviors.

Many researchers chiefly interested in human behavior turned for insights to studies of animal behavior. With animals, running experiments is less expensive, the life cycles are shorter, and more important, there are much greater opportunities for experimental control and drastic manipulation. As an added bonus, the researcher is less frequently engaged in controversy. And many found that the study and understanding of animal behavior as an end in itself provided sufficient intellectual and practical justification.

Much of the earlier animal work was aimed simply at demonstrating genetic control of behavior, frequently by comparing inbred strains. This work was criticized on the grounds that many strains of animals that have been produced by inbreeding are not normal animals (Chapter 21). This suggested that evaluating the performance of the offspring of crosses between such lines is more satisfactory. It was soon found that differences between different genetically uniform lines were commonplace for a great variety of behaviors, and thus this missionary period of demonstrating that behavior is under some genetic influence is drawing to a close.

Only a small sampling of what has been learned about different characters in different animals is possible here. In all the examples cited, such expressions as "genetically determined" or "inherited" do not imply absence of the environmental component of variation, but merely presence of a genetic one. The inheritance of temperamental differences in crosses between dogs has been extensively studied, and such traits as wildness *vs* tameness, the tendency to bite, barking capacity, and differences in obedience to commands, have been found to have genetic components.

In rats, the ability to learn to run through a maze has been established as genetically determined. Selection for susceptibility to morphine addiction in the same species was also successful, indicating a genetic basis for this trait. Similarly, differences in preferences for alcohol have been found in mice—a behavioral phenotype that may in part be based on genetic differences affecting the level of activity of alcohol dehydrogenase, a liver enzyme that takes part in the metabolic breakdown of alcohol. A strain of rats has been selected that voluntarily drinks to intoxication.

Cuckoos show a complex behavioral pattern, also apparently inherited. They deposit their eggs in nests of other species of birds, who unwittingly incubate them and raise the young. By natural selection, each variety of cuckoo has "learned" to recognize nests of species that lay eggs similar in color and size to its own (Figure 12.1). Such species accept the cuckoo's eggs as their own. Beyond that, cuckoos lay their eggs at a time that permits their young to hatch earlier than those of the host and thereby to acquire a selective advantage over their foster sibs. Young cuckoos have even been known to push the unhatched eggs of their foster parents out of the nest.

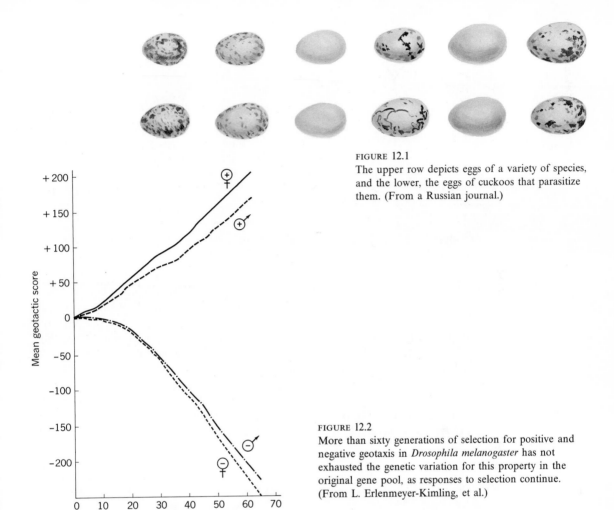

FIGURE 12.1
The upper row depicts eggs of a variety of species, and the lower, the eggs of cuckoos that parasitize them. (From a Russian journal.)

FIGURE 12.2
More than sixty generations of selection for positive and negative geotaxis in *Drosophila melanogaster* has not exhausted the genetic variation for this property in the original gene pool, as responses to selection continue. (From L. Erlenmeyer-Kimling, et al.)

Many genetically determined behavioral characters of insects have been studied. In *Drosophila,* strains differing in mating activity, in duration of copulation, in geotaxis (tendency to fly up or down when offered a choice in a vertical maze-like apparatus), and in phototaxis (tendency to travel towards a light) have been established by selective breeding. Figure 12.2 shows the results of one such experiment.

There is a disease of honey bees, American foulbrood, that usually destroys infected colonies. Certain strains, however, are resistant to the disease. In such strains, workers uncap the wax cells containing infected larvae and remove them from the hive. This behavior appears to be controlled by two recessive genes, one for uncapping and the other for removing. Hence, it is possible to have four kinds

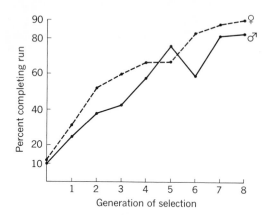

FIGURE 12.3

Selection progress for speed of running through a T-shaped maze by flour beetles of the species *Tribolium castaneum.* The ordinate shows the percentage of beetles completing the run within a two-hour period in each generation of selection. (From I. M. Lerner and N. Inouye.)

of behavior, one resistant (both genes homozygous recessive) and three susceptible (only one or neither of the genes homozygous recessive).

Two behavioral characteristics have been selectively bred into strains of the flour beetle *Tribolium*. One of these is flying ability. Normally these beetles crawl, but when starved some will undertake short hops, or flights. By using as parents of successive generations beetles that were the first of a group to engage in this activity, the percentage of fliers—defined as those who hop within a 24-hour test period—was raised from one or two to 50 in just six generations.

The second *Tribolium* characteristic studied is the ability of the insect to complete a run through a maze-like apparatus in a short period of time. Figure 12.3 shows selection progress for this characteristic. Its biological basis is not fully understood, but it is probably the behavioral manifestation of a physiological trait. *Tribolium* normally have a pair of glands that excrete a mixture of highly

obnoxious substances when the insects are excited, starved, or crowded. These substances act as predator repellents and also serve as a population-control device by causing the beetles to disperse, thus decreasing mating frequency. Although the insects bred in the selection experiment were chosen because of their speed, there is evidence that the basic physiological trait selected was for sensitivity to the unpleasant excretions. In the maze, beetles were starved and crowded, and the fast runners were merely those trying hardest to get away from the crowd. It is suggestive that beetles from the fast strain do not bother to run at all if placed singly in the maze, and only show their speed when other beetles are present.

The study of animal behavior is expanding in at least two important dimensions. One is the observational dimension, which recently gained considerable respectability when the 1973 Nobel prize was awarded to von Frisch, Lorenz, and Tinbergen for their pioneering studies of general animal behaviors such as communication, social organization, and aggression (Box 7.B). Figure 12.4 is an example of species-specific social behavior in which the outcome of a boundary clash is very much dependent on which gull is on its own territory.

The second dimension is that of greater experimental manipulation and analysis, where techniques of molecular biology are being applied (frequently *by* molecular biologists, many of whom are really physicists who paused at genetics on their way to the new frontiers of behavior) to such organisms as fruit flies, round worms, and bacteria, and even to cell cultures. A particularly powerful technique with such a complex organism as a fruit fly is to create a genetic mosaic (Figure 8.1), so that the normal and abnormal behaviors can be investigated relative to which groups of cells contain the normal or abnormal genes. Not all mosaics have the same distribution of cells, so that by studying enough of them, an anatomical map of the sites and modes of gene action affecting behavior may be

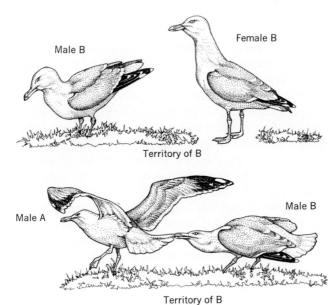

FIGURE 12.4
Boundary clash begins with two pairs of gulls facing each other across the boundary between their territories. Display behavior (*top row*) is a result of conflict between "anger" and "fear." Male B is "choking" in response to upright display of male A; female A utters mew call, while female B adopts the "escape" version of upright posture. When male B invades the territory of male A, A attacks and B flees (*bottom left*). Conversely, when male A invades the territory of B, B attacks and A flees (*bottom right*). (From Tinbergen, N., "The Evolution of Behavior in Gulls." Copyright © 1960 by Scientific American, Inc. All rights reserved.)

constructed. This technique becomes even more sensitive if the enzymes are active within one range of temperature, but become inactive at higher or lower temperature. In this way, the environment can be manipulated in order to activate or inactivate the enzyme in particular regions of the animal's body, making the development underlying the behavior accessible for study.'

12.4 PERSONALITY AND NORMAL HUMAN BEHAVIOR

Although the distinction between normal and abnormal behavior is subject to considerable interpretation, it does seem they are different, and may be under different types of genetic control. At least some abnormal behavior and mental retardation has causation in single defective genes, or in chromosomal imbalance (Section 12.2). Normal intelligence is probably affected by many genes at many loci (Section 12.5), and it is likely that other traits of normal personality and behavior are similarly affected. An important contribution of behavior genetics to the understanding of normal behavior is the demonstration that many personality and behavioral traits are, like most other human characters, inherently variable among individuals in populations.

Compared to the research activity mentioned in Sections 12.2, 12.3, and 12.5, this area is relatively untouched. This is partly due to the chaotic situation that has existed for some time in theorizing about personality differences, and the generally highly individualistic attempts to capture these differences in theory, tests, or ratings. Through all the confusion and inconsistencies, a few generalizations can be made with reasonable confidence. There are considerable hereditary components in personality differences. "Sociability" and "extroversion–introversion" are two traits for which there is convincing evidence for significant genetic components. The ages at which masturbation, orgasm, and other sexual activities begin similarly have significant heritabilities. As examples of studies on the heritability of personality traits and of the traits that are studied, excerpts from two pioneer investigations are presented in Box 12.C.

The intracellular products of the genes that find phenotypic expression in behavioral differences between people are known for only a few major metabolic defects. But pleiotropic pathways of gene action that eventually affect personality are not too difficult to imagine for at least some characteristics. For instance Turner females are not very good at abstract thinking, or finding their way about, or doing mathematical problems; and being an albino in a world of normally pigmented people must affect a person's behavior.

Certain other forms of psychological variability falling within the normal range of expression may derive from subtler developmental effects. It is difficult, for example, to see how differences in the so-called *personal tempo* arise, even though this attribute apparently has a genetic component. It has been found that if a person is asked to tap repeatedly on a table, he or she will do so at a characteristic tempo. The tempos of identical twins resemble each other more than do those of fraternal ones, the similarities within pairs of the latter being of the same order as those among sibs. Musical ability as tested by pitch and loudness discrimination, musical memory, speed of learning music, and even preference for a given composer, has also been found to be heritable on the basis of studies of dizygotic twins and monozygotic twins reared together and apart. Although it is obvious that development of musical ability must depend to a considerable degree on exposure and environment, it appears likely that the genetic endowment imposes a somewhat elastic range of possible achievement.

A comment on what we mean by that last statement: Some people view a person's "genetic endowment" for a trait, particularly one involving some ability such as musical talent, as placing an upper limit on that ability. This limit will be reached only in an absolutely optimum environment. To the degree that the environment is suboptimal, the person will perform somewhere below that genetically imposed upper limit. This is not a very useful conception. Consider instead a bit of metal attached to a rubber band, which is attached to a flat surface. The metal can be moved a short distance from its point of attachment easily, by approaching it with a weak magnet from one direction or another. To move it a greater distance from its point of attachment, in whatever direction, takes a stronger magnet. Once it is, say, a foot from the point of attachment, moving it an additional inch away from the point of attachment requires a greatly increased magnetic force, compared to the force needed to move it away an inch when it is near the attachment. The point of attachment may be viewed as the "genetic endowment," with small departures from that point (the "expected ability"), occurring in response to moderate variations in the environment. But to have a really large departure from the "point of attachment" takes an unusual environmental input. However, another bit of metal attached to a similar rubber band on the same surface, with its point of attachment only an inch or so away from the point that the first could reach only with an unusual environmental input, can be attracted to that same point by a weak magnet, or a small environmental input. A "magnificently endowed" athlete can smoke and still make the team, while a more average person must train rigorously and dare not smoke to make the same team. (It is possible that in characters with some threshhold—possibly schizophrenia is an example—the point of attachment has a genetic weakness such that if given sufficient pull, the rubber band will detach and the metal will move to a very abnormal position. To return it to normal requires a good psychiatric repairman, probably aided by an understanding of the nature of the rubber band, and of the defect that caused it to detach.)

12.5 INTELLIGENCE

One may escape attempting a rigorous definition of "intelligence" by substituting the term "IQ" for it, and then specifying that "IQ" refers to whatever it is that Intelligence-Quotient tests measure. But before considering such an action, let us pause, and consider the trait of "intelligence" as a population biologist might —in much the same way that we considered the heritability ratio (Section 10.3). First, intelligence should be viewed in the context of some population. Second, it should be viewed in the context of some real, operational environment. The discussion of intelligence as some abstract absolute, and speculations on what might measure or affect it, generally leads to long and sometimes heated arguments, and is recommended for those who enjoy such things. It has elements in common with the late-evening arguments of callow humans dissecting the nature of God.

BOX 12.C HERITABILITY OF PERSONALITY TRAITS

Of the numerous studies on the inheritance of personality traits, two by the American psychologist I. I. Gottesman have been chosen to illustrate the kind of results that have been obtained. From these studies, based on comparisons of correlations between monozygotic twins on the one hand and dizygotics of like sex on the other, he has devised indices of heritability. (His method for determining them is not exactly equivalent to the methods given in Section 10.3. Nevertheless his indices do provide an estimate of the proportion that genetic variation forms of the total phenotypic variation, even though they do not correct for the possibility that the differences in the environments within pairs of identical twins are not equivalent to those within pairs of fraternal twins. The statistical significance of his estimates was determined by appropriate tests, which need not concern us here.)

The first study is based on 34 pairs each of one-egg and two-egg twins from Minnesota, who were administered a battery of tests used by psychologists to measure differences in personality. The table shows some selected results from the standard Minnesota Multiphasic Personality Inventory: two significant heritabilities and three that were not statistically significant are shown.

Test	Correlation between twins		Heritability index
	MZ	DZ	
Significant			
Social introversion	0.55	0.08	0.71
Depression	0.47	0.07	0.45
Not significant			
Dominance	0.46	0.21	0.24
Anxiety	0.45	0.04	0.21
Ego strength	0.25	0.47	0

The same twins were given another series of tests, known as the High School Personality Questionnaire. Here the children were rated for different attributes, as shown in the following tables. Significant and nonsignificant heritabilities are given.

The film *Walkabout* deals effectively with intelligence in a populational and operational-environment context. The protagonist, an aboriginal boy performing the walkabout rite, which marks his passage into manhood, clearly possesses considerable intelligence in his operational environment, the Australian bush. The two children who are abandoned in the bush appeared quite bright in their operational environment, a middle-class neighborhood in Sydney. But the Sydney children fare badly in the outback, nor does the aboriginal do well when exposed to even a small part of their environment.

IQ tests were constructed to measure aptitude and predict success in a white

Low score	High score	Heritability index
Significant		
Sober, serious	Enthusiastic	0.56
Group dependent	Self-sufficient	0.56
Confident, adequate	Guilt prone	0.46
Not significant		
Stiff, aloof	Warm, sociable	0.10
Tough, realistic	Esthetically sensitive	0.06
Phlegmatic temperament	Excitability	0
Relaxed composure	Tense, excitable	0

Gottesman's second study, of children in the Boston area, analyzed 79 pairs of identical and 68 pairs of fraternal like-sexed twins. The test used was the California Psychological Inventory. Selected results are presented. Two categories of traits are shown with examples of both significant and nonsignificant heritability estimates.

Personality aspect	Test	Heritability index
Significant		
Introversion-extroversion	Sociability	0.49
	Dominance	0.49
	Self-acceptance	0.46
	Social presence	0.35
Not significant		
Introversion-extroversion	Capacity for status	0.25
	Sense of well-being	0.13

Personality aspect	Test	Heritability index
Significant		
Dependability-undependability	Good impression	0.38
	Socialization	0.32
Not significant		
Dependability-undependability	Tolerance	0.27
	Self-control	0.27
	Responsibility	0.26

middle-class academic environment. They do this job well, and also correlate with many other dimensions of occupation and success in that culture. They are not so good at predicting the success of a stockbroker operating in the very highest economic strata of society, nor do they correlate well with ability to make it in a ghetto. Many people criticize and even fear the IQ tests, sometimes irrationally, but often with cause. They particularly criticize their misuse, either as an attempt to measure some abstract, environment-free, God-like conception of "intelligence," or as a tool to suppress a population that does not occupy the operational environment of the IQ test.

If success in the operational environment of American and European middle-class cultures is of interest, then so is IQ. A fascinating (some say immoral) question is the degree of inheritance of IQ, and the degree of its variability in response to environmental variables. This question has received a great deal of attention, with many studies published, cited as evidence, and generally criticized by other investigators who favor other conclusions, or hope for no conclusions, or seriously question the particular methodology of the study. Many of the criticisms are valid, because IQ is tough to study in the human organism, mostly because of the great difficulty of separating genetic and environmental causation. But in spite of these individual criticisms, the mountain of evidence taken together creates a generalization that no longer seems escapable: in and near the operational environment of the white middle-class culture, IQ has a strong underlying genetic component. One line of evidence is provided by Figure 12.5. The average

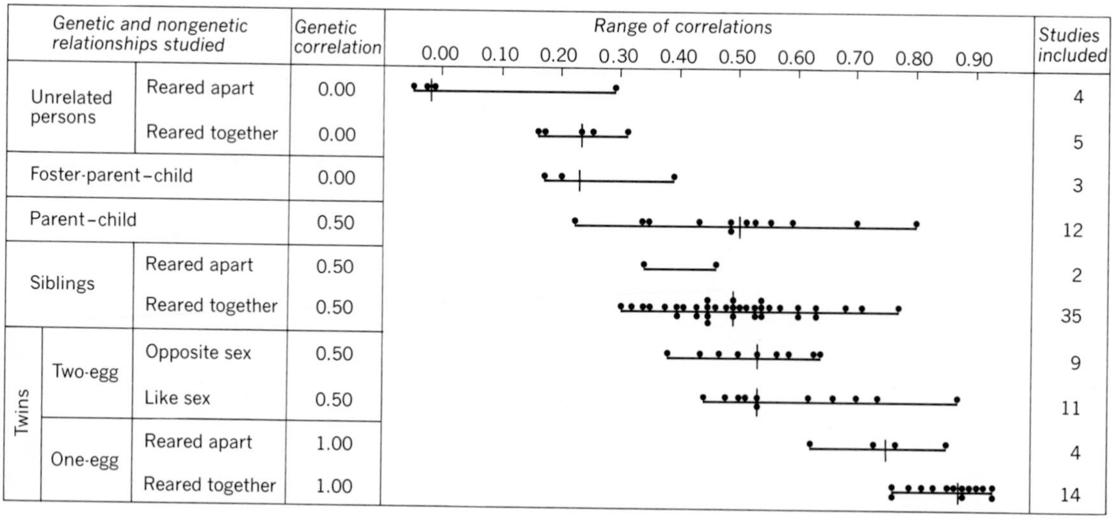

FIGURE 12.5
A summary of correlation coefficients compiled by L. Erlenmeyer-Kimling and L. F. Jarvik from various sources. The horizontal lines show the range of correlation coefficients in "intelligence" between persons of various degrees of genetic and environmental relationship. The vertical lines show the medians.

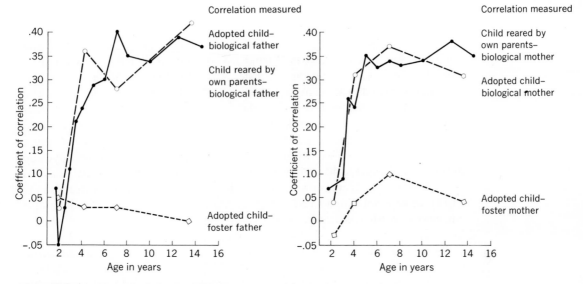

FIGURE 12.6
Correlations between the IQ of children and the educational level of biological and foster
parents. (From M. Honzik, and M. Skodak and B. M. Skeels.)

r values increase as the genetic relationship between the individuals compared increases. A similar, but apparently smaller, effect is shown for increasing similarities in environment.

Evidence from studies on adopted children also indicates the existence of a large genetic component. Thus, Figure 12.6 shows that the education of the biological father and mother correlates with the IQ of their children, whether the children are reared by them or by foster parents. On the contrary, the education of the foster parents has little bearing on adopted childrens' IQ's.

The notion that there can be a single correct answer to the question "what is the heritability of IQ?" is not reasonable. Besides being affected by the genetic compositions and operational environments of the populations tested, heritability estimates may vary because in reality different characters are being tested. Of particular importance are the different skills tested and the weights they receive in the score, how they are tested, and the age of the subjects. Instructive in this regard are some recent data by Sandra Scarr-Salapatek, who estimated heritabilities of four tests of cognitive abilities in adolescent twins drawn from the American black and American white populations:

Test	Black h^2 (160 pairs)	White h^2 (211 pairs)
Raven Standard Progressive Matrices	.59	.87
Peabody Picture Vocabulary Test	.28	.49
Columbia Test of Mental Maturity	.42	.58
Revised Test of Figural Memory	.61	.71

Partitioning of intelligence is nothing new, as people have long discussed such matters as a "mathematical genius" having little "common sense." There is some additional support for the proposition that factors of intelligence may vary independently due to genetic causes. Turner's syndrome (Section 8.5) individuals are generally above average on the Wechsler Adult Intelligence Scale, but are substantially impaired in space-form perception. Within the chromosomally normal population, a particular type of spacial visualization appears to be X-linked; in particular, males who do this well are much more common than females who do it well, and the frequency of males who do well is the square root of the frequency of females who do well. For example, if 0.16 of a population's females have good spacial visualization, 0.40 of that population's males have good spacial visualization (Section 13.1). Increasing attention is being given to children with IQ's in the normal or superior range who are not culturally or educationally disadvantaged, but who nevertheless exhibit some specific learning disability. Such children constitute about 5 percent of the school-age population. One such disorder is *dyslexia,* or word-blindness, for which there is evidence of a strong hereditary factor.

The possibility that individuals have different strengths and weaknesses within the complex trait called intelligence has important implications for educational philosophy and social policy. Arthur Jensen and others have argued that differences in overall intelligence, and in different dimensions of intelligence, are real and strongly influenced by heredity. They have urged that schools educate different children differently, and develop their different strengths rather than attempting to cram them all into the same idealized middle-class mold. Unfortunately, some of the evidence they emphasize was based on differences *between* populations rather than differences *within* populations. Most notably, differences in IQ and other facets of mental performance between different socio-economic classes and between different races were stressed. Thus, what might seem a reasonable educational philosophy, if applied within a single population such as the white middle class, was viewed by many as a suppressive threat if applied in a heterogeneous society made up of different classes and races. Before we briefly look at IQ, class, and race, we should deal with a second part of our fascinating, but perhaps immoral, question on the heritability of IQ.

The variability of IQ in response to variables in the operational environment can be quantified by e^2, or 1.0 minus the estimated heritability ratio. As heritability estimates from different populations or on different traits seem to vary from about .3 to .9, clustering around .5 or .6, e^2 must vary from about .1 to .7, clustering around .4 or .5. Thus, in most of these estimates, there appears to be evidence for a substantial amount of environmental influence on the IQ or other mental abilities of a normal individual encountering a normal set of environments.

We have already reviewed the salutory effects of a distinctly abnormal environment, in the form of dietary therapy devised for galactosemics and phenylketonurics (Section 12.2). The effects of abnormal nutrition on the mental attributes of normal individuals is at present much discussed. Overdosing with vitamin D during pregnancy frequently affects the developing fetus, producing a syndrome that includes several physical abnormalities, behavior characterized as placid and friendly, and mental retardation with IQ in the 50–80 range. Many experiments with rats and other animals demonstrate that preweaning protein deficiency or amino acid imbalance results in life-long learning disabilities, perhaps even carried to second-generation performance of the children of well-fed adults, due to inadequate placental development tracing back to early malnutrition.

The animal experiments suggest that good nutrition during the first three years of human life is crucially important to good mental development in genetically normal children. The common observation that children reared in poverty do poorly on IQ tests lends some support to this suggestion. Among underprivileged children in rural Guatemala, height and mental achievement were positively correlated. Among well-fed upper-class urban children, height and mental development were unrelated. Presumably, in the rural group, both height and mental development were affected by the marginal diets variously available to children in the village, while differences in these two characters were not dependent on variation within the range of good nutrition supplied the upper-class children.

While the hypothesis of nutritional effects on human brain development and function is sensible, it has not been established by rigorous experiment. It has been challenged by the results of a Nazi reprisal during World War II (Box 12.D). These and other observations make it appear that *chronic* malnutrition of the mother, rather than a period of *acute* starvation of a generally well-fed mother, leads to mental retardation of the offspring. Remarkably little is known about either the extent or the effects of chronic undernutrition, and the same may be said about chronic malnutrition among affluent but nutritionally illiterate people. A diet of cola and potato chips may be as debilitating as one that is quantitatively inadequate. Much of the available evidence is well analyzed in E. Schneour's *The Malnourished Mind* (1974).

Besides nutritional and physical elements of the environment, there can be mental and psychological components as well, and these may be *the* important components affecting human mental development. The work of D. Krech, M. Rosenzweig, M. Diamond, and D. Bennett, who studied rats provided with a mentally impoverished or enriched environment, has demonstrated not only different learning abilities among such rats, but physical and biochemical changes in the rat brains associated with these enriched or deprived environments. We have much evidence that dull children tend to come from homes with a stultifying mental environment, while bright children tend to come from homes that are mentally enriched and supportive. But it is not clear whether those environments are created by parents and sibs who share genes for dullness or brightness with these children, or whether these conditions are strictly environmental and tend to be perpetuated in families and other social groups.

A recent experiment, still not completed nor well reviewed, was done by R. Heber with 40 children from a lower-class black community in Milwaukee. From birth, 20 were given daily intensive attention by the project staff for a period of six years. The other 20 grew up in the community as controls. The IQ's, measured at age 6, averaged 126 for those given the enriched environment, and

In September 1944, British, Polish, and American paratroops landed between the British front lines and Arnhem, in an effort to secure a bridgehead across the Rhine and take the land war to Germany. The Dutch underground disrupted the German counterattack. But the effort to secure the bridgehead failed, and the Nazis in reprisal imposed a transport embargo on western Holland. Soon no food was reaching the large cities, and the people there starved until May when the cities were liberated by the Allies.

Records kept during the Dutch famine have several useful features: Extensive and reliable data were maintained on such things as birth records; the famine was sharply circumscribed in both time and place; and the type and extent of nutritional deprivation were known with a precision unequalled for any other large human population before or since.

A study was made of 125,000 Dutch males who had been born during, or close to the time of, the famine; at the time of the study they were being examined for military service and were 19 years old. The data included information on mental retardation, IQ (measured by the Dutch version of the Raven Progressive Matrices—the higher score signifies poorer performance), and birth date and place. The latter allowed determination of whether each boy was in a famine city (under Nazi rule) or a control city (behind Allied lines or with a food supply nearby) at the time, and assignment of each to a famine-exposure birth cohort. The pregnant women in the famine cities could not be adequately fed, as indicated by birth weights. The boys were further divided roughly by socioeconomic class, based on the occupational status of their fathers—manual or nonmanual worker.

Three conclusions seem clear from these data: starvation during or just preceding pregnancy had no detectable effects on the mental performance of surviving male offspring during their late adolescence ; mental performance of surviving males had no clear association with lowered birth weights recorded in hospitals during and after the famine; and, the association of social class with mental performance was strong.

One should be cautious in placing too much weight on a single set of data. The circumstance of starvation imposed on a large human population was repeated during the more recent Biafran rebellion in Nigeria. The Biafrans, including their young children, were starved to the point that many children developed the protein-deficiency disease kwashiorkor. For the sake of the Biafran children, we hope that the absence of enduring effects, such as found by Stein, Susser, Saenger, and Marolla also obtains for Biafra's population. It may be that the human fetus accords greater homeostatic protection to its developing mind than a rat allocates to its developing learning abilities.

96 among the controls. The 20 enriched children were surely not raised in a normal operational environment, but if the abilities predicted by a high IQ are indeed important, then perhaps this Milwaukee experiment gives us some clues about what a normal operational environment for children should be.

The question of IQ performance between socioeconomic classes, or between races, is often badly stated. It comes out sounding as if the issue were whether or not two groups have the same genetic endowment for IQ or some other mental

Socioeconomic class	City group	Birth cohorts				
		Conceived and born shortly before the famine	Conceived before, born during the famine	Conceived before, born after the famine	Conceived during, born after the famine	Conceived and born shortly after the famine
Average birth weight (grams)						
Both classes	Famine	3470	3250	3040	3210	3290
	Control	3290	3360	3320	3430	3360
Severe mental retardation (number of individuals per 1,000)						
Both classes	Famine	2.9	3.2	4.0	4.5	4.0
	Control	3.0	4.1	4.7	3.6	3.7
Mild mental retardation (number of individuals per 1,000)						
Manual	Famine	41	45	45	45	53
	Control	68	64	47	60	66
Nonmanual	Famine	11	11	10	7	11
	Control	8	12	7	11	10
Raven Progressive Matrices Test (mean scores)						
Manual	Famine	2.7	2.7	2.6	2.6	2.7
	Control	2.8	2.8	2.8	2.8	2.8
Nonmanual	Famine	2.2	2.1	2.1	2.1	2.2
	Control	2.2	2.1	2.1	2.1	2.1

THE DATA ARE FROM Z. STEIN, M. SUSSER, G. SAENGER, AND F. MAROLLA.

attribute. Absolute genetic equality is not a reasonable expectation, and even in the exceedingly unlikely event that two groups were precisely equal, they could not long remain so. Births and deaths, coupled with the dynamic nature of genetic recombination and drift, would result in their becoming different. A more reasonable approach is to accept that there must be differences, and ask the more pertinent questions as to *direction* of the difference, and *significance* of the difference. *Significance* may mean the statistical validity of the difference, but should also evaluate the magnitude of difference in terms of biological significance, or educational, or economic, or social significance. It is quite possible to have a difference between two groups that, because of the power of the test employed, is statistically highly significant, but of no serious biological (or educational, or economic, or social) significance. The converse may also be true.

It is clear that IQ's of both adults and of their children are highly related to occupational and socioeconomic class. Boxes 12.D and 12.E present examples of such evidence for the children born into these classes. For adults in occupations of high status, IQ's have a high mean, a truncated range, and a variance of about 10 (see Box 10.A for a description of variance and mean). In occupations of medium status, the adult IQ mean is nearer average, with a broader range, and a variance of about 15. In occupations of low status, the adult IQ mean is also low, but includes the entire range and has a variance of about 20. Thus people with high IQ's may choose or be forced into low-status occupations, but the reverse is much less common.

R. Herrnstein has speculated on mobility in a "meritocracy," in which people would attain positions because of their abilities, rather than family connections,

---

BOX 12.E OCCUPATION, IQ, AND FAMILY SIZE

Demographic statistics relating socioeconomic class, IQ, and family sizes are abundant. Here, we shall give a composite picture from several sources. The data are not strictly comparable, and for the sake of compactness, some of the occupational categories are not described in the original terms. They are, however, sufficiently accurate to make the general picture emerging reliable.

The association of intelligence quotient with socioeconomic status is remarkably uniform in the three countries.

In the Chicago study, done in 1935 before a careful matching of an adopted child's background to its adopting family was common practice, a comparison is made between the IQ of the biological children in the adopting families and the IQ of the adopted children.

Occupational status of parents	U.S.	USSR	Chicago Adopted	Chicago Biological	England	Average size of English family
Professional	116	117	113	119	115	1.73
Semiprofessional	112	109	112	118	113	1.60
Clerical and retail business	107	105	111	107	106	1.54
Skilled	105	101			102	1.85
Semiskilled	98	97	109	101	97	2.03
Unskilled	96	92	108	102	95	2.12

Average IQ of children

inherited wealth, or racial and religious restrictions. As such environmental components as nutrition, culture, and education would be equalized, an even greater proportion of the remaining variation in mental ability would be genetic. Under such conditions, revolutions would be less possible, because the ruling class would not include people of low ability who had inherited power, and the downtrodden classes would be without leaders because persons from these classes who were capable of leadership would rise and join the ruling class.

The study of IQ and race is both emotionally charged and refractory. Mobility between races is clearly more difficult than mobility between socioeconomic classes, and thus that cause of IQ stratification is less important. But there are serious confounding variables that include nutrition, housing, education, culture, language, and, perhaps most important of all, the psychological impact of racism. Because of the difficulty of separating genetic from environmental causation, many have persuasively argued that such studies should not be done at all. But with or without adequate studies, information on racial differences in IQ performance is widely available, and is being acted upon.

In recent years, chinese and jewish people in the United States have performed well on IQ tests of various types. These tests are supposed to predict success in academic pursuits, and thus it is not too surprising to find over twice as many jewish professors and five times as many chinese professors as the proportions of these groups in the U.S. population would predict. This may, perhaps, be accounted for by the entry into the United States of jewish and chinese intellectuals fleeing suppression in Europe and Asia. Interestingly, white Americans who are not jewish are on university faculties in almost exactly the same proportion as they are in the population at large. Whatever the mixture of genes and culture and environment that produces intellectuals, an unusually high proportion of young jewish and chinese Americans develop into intellectuals. An important social question today asks whether young black and native American intellectuals, among others, whose elders are underrepresented among professors, should be favored in the competition for faculty positions, or whether absolute equality of opportunity is enough.

Another issue to be considered is the possible erosion of intellectual faculties of future human generations through differential reproductive rates of persons differing in intelligence. Thus, heritability of IQ in the "breeding value" sense of heritability becomes of some concern. Both in the past and at the present time, fears have been expressed that inequalities in birth rates among people in different occupational classes, together with differences in the average IQ of families of different sizes, may erode the level of intelligence in the human population. This argument is based on the kind of information shown in Boxes 12.D and 12.E. The fact that people of low economic status produce more children than those of higher status may have a selective effect in the direction of decreased intelligence. Irrespective of status, children of larger families tend to have lower IQ's. In one Scottish study, a negative correlation between family size and average IQ of -0.28 has been reported; the mean Stanford-Binet score of children who had no sibs was 113, that in sibships of six, 91.

There are at least two reasons why we may not be declining in IQ as quickly (or at all) as the data suggest we should be. One is that the lower IQ performance of children from large families may mean simply that large families provide a less satisfactory environment for the development of IQ, rather than indicating that parents with lower IQ's tend to have more children. Indeed, there are some tenuous indications that where average family size is going down, average IQ performance is going up.

The other reason is that simple correlations, or gross associations with socio-economic class, don't tell an adequate story, and may even miss some important data. Figure 12.7 indicates that the picture is much more complicated than just lower-IQ people having more children. This illustration is approximate, is for the western European and North American white populations, and is surely subject to change as birth control methods gain understanding and acceptance, and as in-centives and disincentives to having large families come and go. At present, several things are clear, although the causes may not be. People with very low IQ's have few children, possibly because they cannot find mates, and possibly be-cause it is clear they cannot adequately support them. People with near average IQ's are below average in reproductive rate, and this is particularly pronounced for IQ's in the high 90's. These people may regard their success in the competitive societies in which they find themselves as marginal, and may realize that large families could tip the balance against them. People with IQ's in the 70–90 range tend to have large families, perhaps in ignorance of possibilities for limiting family size, or perhaps in the hope their children will later take care of them. And people with very high IQ's tend to have very large families, perhaps because they can afford them. Whether the overall effect of all this is gain, loss, or steady-state in population IQ is uncertain. A further complication is that many studies do not include information on the association of IQ with differential death rate, differential celibacy rate, percentage of barren mariages, or of differential genera-tion time, i.e., age of parents at reproduction. Assortative mating also distorts both data and predictions. One study found that the IQ's of spouses correlated higher than any other measured behavioral trait.

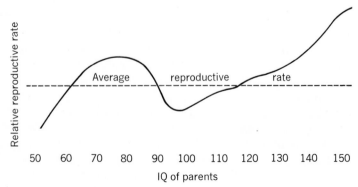

FIGURE 12.7
Data drawn from several sources, indicating relative reproductive rates of parents classified by IQ.

A logical test is to monitor IQ performance over time. This has two difficulties. An improvement or degeneration in the environmental components of IQ could mask genetic changes. More important, since most tests are somewhat culture-bound, the same question does not mean the same thing at different times. For instance, the meaning of the word "Mars" has changed from a planet to a candy bar and back to a planet over three generations of testing. To remedy this, the IQ test is periodically updated and standardized to a population mean of 100.

Finally, the heat generated by the IQ controversy has puzzled some observers. It appears that, at least in academe, human value is frequently measured or even equated with IQ performance. We think that tacit assumption needs careful evaluation.

12.6 INSTINCT

It is clear that some types of animal behavior are learned, but that most animal behavior is largely instinctive. It is also clear that in humans a much greater pro-portion of behavior is learned than in any other known animal. It is not clear how much of some kinds of human behavior may vary within or between popula-tions due to genetic differences between people or groups of people. Nor is it clear how much of some kinds of human behavior may be species-specific, or instinctive, i.e., part of the evolutionary baggage of *Homo sapiens.*

One of the major concerns of humankind is violent, or destructively aggressive, human behavior. Several books, including *On Aggression* (1963) by Konrad Lorenz, have done much to raise questions about the extent of inborn tendencies to defend territory (be it intellectual territory, owned property, real estate, or nation), to establish social pecking orders, or to respond to submission signals. The evidence is strong that many diverse and interesting forms of such behaviors are part of the evolutionary baggage of many animal species, including insects, birds, fish, and mammals (but not all species of insects, birds, fish, or mammals). Humans, singly or in groups, can be observed engaging in behaviors that bear considerable similarity to some of the patterns of behavior that are instinctive in other animals (Figure 12.8). What is not clear at present is to what degree human culture can reinforce or override such behavior patterns, if indeed there is a genetic basis for them at all in the human species.

Many people dislike the notion that humans are naturally aggressive, or that they enjoy destroying other creatures. Yet we all know people who use a light fishing tackle in order to prolong the fish's struggle, to maximize the personal sense of mastery and skill. Human killing differs from that of most carnivorous mammals in that many of the victims are of the same species as the killer. What-ever the origin of this behavior, almost every human society has regarded killing members of certain other human societies as desirable, and the folklore of almost every society has heroes who achieved their status by destroying human enemies.

It may not be that these behaviors are inevitable, but it does appear that they are easily learned by most humans, while other skills and concepts are more difficult to acquire. Perhaps therein lies the basis for human behavioral instinct.

If there are genetic bases for aggressive and destructive forms of behavior, it is surely not clear at present whether there is significant genetic variability for such behavior, either between people within populations, or between populations.

FIGURE 12.8
Norway rats (*Rattus norvegicus*) fight in a species-specific manner whether they are raised in isolation or with a group of other rats. The aggressor approaches, displaying his flank and arching his back (**A**). Then, standing on their hind legs, the two rats wrestle. They push with their forelegs (**B**) and sometimes kick with their hind legs (**C**). If one rat is forced to his back as they tussle (**D**), he may give up; otherwise the tournament phase ends and the real fight begins with a serious exchange of bites (**E**). (From I. Eibl-Eibesfeldt, "The Fighting Behavior of Animals." Copyright © 1961 by Scientific American, Inc. All rights reserved.)

A dog's response to another dog's submission signal (rolling on the back and exposing the belly) is normally to cease fighting. A typical human response to submission signals, be the combat intellectual or physical, seems to be similar. Is this learned, or is it inherent? When U.S. soldiers killed the villagers at Mai Lai, the nation and the world were horrified. When U.S. pilots killed other villagers with napalm and fragmentation bombs, the outcry was distinctly subdued by comparison, and no pilots were brought to trial. An important difference was that the soldiers at Mai Lai were in a position to perceive submission signals, and killed in spite of them, while the pilots initiated homicide-at-a-distance. The ability to kill at a distance, while useful for hunters, may have made warfare more lethal than it would otherwise be. Closer to home, a 1968 Chicago study found that many more serious assaults were made with knives than with guns, but that more of the gun assaults were fatal. Knives surely kill as well as guns if the victor in a violent confrontation does not break off the attack. But knives are employed at close range, where submission signals can be seen and heard.

13
POPULATION GENETICS

Most of the discussion thus far in the book has been concerned with the inheritance of differences between individuals, although the concept of heritability is one applicable to the population level. We are now ready to extend the discussion to genetic properties of interbreeding groups of individuals, the subject matter of population genetics.

13.1 THE HARDY-WEINBERG EQUILIBRIUM

Shortly after the rediscovery of Mendel's paper, many of the "new geneticists" began to give lectures describing the Mendelian explanation of inheritance. One of these was R. C. Punnett. Punnett, however, was confused by the knowledge that although blue eyes were recessive, there were many blue-eyed people in England. He was unable to account for this, and asked a fellow cricket-player, the mathematician **G. H. Hardy** (1877–1947), about it. Hardy quickly came up with the explanation. He recognized that Mendelian inheritance is binary (i.e., each locus in an individual has two identical or dissimilar alleles, which segregate and then recombine in the offspring). Because it is binary, he noted that a binomial expression is appropriate for characterizing the distribution of the alleles at a given locus in a population, and the algebraic expansion of this binomial expression gives the frequencies of the genotypes for that locus, in that population. This became known as the Hardy-Weinberg law (Box 13.A), as both Hardy and **W. Weinberg** (1862–1937), a German physician, independently arrived at this conclusion in the same year, 1908. Actually, the American geneticist and breeder W. E. Castle had published this conclusion in 1903, less than three years after Mendel's paper was rediscovered. But he considered it too obvious for special notice, and did not actively push for its acceptance. Thus the basic theorem of population genetics commemorates the work of Hardy and Weinberg. (But Castle bred the white and black "hooded" rat, popular as a household pet, and this contribution may directly affect more peoples' lives than his "trivial" law.)

The consequences of this law make it clear that allele frequency, as well as a dominant-recessive mode of gene expression, affect the frequency of a character in a population. Thus, if the frequency of the gene for blue eyes is 0.8 (as in Box 13.A), the frequency of the homozygous recessive genotype is 0.64. Note that the frequency of the recessive genotype is less than the recessive gene's frequency, which made Punnett's intuitive worry appropriate, but it still explains how the majority of English people could have blue eyes. (Actually, the inheritance of eye color is more complicated than this, but it is approximated by inheritance of a recessive gene for blue at a single locus, and it *is* the character that led Punnett to pose the question to Hardy.)

In Section 7.1 we referred to random fertilization, in which chance governs the union of the different kinds of sperm with the different kinds of eggs. A similar concept, used in Box 13.A, refers to matings at random between individuals having different genotypes. If matings are entirely governed by chance, the mating system is described as **random mating** or *panmixia*. It is a rare population in which it is really true that each individual has an equal chance of mating with any other individual. It is certainly not true in human populations, in which members of the same sex cannot mate to produce offspring, and the probabilities of matings between opposite-sex individuals are affected by distance, socioeconomic class, age, and many other factors. But the theoretical requirement of random mating is satisfied for purposes of genetic analysis if the different genotypes are well distributed throughout the population, and if matings occur without regard to genotype. For instance, most humans choose mates without regard to ABO blood type, and thus predictions of the frequencies of ABO alleles and genotypes (based on the population being panmictic and in Hardy-Weinberg equilibrium) are usually very close to the frequencies actually observed (Box 13.A).

There are often meaningful departures from random mating, particularly among human populations. These departures have been classified into four types that may occur in different populations, sometimes in combination with each other.

In **assortative mating,** like phenotypes tend to mate together in more instances than would be expected from their frequencies.

In **disassortative mating** the opposite is true, that is, there are more matings between unlike phenotypes than expected on a chance basis.

In **inbreeding** or **consanguineous mating,** there are more matings between relatives than under random mating; this departure might be described as genotypic assortative mating, or just **genotypic assortment.**

In **outbreeding,** or **crossbreeding,** a common technique in animal breeding, individuals of different breeds or from different populations are mated to produce **hybrids.** This is **genotypic disassortment.**

A concept that we have been using in this and several previous sections is that of allele frequency in a group (or population) of organisms of the same species. Different populations of the same species frequently have different characteristic allele frequencies. For example, in Israel, which is now populated by many

---

BOX 13.A CONSTANCY OF ALLELE AND GENOTYPE FREQUENCIES

Consider the gene pool of a population that has only two alleles at a particular locus; the frequencies of these alleles in the population are given as p and q, and $p + q = 1$. The genotypic frequencies in a filial generation produced from the random mating of a parental generation having frequencies p and q for the two alleles can be obtained by expansion of the square of this binomial, $(p + q)^2 = p^2 + 2pq + q^2$. For example, in a parental population consisting entirely of Aa individuals, p and q are both equal to $\frac{1}{2}$. In the F_1 population produced by their random mating, the gene frequencies will still be $\frac{1}{2} a$ and $\frac{1}{2} A$, but rather than all individuals being of genotype Aa, the genotypic frequencies will be $(\frac{1}{2} A + \frac{1}{2} a)^2 = \frac{1}{4} AA + \frac{1}{2} Aa + \frac{1}{4} aa$.

Now let us examine a gene pool in which the gene frequencies are $p_A = .8$ and $q_a = .2$. These, in the absence of selective advantage of one genotype over another and of mutation, are also the gametic frequencies. The genotypic frequencies in the next generation following random mating are readily obtained from the following Punnett square by multiplying the appropriate gamete frequencies in the margins. The population thus consists of 64 percent AA's, 32 percent Aa's, and 4 percent aa's. Note that there is only one way to get

		$p_A = .8$	$q_a = .2$
Eggs	$p_A = .8$	$p_A p_A$ .64	$p_A q_a$ .16
	$q_a = .2$	$p_A q_a$ .16	$q_a q_a$ .04

the homozygote AA: by having an A-bearing sperm (frequency .8) fertilize an A-bearing egg (frequency .8). But there are two ways to get the heterozygote Aa: An uncommon (frequency .2) a-bearing sperm can fertilize one of the more frequent (.8) A-bearing eggs, or a common (frequency .8) A-bearing sperm can fertilize one of the uncommon (.2) a-bearing eggs. Thus the coefficients of the binomial expansion make biological sense, with homozygous genotypes having a coefficient of 1 and heterozygous genotypes 2.

In the next generation, six types of matings (i.e., combinations of parental genotypes) are possible. Under random mating, the frequency of each type of mating depends on the frequencies of the genotypes in the population. The table below shows the types and frequencies of expected matings, with brackets emphasizing reciprocal crosses. The sums of the expected offspring fre-

immigrant groups previously isolated from each other, examination of the glucose-6-phosphate dehydrogenase (*G6PD*) locus shows the following frequencies of the sex-linked recessive allele causing deficiency of this enzyme among Israeli males of various origins:

Kurdish, .60	Yemenite, .05
Persian and Iraqui, .25	North African, .02
Turkish, .05	European, .002

In the usual notation, frequencies of recessive alleles are designated by q, and dominants by p. Remember that in a two-allele situation, $p + q = 1$, and $p = 1 - q$. For the Israelis of Turkish origin, q of the *G6PD* gene is 0.05 and p, 0.95.

The notation can be extended by multiple-allele situations. Thus, if we consider that at the ABO locus there are three alleles (an oversimplification to be corrected

quencies reconstitute the same array of genotype frequencies as in the parental generation, showing that an equilibrium exists in this population.

Mating		Frequency		Offspring		
♀	♂	♀	♂	AA	Aa	aa
$AA \times AA$		$.64 \times .64$	$= .4096$	.4096		
$AA \times Aa$		$.64 \times .32$ $.32 \times .64$	$= .4096$	.2048	.2048	
$AA \times aa$		$.64 \times .04$ $.04 \times .64$	$= .0512$		.0512	
$Aa \times Aa$		$.32 \times .32$	$= .1024$	.0256	.0512	.0256
$Aa \times aa$		$.32 \times .04$ $.04 \times .32$	$= .0256$		.0128	.0128
$aa \times aa$		$.04 \times .04$	$= .0016$			.0016
			Total	.64 p^2	.32 $2pq$	.04 q^2

The binomial formula can be extended to a polynomial for more than two alleles per locus, and ploidy levels other than diploid:

$$(p_1 + p_2 + p_3 + \ldots + p_n)^c,$$

where n = the number of alleles at the locus, p = the frequency of each allele, and c = the ploidy level of the organism. For instance, in (diploid)

humans, the genotypic frequencies for the ABO blood groups may be calculated by expanding

$$(p_A + p_B + p_O)^2 = p_A p_A + 2 p_A p_B + p_B p_B$$
$$+ 2 p_A p_O + 2 p_B p_O + p_O p_O.$$

Note that the genotypic frequencies are given by the products of the p's and their coefficients, while the identities of the genotypes are supplied by the subscripts. Thus, for allele frequencies $p_A = .1$, $p_B = .4$, and $p_O = .5$, the genotypic frequencies are $AA = .01$, $AB = .08$, $BB = .16$, $AO = .10$, $BO = .40$, and $OO = .25$.

If mating is not at random, allele frequencies may remain constant, but the genotypic frequencies change from generation to generation. If we consider, for instance, complete assortment in our two-allele example, so that only three kinds of mating occur: $AA \times AA$, $Aa \times Aa$, $aa \times aa$, we can see that because of segregation of AA and aa genotypes in the $Aa \times Aa$ matings, the number of Aa individuals will be reduced by half in every generation. In the limit, the population will consist entirely of AA's and aa's, and their frequencies, and clearly those of the alleles, will be .8 and .2 respectively.

FIGURE 13.1

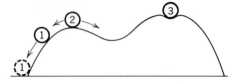

in Chapter 20), *A, B,* and *O*, it is found, for example, that Germans in Berlin have the frequencies:

$$A = .29, B = .11, O = .60,$$

$$p_A + p_B + p_O = 1,$$

$$.29 + .11 + .60 = 1.$$

Next we may consider the concept of **genetic equilibrium.** An equilibrium, as the name implies, refers to a state of balance. We may, however, distinguish between *stable* and *unstable* equilibria, as illustrated in Figure 13.1. Allele frequencies are said to be in equilibrium when they remain constant from generation to generation. *Such equilibrium under random mating and free from disturbing forces implies that genotypic frequencies will also be constant in successive generations* (Box 13.A).

13.2 DISTURBING FORCES

The frequencies of alleles and genotypes in a population will be in Hardy-Weinberg equilibrium if mating is (operationally) at random and if there is no significant disturbance due to selection, mutation, immigration, segregation distortion, or drift. Interestingly, it is these same disturbing forces that change allele frequencies, and provide the driving forces of evolution (Section 4.3). By definition, then, a population in Hardy-Weinberg equilibrium is not evolving. In fact, however, the changes in allele frequencies due to one or more of these forces may be so slight that departures from H-W equilibrium cannot be confidently detected, but nevertheless over many generations, evolution may be progressing very effectively.

One appropriate definition of selection is: the differential reproduction of genotypes. Different genotypes contribute alleles to the gene pool of each successive generation not precisely in proportion to their number, but relative to the selective advantage they enjoy. If selection is of the stabilizing kind, an equilibrium may still persist. For example, if only the heterozygotes from a population (in which the initial frequencies were $p_A = q_a = 0.5$) reproduced, the gene frequencies would still remain constant. So would the genotypic frequencies of 0.25, 0.50, and 0.25.

Under directional selection, this would not be the case. The frequency of the allele that conferred a selective advantage on its carrier would increase until it became the sole component of the gene pool. The rate of this process would depend on whether the allele selected was dominant or recessive, and on the degree of the advantage. Box 13.B illustrates these points.

Situations can arise in which there are combinations of directional and stabilizing forms. Figure 13.2 shows the results of an experiment, begun in 1940, in which selection was directed to increase the number of blades on a chicken comb. The original population consisted of a mixture of birds with one, two, and three blades. By 1944, birds with five blades had made their appearance; by 1946, the flock had become a stable mixture of three-, four-, and five-bladed individuals. Apparently, the directional force of artificial selection was being defeated by the stabilizing force of natural selection. What happened may be described as resulting from **genetic homeostasis,** the property of a population to regulate or stabilize its gene-pool contents. Various examples of this property have been demonstrated in laboratory and natural populations.

The equilibrium in the chicken experiment and the one in which the heterozygotes are favored are dynamic. In the latter, selection favors reproduction by the heterozygotes, but recombination continually produces homozygotes of both alleles, while in the former, alternative forms of selection favor alternative

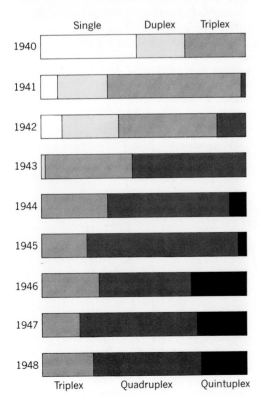

FIGURE 13.2
Changing population composition resulting from directional and stabilizing selection for the number of blades on the comb of chickens. (Based on the data of L. W. Taylor.)

genotypes. The equilibria in both these cases are maintained by a balance of forces, rather than the absence of them.

Mutation is the second of the forces that may interfere with the maintenance of the genetic equilibrium. If the rate of mutation from A to a is m per generation, then the frequency of A will drop from p in the foundation population to $p - mp$, while that of a will increase from q to $q + mp$. If this process is recurrent and continues indefinitely, eventually all A's will mutate to a, which will become fixed at the frequency of 1.

BOX 13.B RATE OF SELECTION PROGRESS

Natural selection usually operates in such a way that different genotypes have *on the average* different numbers of offspring. In artificial selection used in plant and animal improvement, certain genotypes (or phenotypes) are prevented from reproducing. For continuously varying characters, the distribution is **truncated;** all individuals falling on one side of a demarcation are discarded, and those on the other become parents of the next generation. Among the selected group, however, natural selection can still operate. For help in visualizing the changes under selection in the allele frequencies of the gene pool, some examples of truncated selection are shown.

In each of the three graphs, two situations are represented. In one, the **selection intensity** is 10 percent for each sex, meaning that each selected pair is allowed to produce 20 offspring to maintain a constant population size. This value is unrealistic for humans, but common among chickens. In the other, a *lower* intensity of 50 percent is applied with each pair giving rise to four descendents.

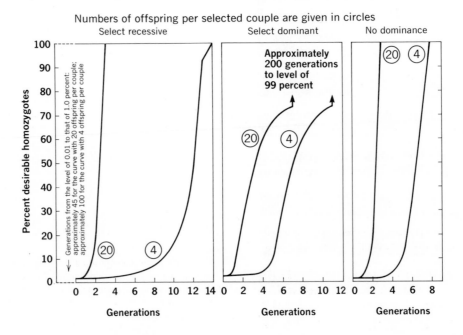

Numbers of offspring per selected couple are given in circles

Mutation pressure, however, can operate in both directions: *a* may show **reverse mutation** to *A* with a frequency of *n*. This will produce, in a single generation, frequencies of

$$p - mp + nq \quad \text{for } A$$

and

$$q + mp - nq \quad \text{for } a.$$

Under these simplified circumstances, an equilibrium will be reached if the

When there is no dominance, selection progress is fairly rapid, as shown in the right graph, because all three classes are distinguishable.

The left graph depicts the situation in which the recessive homozygote is the desired genotype. When its frequency is very low at the start, selection progress is slow, because to keep the size of the population constant, heterozygotes and dominant homozygotes must be permitted to reproduce. However, once enough recessive homozygotes are available for use as breeders, progress is exceedingly rapid. The gene pool then proceeds to **fixation,** that is, finally only one allele is represented in it. The frequency of this allele becomes 1.0; that of the other allele, 0. Differences in reproductive rates are important in this situation. If 3 percent of a population is homozygous recessive and if the population size can be maintained with only 10 percent of its members reproducing, the recessive allele may be fixed after only two generations of selection. If, however, 50 percent of the population must reproduce in order for population size to be maintained, eleven generations of selection are required for fixation of the recessive allele.

Selection for a dominant homozygote, shown in the middle graph, has a different pattern. Progress is reasonably rapid at first because even at a low frequency of *A*, there are enough heterozygotes available to use as parents. When the frequency of *A* becomes high, selection slows down. With complete dominance, *AA* and *Aa* are undistinguishable, and *aa* individuals thus keep appear-

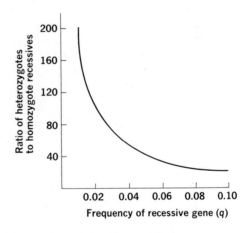

ing in the population for a long time. For instance, in black breeds of cattle, recessive red-colored animals are still occasionally found after many generations of selection against them. Animal breeders, who want to fix a desired genotype in as few generations as possible, must resort to progeny testing to distinguish *AA* from *Aa* individuals. Those identified as *AA* are then used to produce the succeeding generation.

Finally, in connection with the ineffectiveness of selection against recessives, the graph above shows the proportion of heterozygotes in a population as a function of gene frequency. The ordinate gives the ratio of the number of heterozygotes to that of recessive homozygotes at various low values of *q*. At $q = 0.5$, the ratio is 2.0, and at 0.9 it is 0.2.

losses in the numbers of A by mutation are balanced by the gains from reverse mutations. Because, however, mutation from the functional form of a gene to less-functional or non-functional forms can occur in many ways, while full function can be restored by a reverse mutation in only a few ways, mutation pressure in the absence of counterbalancing selection can lead to near fixation. Degeneration of organs that have lost their adaptive value (for instance, the evolution of blindness in cave-dwelling fish) can be accounted for by this process. (However, it is also possible that there is a selective advantage for cave-dwelling fish not having functional eyes.)

In fact, the frequency of mutation is typically so low that—even in the absence of balancing back mutation or selection—observed H-W frequencies are not affected by it. Mutation thus cannot generally be detected by noting mutation-caused departures from H-W equilibrium.

Immigration, segregation distortion, and drift, already discussed in Section 4.3, are complex and difficult to analyze. But the important point is that whatever new alleles are brought in by mutation or immigration, and whatever changes in allele or genotypic frequencies arise from segregation distortion and drift, these alleles and the genotypes that include them are then subject to selection pressures that determine the subsequent composition of the gene pool (Box 13.C).

The results of the relationships described in Boxes 13.B and 13.C lead to the conclusion, surprising at first sight, that even extreme selection against a rare recessive allele can be an ineffective process. Let us consider as an example PKU and use numerical values that should be viewed only as rough approximations. If we assume that the white population of the United States is in equilibrium at this locus, and that the incidence of the disease is 4 in 100,000, then $q^2 = 0.00004$, $q = 0.0063$ and $2pq$, or the proportion of heterozygotes, is $2 \times 0.9937 \times 0.0063 = 0.0125$. The mutation rate, assuming complete lethality, is equal to q^2 or 40 per million zygotes in each generation. Taking the population at 200 million and assuming that there are 5 million births a year, we arrive at the numbers:

number of PKU babies born annually	$= 200$
number of carriers of newly arisen mutations in ten million gametes (double the number of PKU zygotes)	$= 400$
number of heterozygotes in the population accumulated from previous mutations	$= 2,500,000$

Thus, a vast majority of the deleterious alleles is carried by heterozygotes. Failure of the recessive homozygotes to reproduce removes only 400 alleles of the two-and-one-half million. The folly of any program to purge the human gene pool of detrimental alleles by sterilizing homozygotes (or in any other way preventing them from producing offspring) becomes immediately apparent from these computations.

For illustrative purposes we shall consider an extreme case, that of a recessive allele, which in a homozygous state is lethal or renders the carrier sterile. Let us indicate the relative fitness of three possible genotypes as

AA	Aa	aa
1	1	$1 - s$

where s, the selective disadvantage, is equal to 1. In other words, the fitness of AA is equal to that of Aa, and the fitness of aa is zero.

At birth the respective genotypic frequencies are

AA	Aa	aa	Total
p^2	$2pq$	q^2	1

The a alleles are lost from the population at the rate q^2 per generation. (Note that the remaining population is made up of $1 - q^2$ of the original population, and that the frequencies of the surviving alleles are slightly modified because $1 - q^2$ is not quite 1.00. This can be taken into account for both selection and mutation, but we will ignore it in this example.) An equilibrium is established when the losses are balanced by replacements, i.e., when m new a genes are created by mutation of A to a per generation. Thus, at a mutation-selection equilibrium, $m = q^2$ when the fitness of the homozygote is zero. If the fitness of the homozygote is greater than zero, but less than one, the more general expression can be used:

$$\frac{m}{s} = q^2$$

Thus the equilibrium value of q for a is $\sqrt{m/s}$, or for a recessive lethal gene in which $s = 1$, $\sqrt{m}$.

The remarkable property of this relation is that it permits the computation of the rate of mutation. When a population is in genetic equilibrium, the rate of mutation to a recessive lethal allele is equal to the proportion of homozygous recessive genotypes found in a given generation. In practice, of course, determination of mutation rates is not so simple, because (among other difficulties) it is not always possible to determine whether a population is, indeed, in equilibrium. Nevertheless this method has been used to estimate human mutation rates (Chapter 16).

If heterozygotes could be identified and removed from the breeding population, such programs (negative eugenics) could work. For instance, heterozygous carriers of the gene for galactosemia (Section 12.2) can be reliably detected. Galactosemia occurs in newborn English and U.S. children at the rate of about 25 per million. About 10,000 people per million are heterozygous carriers, and could be identified by a mass screening program. We could then effectively eliminate galactosemia by preventing the identified carriers from reproducing. But does eliminating the possibility that 25 galactosemic babies will be born per million births justify the destruction of the reproductive rights of 10,000 people per million? Even if we decide it is worth it, the problem is that all of us are carriers of, on the average, several detrimental alleles. Thus eugenics based on heterozygous screening on a scale large enough to be of significance would not lead to improvement but rather to termination of the human species. Comparison of eugenic proposals with other methods of management of human genetic resources will be made in Chapter 22.

Another class of deleterious alleles is kept in the gene pool not by virtue of mutation pressure, but because of overdominance (or pseudo-overdominance, which might result from the close linkage with a favorable allele or alleles at neighboring loci). Overdominance is described in Section 7.4 for the hemoglobin alleles. If the selective disadvantage of one homozygote, relative to the heterozygote, is indicated by s, and of the other by t, so that the relative fitnesses of the three genotypes are

$$\begin{array}{ccc} AA & Aa & aa \\ 1-s & 1 & 1-t, \end{array}$$

it is readily demonstrable that the equilibrium values are $p = t/(s + t)$ and $q = s/(s + t)$. If both kinds of homozygotes died before reaching maturity, or were sterile, or for any other reason produced no offspring ($s = t = 1$), the gene frequencies of A and a would be 0.5, as noted in the second paragraph of Section 13.2. The main point is that a severely deleterious or lethal allele can be maintained in a population at high frequency when the heterozygote exceeds both homozygotes in fitness.

As evolution proceeds, a disadvantageous mutant gene may become advantageous because of changes in either external or genetic environments. An example is the gene for hornlessness in cattle. In natural populations horns originally served a useful function, perhaps in fighting and defense. But after cattle were domesticated, horns led to economic losses by causing injuries to animals kept in close confinement. The breeders then took advantage of the presence of a rare dominant allele for hornlessness in the gene pool and, reversing the direction of selective advantage, produced polled breeds. Similar situations, of course, occur in nature.

The reserve of mutations in populations is an important factor in evolutionary change and, despite the deleterious effects that mutant genes tend to produce, bestows an overall advantage for the species in the process of adapting to changed conditions. This will be discussed in more detail in Section 13.3.

13.3 POPULATION STRUCTURE

It is useful to divide the genetic variability within a species into four levels, and to consider the evolutionary forces and genetic mechanisms that lead to this variability.

First is the level of major geographic variability, called *provenance* (from the French for origin) variability in some studies and racial variability in others. At this level, groups are separated by great distances, and thus isolation between them is important, and migration is relatively ineffective in reducing the differences between them. A mutant allele that becomes established in one may be slow to move to other groups. The environments of the different places are usually

FIGURE 13.3

A provenance study of the western North American species *Pinus contorta* (called lodgepole pine in the mountains and shore pine near the coast) in a study plot in the Kaingaroa National Forest in central North Island, New Zealand. None of these trees are native to New Zealand. All trees in a block are from the same area of the *P. contorta* range, and all are exactly the same age (10 years). Note differences from block to block. *Left:* The three blocks starting from the near corner and moving to the truck are from an elevation of 3,700 feet in Oregon's Cascades, from 5,500 feet in northeastern Washington, and from 3,200 feet in central Oregon. From the right edge to left in the next tier of blocks are trees from the central California coast (200 feet), from a successful New Zealand plantation, and from the central Oregon coast (50 feet). *Right:* This picture was taken facing toward the truck in the upper right part of the study plot—from the place marked by an arrow. The foreground trees are from 4,000 feet near Jasper, Alberta, the tall trees on the right from 2,000 feet near Prince George, British Columbia, and the tall trees on the left from near sea level at Long Beach, Washington. The site is highly uniform, as indicated by the border row in the picture on the left. (Courtesy of H. G. Hemming, New Zealand Forest Service photo.)

significantly different, causing selection to favor different alternative states of some characters and, thus, the alleles that govern them. Figure 13.3 shows an experiment investigating this level of variability; in this type of study, which is called a common-garden experiment, seedlings from many different provenances are brought to the same location where they can develop under similar environmental conditions.

Second is the level of local population differences. Local populations may be characterized by intermittent migration, which would make it unlikely that any particular new mutant allele could become frequent in some while remaining absent from others. Strong selection differences could, however, maintain genetic differences in the face of occasional immigration, and recent experiments with

BOX 13.D S. S. CHETVERIKOV (1880–1959)

After the rediscovery of Mendel's laws, Darwinism suffered a temporary eclipse; natural selection was relegated to a secondary role in evolution, and, at least according to one school of thought, mutation assumed the principal role among evolutionary forces. R. A. Fisher, Sewall Wright, and J. B. S. Haldane, independently of each other, formulated the basic tenets underlying the synthesis of Darwinism and Mendelism, for which they also provided a quantitative basis. Only recently has the fourth of the co-founders of the current synthetic theory of evolution, the Russian naturalist S. S. Chetverikov, been given credit for his role in its formulation.

There are two reasons why Chetverikov was so long neglected. First, his now classical 1927 essay on evolution from the genetic standpoint was originally published in Russian and did not become fully available in English until 1961. And, second, the bizarre and tragic period in the history of Soviet genetics under Stalin, which is described in Chapter 23, not only robbed Chetverikov of honor, but led to his demotion, dismissal, and exile, and barred him until just before his death from any possibility of further intellectual or experimental contribution to science.

Chetverikov's essay included most of the basic ideas on which current evolutionary theory rests. He recognized the existence of the great store of genetic variability supplied by mutation in natural populations; he understood clearly how particulate Mendelian inheritance makes it possible to maintain this variation; he visualized the concepts of allele frequency and the gene pool; he realized the roles of such evolutionary forces as isolation and drift; he emphasized the importance of such phenomena as pleiotropy of gene action, and of polymorphisms in populations.

Many of his notions were in the air at the time. Thus, in America, F. B. Sumner studied variation in wild mice, and G. H. Shull and E. M. East in corn. Chetverikov was, by training and profession, a lepidopterist, and had the largest butterfly and moth collection in the USSR. Yet like many other geneticists, he used the fruit fly for his important

experiments. It was Chetverikov's simple drosophila experiment that was probably the first one based on strict Mendelian considerations. Capturing inseminated females in the wild, he inbred their offspring to find a great reservoir of recessive alleles carried by them. Chetverikov laid the foundations of a great school of Russian population geneticists, in which his students, N. Timofeev-Resovsky, B. L. Astaurov, N. P. Dubinin, and others, played a prominent role until the government fiat practically extinguished genetic research in the USSR.

His personal biography was tragic. Dismissed from his university position, he was successively a consultant to a zoo in the Ural mountains and a mathematics teacher in a junior college. Reprieved for a while, he was able to do some work on selection in a species of silkworm, but when genetics was formally outlawed in 1948, he lost his post again. He ended his days blind and in poverty, and it was only just before he died in 1959 that relaxation of the attitude of the Soviet government towards genetics brought him some measure of recognition.

THE PHOTOGRAPH IS REPRODUCED COURTESY OF B. L. ASTAUROV.

plants indicate that this frequently happens when the environmental differences are great enough. But the force most commonly responsible for local population differences is drift. There are two situations in which drift may operate. In the first, a small population is isolated over many generations; an example of a group of populations in this situation is provided by small isolated Swiss villages. In the second situation, there is a population bottleneck; for example, a large population of loblolly pine may become established on an abandoned field from just a few parents growing at the edge of the field. In either of these situations, the drift-caused differences tend to be not particularly adaptive, and are probably transient, as over time some immigration from neighbor populations will likely occur.

Third is the level of family differences within populations. Mutation is not a major contributor to this level of genetic variability, which it should be recognized, does not have the time dimension of many generations that the first two levels possess. The mating system plus the amount of genetic variability maintained within a population are of overriding importance. If assortative mating, or consanguineous mating, is common, the differences between families will be greater. They will also increase if immigrants join the breeding population, either as family units, or as mates of local individuals. Segregation distortion, if present, would also add to differences between families.

Finally, the variability within families, which in most studies is found to be a very large reservoir of genetic variability, depends on the average heterozygosity of the parents, and the mating system. For instance, the seedling offspring of a single fir tree are least variable if only one other tree serves as the pollen donor. They are more variable if a pollen mixture from most or all of the trees in that stand (population) is effective in fertilization. They are even more variable if a significant proportion of the effective pollen blows in from distant populations. We will use the rest of this section to discuss the important question of the maintenance of genetic variability within populations.

The discovery of mechanisms for maintaining genetic variability—which include mutation pressure and the balance between the reproductive rates of heterozygotes and homozygotes—dealt a death blow to a typological view of the species. In the presence of polymorphisms at a great number of loci, the notion that there is a *type* for each given species became entirely untenable and gave way to the statistical interpretation of evolution. Mutation pressure that is opposed by natural selection can usually maintain disadvantageous alleles at only very low frequencies, such as those reported for the alleles for PKU and galactosemia in Section 13.2. A **genetic polymorphism** is defined as the presence of two or more alleles in a population at frequencies greater than can be explained by a mutation-selection balance.

One of the first to recognize the existence and the significance of the stores of expressed and concealed genetic variability in natural populations was the Russian naturalist and geneticist **S. S. Chetverikov** (1880–1959). He may be considered the founder of experimental population genetics (see Box 13.D). Ever since his 1926 experiment, evidence has been accumulating that the capacity of populations for storage of direct genetic messages and messages concealed by dominance and epistasis is tremendous. For some characters, it has even been claimed that the total variability of a population could be reconstituted from the alleles contained in very few individuals (but see the founder principle mentioned in Sections 4.3 and 19.3). Recently, a great spurt of activity in the study of population structure has been initiated by the discovery of many biochemical polymorphisms in *Drosophila* by the American geneticists R. C. Lewontin and J. L. Hubby. The polymorphisms appear to exist at a high proportion of loci producing particular proteins, including specific enzymes, in both wild and laboratory populations. Biochemical polymorphisms appear also to be widespread in human populations. Thus the English geneticist H. Harris found that three out of ten arbitrarily chosen enzymes were polymorphic. It has been computed from blood-group data that humans are heterozygous for about 16 percent of their loci, a figure remarkably close to one found experimentally in *Drosophila*.

The genetic population structure of characters exhibiting the heterozygote-superiority, or overdominance, mechanism is called **balanced polymorphism**—that is, populations contain alleles for the locus in various frequencies depending on the s and t values (see Section 13.2). **Transient polymorphism** is also known, in which the presence in a population of two or more alleles for a locus at intermediate frequencies is only temporary. For instance, many African populations are polymorphic for the alleles specifying hemoglobins C, A, and S. The Hb^C allele, as judged from geographical distribution, arose from a mutation perhaps only once in the last 1,500–2,000 years, and spread in malarial areas. Upon reaching places where hemoglobin S provided a defense against malaria, the Hb^C allele, being less debilitating, began to displace the Hb^S allele. If this process were to continue, the present-day polymorphism for the three alleles would be reduced to two, Hb^C and Hb^A. Chances are that now, with malaria being brought under control, Hb^A may replace both.

Mechanisms for maintaining polymorphisms other than heterozygote superiority are also known. They include alternation of selective advantage between alleles in the course of a life cycle or from generation to generation, differences in preference for ecological subniches between the different genotypes, and frequency-dependent selection.

In addition to balancing mechanisms affecting characters controlled at a single locus, there are others, perhaps more complex, that deal with polygenically determined traits. Their complexity lies not only in the fact that many loci are involved, but also in their usual remoteness in the developmental pathway from the original gene products. This means that whatever balance is attained, it involves a great many intermediate biochemical and physiological interactions. While genetic variability is thus tolerated, and perhaps even selected for, not any old combination of alleles will do, and the resulting genetic homeostasis has been demonstrated for various metric properties with several species.

In the selection of chickens for egg production, large egg size is one of several economically desirable objectives. Hence, breeders exercise artificial selection to obtain hens that lay larger eggs. But the biologically balanced phenotype, as a result of prolonged natural selection, includes a smaller egg than the breeder wants. The highest reproductive fitness in the selected flock is found among hens

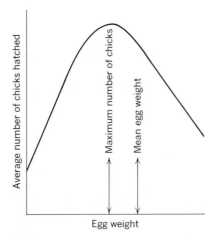

FIGURE 13.4
The curve shows the relationship between the size of eggs laid by pullets and the relative fitness of the pullets as judged by the number of chicks they produce on the average.

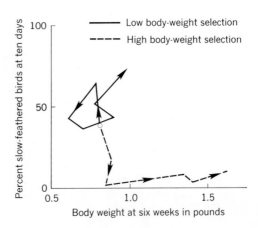

FIGURE 13.5
Correlated response in feathering rate to selection for low body weight. (Based on the data of G. F. Godfrey and B. L. Goodman.)

laying smaller than average-sized eggs, the flock's average egg size being maintained by the pressure of artificial selection. If the breeder suspends artificial selection for this trait, it reverts to a lower average egg size by natural selection (Figure 13.4).

The final point that needs to be made about balance regards **correlated response.** Darwin recognized that selection exercising its pressure on some characters unavoidably modifies others. He was viewing this phenomenon on a phenotypic basis and assumed, entirely correctly, that there is an interdependence of structures and functions in a harmonious living being. Today we can also understand this phenomenon both on the level of the genotype because of pleiotropy and linkage, and on the level of the gene pool as a form of coadaptation.

We refer often to chickens, which provide many useful illustrative examples because the genetics of the fowl has been intensively studied, in part for economic reasons. Figure 13.5 shows selection progress for high and low body weight in an experiment in which the rate of feathering of the chicks was not used as a criterion of selection. In the line selected for high weight, feathering rate remained rapid. But in selection for low weight, which was not very successful since the modification was "anti-adaptive," a correlated response of slowing down the rate of feathering was observed. Evidently, the small birds could not support metabolic activities necessary for rapid feathering. It might be noted that a common form of correlated response to prolonged artificial selection of metrical traits is a drop in reproductive fitness. This is to be expected if coadaptation cannot keep pace with extreme changes induced by selection, because natural selection operates on the totality of fitness components, and sudden or rapid changes in one of them unbalances the rest.

14

SELECTION

We have already examined a number of instances of natural selection, such as for industrial melanism in moths and for hemoglobin variants in humans. Much evidence has been compiled that natural selection is operating on a number of traits in various organisms including our own species. An example of a dynamic interaction between selective forces is provided by the Australian experience in attempting to control the rabbit population. In 1859, a gentleman farmer in southern Australia imported twelve pairs of wild rabbits from England in order to be able to satisfy his passion for hunting. Within a few years, the rabbit became a major agricultural pest on the whole continent. Many ways of eradication were tried without success. In 1950 a virus producing myxomatosis, a disease fatal to rabbits, was deliberately released in the countryside and rapidly spread throughout Australia. It caused such a widespread epizootic, that in some areas the census numbers of rabbits went down from 5,000 to 50 in a matter of six weeks. However, year after year, fewer and fewer rabbits died of myxomatosis, so that this solution did not turn out to be a permanent one. What happened was that the epizootic became a powerful selective force for resistance of the rabbits to the disease. At the same time, some of the viruses mutated to less virulent forms, permitting them to multiply without killing the hosts. Survival of the more resistant rabbits was advantageous not only to the rabbits but to the parasite as well. This is an example of the general phenomenon of interspecific coadaptation (if viruses can be considered to be species), characteristic of host-parasite relations.

14.1 NATURAL SELECTION IN HUMANS

In the human species evidence for natural selection at various loci is more tenuous and speculative, largely because of difficulty in obtaining data. Nevertheless, many signs suggest that much human polymorphism is maintained in populations by a variety of selective processes, including that of heterozygote superiority, which seems probable at the locus controlling the MN blood group. Also of interest is the more commonly known ABO blood-group polymorphism (see Chapter 20).

It has been found that people differing in genotype at the ABO locus show different incidences of a number of diseases. A recent (1970) survey lists 31 such associations of ABO blood types and various diseases that have been found to be statistically significant, most from data on over 400 patients, and several from observations on thousands of patients. A few examples are given below:

Disease	Blood groups contrasted	Ratio of incidence[1]
Ventriculi cancer of the intestine	A/O	1.22
Malignant salivary gland tumors	A/O	1.64
Cancer of mouth and pharynx	A/O	1.25
Multiple primary cancer	A/O	1.43
Duodenal ulcer	O/A	1.35
Tertiary syphilis	B/O	1.51
Acute hepatitis	A/O	1.27
Cirrhosis of the liver	A/O	1.50
Smallpox (sib comparisons)	(A+AB)/(B+O)	6.09
Influenza A_2	O/A	1.49

[1]As explanation, in the first line, 1.22 indicates that the incidence of ventriculi cancer of the intestine is 22 percent higher among *A* blood-group people than among *O* blood-group people.

A connection has also been suggested between the geographical distribution of the ABO genes and the previous epidemiological history of the various areas. Where plague was once common, there seems to be a relative shortage of O individuals, whereas areas known to have had severe outbreaks of smallpox show a similar deficiency of people belonging to the A group. A possible explanation, by no means agreed upon by all immunologists, is that in these cases the causative agents of the disease (the bacterium *Pasteurella pestis* for plague, and the *Variola* virus for smallpox) have immunological properties similar to the respective blood group antigens. Those O individuals having the one disease, and the A having the other, are not able to recognize the infective agent as a foreign antigen and to produce sufficient antibodies to combat it.

Still another example of possible natural selection, this one probably operating on more than a single locus, may be found in heart disease. The genetic basis of coronary artery disease is not understood as yet, but the disease is known to be common in families with a history of defective lipid metabolism and among diabetics. Recently, the incidence of the disease seems to be increasing. In part this may be explained by increased longevity. More people are now surviving to the age when arteriosclerosis tends to occur. And in part this may be connected with eating habits. There is a correlation between incidence of the disease and amount of animal fats in the diet.

A New Zealand physician has suggested that susceptibility to heart disease may have had a selective advantage at one stage of human society, because groups not having to support people past their reproductive age would have an advantage in periods of food scarcity. Haldane had earlier proposed the existence of such selection for liability to cancer and other diseases of old age, although such intergroup selection is generally not very efficient. When tools were invented and fire domesticated, our ancestors became largely carnivorous. A rich reservoir of previously unexploited food resources was opened for the hunting societies of the Stone Age and with the change in diet, selective forces for heart disease susceptibility came into operation. With the invention of agriculture, the diet once more became predominantly vegetarian, as it still is in many primitive societies or in underdeveloped areas. In recent years the animal-fat consumption in the more prosperous countries has again been rising, with a consequent increase in incidence of the disease among genotypes predisposed to it, perhaps by stone-age selection. Speculative and unsupported by concrete evidence as this hypothesis is, it is worth mentioning as another possible instance of interaction between our biological and cultural evolution.

In general, in spite of our increased life-span (Box 15.A), considerable opportunity still exists for natural selection to occur in humans. Statistical studies verifying this have been carried out. In the past, much of selection operated by eliminating disadvantageous genotypes in the intrauterine, early postnatal, and adolescent stages in life. There also was and still is some prezygotic selection produced either by segregation distortion, or by competition between sperm, or by sperm-egg incompatibility. Today selection pressure is largely directed towards differential fertility. On the average, about one-third of the members of one generation contribute two-thirds of the genes in the gene pool of their great grandchildren's generation. Even more extreme selection opportunities exist in certain populations. In one study on Navajo Indians in New Mexico, one of the founders of the population was found to have, after eight generations, 411 descendents, forming 14 percent of the total group living today. Indeed, six persons among the founders accounted for more than half of all descendents. Clearly, considerable natural selection can be accommodated with reproductive patterns of this type.

A more quantitative approach to estimating the opportunity for selection is to compute the ratio of the variance in number of children per family to the square of the average number of children per family. In the United States, this ratio went up from about 0.25 for the cohort of women born in 1839 (when the average number of children was 5.5) to 1.14 in the 1901–05 cohort (when the average number of children was 2.3). This ratio is going down again, as there is recently a strong tendency for families to have more nearly the same number of children. Since Darwinian fitness is associated with differential reproduction, there is much greater opportunity for natural selection when this ratio is high than when it is low.

Another trend currently affecting selection in humans is a reduction in inbreeding, associated with greater physical and social mobility (Chapter 21). As a consequence, the frequency of rare recessive genes coming together in the homozygous state, and thus becoming more exposed to selection, will be reduced for several generations. A trend that appears to be partly balancing this is increased assortative mating for a number of traits, such as height. The most striking of the traits for which there is assortative mating is intelligence, and there may be a resulting substantial increase in the variability of intelligence in the normal range. The American public university system is a major factor enhancing this trend, as it brings students from different places and social strata to a common place, where intelligence is valued, just prior to their reproductive years.

Perhaps the clearest trend is the relaxation of selection for traits associated with survival and, more gradually with medical and family-planning advances, for fertility. Human fertility is increasingly becoming the result of state of mind rather than of physiology.

14.2 RELAXATION OF SELECTION

An example of the effects of the presence or absence of the selecting agent is provided by natural selection for resistance to bubonic plague in rats. Rats, in India, taken from areas where plague epidemics had recently occurred, showed only a 10 percent mortality when infected with the disease in the laboratory. Rats taken from areas free of plague, however, had a mortality of 90 percent. This suggests, since plague had probably visited these latter areas at some time in the past, that the alleles for plague resistance may be disadvantageous in the absence of plague. Thus, during the many generations in which rats had reproduced their populations since the last plague epidemics, it appears that the alleles for resistance may have dropped to relatively low levels.

When tuberculosis was a major cause of human mortality, resistance to it had a selective advantage. Today, this selection pressure no longer exists over the larger part of the world. If the alleles for resistance have no other effect, their frequencies may be expected to drift. If they are disadvantageous for some reason, they would tend to be reduced, or even eliminated, if tuberculosis remains controlled and rare.

Figure 14.1 shows results from a selection experiment on chickens, in which crooked toes (the extreme form producing club feet) were selected for. As may be readily seen, after three generations of selection it was still possible to reverse the direction of selection and return a sub-line to the initial level. But after eight generations, the line selected for crooked toes had nearly 100 percent incidence of the defect, and reversal would then have been difficult or perhaps impossible.

The question has often been raised whether the various preventive and therapeutic measures now available have a **dysgenic** (that is, detrimental) effect on

BOX 14.A SELECTION FOR EGG PRODUCTION IN THE FOWL

The University of California at Berkeley carried out a long-term experiment designed to explore the optimum methods of selection for an economically useful character, the number of eggs laid by a flock of hens. The foundation flock in 1933 averaged a little over 120 eggs per year per pullet, a figure roughly comparable to the level of commercial egg production of the day. A few sires were added from other populations in the early part of the experiment, but after 1941 the flock was *closed*.

THE EARLY DATA, TAKEN FROM I. M. LERNER, HAVE BEEN SUPPLEMENTED BY COURTESY OF D. C. LOWRY.

In each generation, the flock is composed of spring-hatched pullets.

The actual methods of selection changed as more information on the theory and empirical results of selection was obtained. Basically, however, selection emphasized the performance of families of full-sibs and half-sibs. Since egg production is a *sex-limited* character, expressed only in females, sires were selected in the early part of the experiment on the basis of progeny and sister tests. In later years sister tests were used exclusively. One of the shortcuts found in the course of the experiment was that the criterion of selection need not be production for the whole year. Spring-hatched pullets start laying late in the summer or early in the fall. By using the record of egg production to December 1 or January 1, the interval between generations—and hence the gain per unit of time—could be shortened. Other economic considerations (for example, resistance to specific diseases, egg weight and quality) were also selected for during most of the experiment.

The illustration shows the average egg production of the flock at five-year intervals. It may be readily seen that selection has been extremely effective. The economic significance of doubling the number of eggs laid in a year, from 125.6 in 1933 to 249.6 in 1965, is obvious. Note that by January 1 the birds of the last generation laid as many eggs as the foundation flock took until July 1 to lay. Similarly, production to July 1 in the last generation exceeded the 1943 production figure for the whole year.

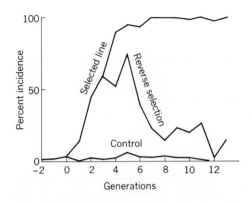

FIGURE 14.1
An experiment in which chickens were selected for the incidence of crooked toes. The selected line has become nearly fixed for this trait. Selection in the reverse direction, initiated in a subline taken from the selected line after the incidence had been greatly increased, was able to bring down the incidence of the defect to near that of the control line.

the human gene pool. The formal mathematical answer is different for recessive genes from that for dominant genes. Using crude approximations for illustrations, and assuming that q for the recessive allele causing PKU is 0.01 and that the fitness of the homozygote recessives was zero before treatment was discovered, J. F. Crow computed the effect of restoring normal reproductive capacity (fitness of 1) to all PKU patients. It turns out that permitting these q^2 homozygotes to contribute to the gene pool of the next generation would increase the incidence of the disease at a rate of 2 percent per generation, a rate that would double the incidence over 40 generations, or some 1,200 years. For rare dominant alleles responsible for defects, the situation is different: the incidence of the defect produced may be expected to increase at the rate they had previously been eliminated by selection each generation (Section 13.2). For previously lethal dominants, this could double the incidence in the first generation after bearers of the allele become able to reproduce normally.

Too little is known as yet about human allele frequencies in different generations to estimate accurately the actual effects of reversal, suspension, or relaxation on various traits. Several attempts, however, have been undertaken to gauge the effects, by comparing incidence of various defects in primitive societies, in which they might be expected to have selective disadvantages, with that in more advanced societies in which selection has been relaxed. The frequency of red-green color blindness in populations of eskimos, Australian aborigines, and other hunters and gatherers, is found to be about 2 percent among the males. Groups somewhat removed from hunting and gathering, such as some American indians and African Negroes, have an average incidence nearer 3 percent. Populations still farther removed from being hunters in time or mode of existence, including Europeans, Philipinos, and some Arabs, show a frequency of about 5 percent, with some populations in Britain, France, Switzerland and Russia having about 9 percent. These statistics, as well as similar ones on visual acuity, indeed suggest that selection no longer operates against such eye defects in modern urban societies. Dorothy Parker's "Men seldom make passes at girls who wear glasses" may be a clever poem, but it seems to have little predictive value with respect to the reproductive fitness in modern societies of either males or females, whose defective vision may be corrected by a variety of technological options.

BOX 14.B SELECTION PROGRESS IN MICE AND CHICKENS

D. S. Falconer selected for 6-week body weight of mice for 23 generations. In each of six experiments (noted as A, B, C, D, E, and F in the graphs), an initial line was subdivided at generation one as follows: a group of 8 ♂ and 8 ♀ mice was selected at random, and in each subsequent genera-

tion offspring selected at random from this control line were mated to each other; eight of the heaviest remaining ♂ and ♀ mice were selected and mated, and in each subsequent generation, eight pairs of the heaviest mice of this line were selected and mated; the lightest mice were similarly selected and mated to create the third line. As the experiments progressed, matings failed with increasing frequency as inbreeding within these closed populations increased (although the mating scheme employed tended to minimize inbreeding), and some lines were reduced to three, or even two, breeding pairs on a few occasions. Finally, after 23 generations, the experiment was terminated because of the low reproductive capacity of some of the lines.

We can make rather different conclusions if we look at only a single experiment and time of observation. For instance, by observing the first ten generations of replicate C, we might conclude that selection for low body weight is completely ineffective. The first 12 generations of replicate F indicates a strong asymmetry of response (the heavy-body-weight line averaging 13 grams above the controls, the low-body-weight line only 5 grams below) while the final eleven generations indicate a plateau in both selected lines, with no further gain in the heavy-body-weight line and only a slight further reduction in the low-body-weight line. The B replication indicates a continuing and generally symmetrical response to selection in both lines throughout the experiment. When data from the different replicates are superimposed (page 253), a generally consistent picture emerges. Data points from the control lines fluctuate about the initial values of around 22 grams,

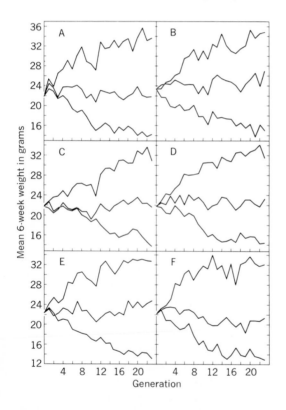

THE FIRST TWO GRAPHS ARE FROM GENETICAL RESEARCH 22:291–321, WHERE THE RESEARCH IS REPORTED IN FULL DETAIL. DATA OF THE THIRD GRAPH ARE COURTESY OF D. C. LOWRY.

largely in response to genetic drift due to the small number of breeding pairs. There is a clear response to selection for both heavier and lighter mice, and this seems to be tapering off in the latter half of the experiment.

Single experiments or single generations frequently do not conform well to genetic selection theory, sometimes failing, sometimes exceeding expectations. Thus, any single selection program is something of a gamble. But the experience of all such programs to date indicates that, on the average, the predictions based on genetic selection theory are borne out remarkably well, and further, that the investment in a selection program to improve a crop or herd is one of the best options available in agriculture. It further affirms, as do Falconer's experiments, that selection carried on over many generations must be based on a large and genetically diverse breeding population, in order to avoid or postpone matings between relatives, and the difficulties associated with such inbreeding.

As previously indicated (Section 13.3, Figure 13.4), complications may arise because of genetic homeostasis, in which artificial selection is opposed by natural selection. The adjacent graph presents as an example the results obtained in a selection experiment for long shanks of chickens. Coadaptation of genes for leg length and reproductive capacity is a slow process. In this example, successful selection for increased length of shank had a setback, because the birds with the longest bones had, as a correlated response, lowered fitness. After a period without consistent gains, however, a new balance was reached, and the increase in shank length resumed.

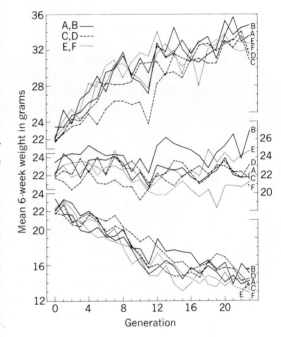

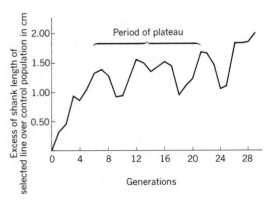

The surgical or orthodontial correction of various mouth and tooth abnormalities, some of which clearly have strong hereditary components, may operate in three ways. Orthodontists and other dentists stress that such correction improves the health of the patient. Thus, the patient is more likely to survive and the alleles specifying the abnormalities are more likely to be passed on. The treatment itself frequently includes X-ray diagnosis, which may increase mutations (Chapter 17). These will normally be of the somatic variety and in the vicinity of the mouth, and only if procedures are sloppy will such X-rays reach the gonads. But most important, if television commercials portray the truth, a good smile is dynamite for attracting a mate. Thus, corrective dentistry's contribution to increased frequencies of the hereditary causes of dental abnormalities in future generations may be great indeed.

Such increases in frequencies of deleterious alleles as those we may observe if we move from data on the most primitive people to those on people whose technology and medical science compensate for genetic disadvantage, have alarmed some geneticists. Some even find in them a threat to continued human existence. But, as has already been shown in Section 13.3, negative eugenic measures can do little to control the frequency of genetic defects. Furthermore, it is estimated that at least 10 percent of all human zygotes carry a new mutation. Relaxation of selection pressures for various disadvantageous traits occurred long before *Homo sapiens* emerged.

It is thus possible that there may be a vicious circle between biology (which is responsible for increases in hereditary defect) and culture (which provides corrective devices that also screen out biology's selective defenses against too many defects). We are already accustomed to corrective lenses for near- or farsightedness. Surgical intervention to correct such abnormalities as a harelip or polydactyly (extra digits) is commonplace. Replacement therapy, by hormones for diabetes, or by other substances for anemia, is available. Dietary circumvention of some metabolic blocks has already been discussed, and a variety of other **euphenic** (that is, improving the phenotype) measures are being developed for the various genetic ills to which humans are subject.

The scenario based on the proliferation of corrective devices, coupled with an understanding of genetic recombination, leads to speculation that there may come a time when only an unusual person will dare stray far from the life-support machines. The problem of human impact on preserved wilderness areas, now a vexing problem for the Park Service and others who administer wilderness areas, will disappear. Only one person out of very many will possess the physical and mental equipment to survive farther in the wilderness than a day's hike from trailhead. Those who today defend being in the wilderness as an uplifting and enriching experience may well wonder at the magnitude of the spiritual loss for future peoples.

It has been suggested that such extreme dependence may not develop, because the cost to society of providing such corrective devices may first become an intolerable burden. But because cultural evolution of human populations is currently so much more rapid than their biological evolution, we can expect to "keep ahead of the game" for a long time to come. Apart from that, bookkeeping of social costs and gains is not a simple matter, and surely cannot be lightly invoked where human welfare is concerned.

Perhaps more important aspects of negative eugenics are the psychological effects on the affected persons and their families. Voluntary restriction of reproduction, where recurrences of grave defects are expected to be high, is an obvious partial solution of the problem, and is considered further in connection with genetic counseling (Chapter 19). The possible role of positive eugenics is discussed in Chapter 22.

14.3 SELECTION IN AGRICULTURE AND MEDICINE

Natural selection proceeds both in the presence and absence of humans. Even though our ancestors may not have selected at all when they gathered seed from plantations for the next crop, there was continued natural selection for adaptedness to the plantation environment, and domestication to plantation varieties thus slowly progressed. But we are primarily interested here in artificial selection—human-directed adaptation of populations or varieties to human needs. This can proceed crudely, as it did through centuries of domestication of farm animals and crop plants. At present, it can proceed with much greater precision, as the principles of selection and inheritance that have been developed in the century since Darwin and Mendel are applied to plant and animal breeding.

A distinction is often made between natural and artificial selection. But if humans and human activities are considered to be a part of the ecosystem, the distinction blurs, and properly so. Domesticated crops and animals respond to selection pressures imposed directly or indirectly by these human activites, as well as by other components of their environment. But the principles underlying both artificial and natural selection are exactly the same, and that is why a knowledge of one has repeatedly led to better understanding of the other. If there is a difference, it is that artificial selection is directed to an end other than reproductive fitness (frequently with cleverness, but not always with intelligence).

An early form of artificial selection, closely mimicking natural selection, was *mass selection,* in which only the most desirable phenotypes were permitted to propagate. This technique of selection, although effective for characters of high heritability, is not an efficient one where heritability is relatively low. In order to increase the accuracy of estimation of genotypic merit, breeders developed *individual-plus-family selection.* In a breeding scheme of this type, in addition to the phenotypic performance of the individual tested, the phenotypes of its sibs, or its progeny, or other relatives, are also taken into account. With the development of methods of population and biometrical genetics, it has become possible to choose from among different methods of selection (mass, full-sib test, half-sib test, progeny test, pedigree, or a combination of these), the ones that would produce the most rapid, or the most economical, or the most prolonged gains in the objective sought.

BOX 14.C HYBRIDS

Reference is often made to "the miracle of hybrid corn" as one of the crown jewels of American agriculture. There is no question that its development is a spectacular achievement (although American indians had already effected notable improvements in their domesticated varieties, and that given to the Pilgrims and other early European explorers and immigrants was well advanced compared to primitive corn). The pictures of corn contrast the oldest known ear of corn (left), dated about 2000 B.C. and found in New Mexico's Bat Cave, with a modern hybrid dent corn ear.

However, some corn geneticists have questioned the wisdom of using a hybrid breeding scheme. Briefly, breeding productive hybrid corn depends on identifying the specific combining abilities of two (or sometimes four) highly inbred, and therefore homozygous, lines. When such lines are crossed, the resulting uniformly heterozygous lines have high predictability and heterotic vigor. This vigor derives from the nonadditive components of genetic variability. The critics suggest that had we employed other breeding schemes, based on the additive component of genetic variability, yields today would be even higher than those we are getting from the hybrids.

In defense of the hybrids—they offer great uniformity, which is important if an entire field is to be harvested at once by machinery. (This genetic input—allowing people to be efficiently replaced by machines—has considerable social importance.) But what is historically more important, there is built into the use of hybrid corn a seed-production requirement that makes it very attractive to private investment capital. After selection is suspended in an additive genetic scheme, the line of corn tends to maintain about the same mean production. Thus, farmers would only have to buy nonhybrid seed once, and their corn crop would repeat its high performance every year if they simply saved and planted seed from their own fields. They would go back to the seed companies only when, through selection experiments, the companies had further improved the line, or developed a new one, to such an extent that buying the new seed would be worth the added cost. Not a very profitable arrangement for the seed companies. The offspring of hybrids, on the other hand, segregate into a great variety of different genotypes, with chaotic loss of uniformity and markedly lowered average performance. **Thus the farmers cannot use seed from their own crops, but must every year buy seed from the seed** companies, who own the inbred lines that produce

THE PHOTOGRAPHS OF CORN ARE FROM P. C. MANGELSDORF, "THE MYSTERY OF CORN." COPYRIGHT © 1950 BY SCIENTIFIC AMERICAN, INC.

the guaranteed-performance, genetically uniform, hybrid corn.

In defense of the seed companies—they need to be able to sell seeds every year if they are to stay in the seed business. If they had not done the work to improve corn, probably the only other candidate for the job would have been the government, as individual farmers would not have been able to support the research necessary—regardless of whether an additive or a nonadditive breeding scheme was being sought. While theoretically

"He's kicked the Corporal!"

"He's kicked the Vet.!!"

"He's kicked the Transport Officer!!!"

"He's kicked the Colonel!!!!"

we would probably be better off if an additive scheme had been developed, improved corn from a nonadditive scheme is better than no improvement at all: our government has a track record of allocating more of its agricultural-support resources to subsidy programs than to research. So it is perhaps fortunate that private industry was able to find a way to do the job. The USSR, with its economic and governmental philosophy, could have been the logical control in this giant experiment. But it turned its back on genetics during the period when the experiment was in progress.

Mules are quite another story—they are the result of interspecific hybridization between female horses and male asses (the Greek word for mule was *hemionus,* i.e., "half-ass"). While substantial hybrid vigor is derived from the cross, the vigor of the progeny also reflects the fact that much care typically goes into the selection of both the horse and ass parents. Male mules are sterile, and female mules are only very rarely fertile.

The mule has more than its share of admirable qualities. It is hard of hide, sure of foot, sound of constitution, and better able to endure extreme climates, thirst, and hunger than a horse. It is courageous, intelligent, and sensitive. When mishandled or misunderstood, as they frequently were when the practice of war in earlier times put them under soldiers untrained in mule handling, mules commonly gave muleish reprisals. During World War I, a cartoonist for the British comic weekly *Punch* depicted (as reproduced here) a cluster of mules becomingly increasingly mirthful as one of their number, who is out of sight, kicks at military personnel of successively more exalted rank.

Finally, hybridization is an important technique for establishing a highly variable base population before other techniques of selection are applied. Many, perhaps most, of our domesticated plants and animals are derived from a broad interpopulational or interspecific base resulting from early hybridization.

In order to make such decisions, genetic architecture experiments are generally performed to investigate such things as the amount of additive and nonadditive genetic variability and the genotype–environment interactions. Data from such an experiment, recently performed by Weyerhaeuser forestry research personnel at Coos Bay, Oregon, are presented below. The organism is douglas-fir, the character, growth in height.

	Variance components			
Type of family	Between families	Location × family	Within families	Heritability
Full-sib	70	− 1	626	.20
Open-pollinated	31	4	671	.18

The data may be interpreted as follows: Full-sib families have about half the population's additive genetic variance, plus a moderate proportion of the nonadditive variance, contributing to differences between families (Box 11.A). The open-pollinated families, which probably approximate half-sib families (many different trees contributing pollen to the wind-blown pollen cloud), have about one-fourth the population's additive (and little nonadditive) genetic variance contributing to differences between families, with the remainder contributing to variation within families. Note the full-sib between-family component is slightly more than twice the open-pollinated component, indicating mostly additive genetic variance with some nonadditive variance. The open-pollinated families are internally more variable than the full-sib families, as would be expected because many trees rather than only one are contributing pollen to produce the offspring. The location-by-family component is small, indicating little genotype-environment interaction in this set of environments. Thus the company may select one set of good trees for general use in these environments. The heritability estimates are moderate, indicating that most of the expressed variation is environmentally caused, and that mass selection will not be very efficient. However, h^2 of .18 indicates substantial genetic variability, and the company is justified in using an additive genetic variance scheme, such as an individual-plus-full-sib family design, to select for faster early height growth.

The literature on the use of selection in breeding practice is vast. One example of the economic significance of animal improvement methods is given in Box 14.A. Complications arise when, rather than a single criterion, multiple criteria of genetic worth have to be used. In selecting for increased milk production, for example, the amount of butterfat and nonfat solids should be considered in addition to the yield of milk. Only after the development of computers did it become possible to put into efficient operation *selection by index,* which takes into account the great number of different variables determining the total genetic worth of individuals.

Breeders of egg-laying chickens now have to combine in an index properly weighted heritabilities, genetic correlations, and economic values of as many as sixteen different traits expressed by the individuals and their relatives: fertility, hatchability, incidence of crippled chicks, chick mortality, pullet mortality, degree of broody behavior, body size, rate of sexual maturity, rate of egg laying, egg size, shell thickness, egg shape, shell texture, internal egg quality, shell color, and frequency of blood spots in the eggs.

Questions frequently asked are: How consistent is response to selection from generation to generation, or from experiment to experiment, and how long can it be effectively continued? These are addressed in Box 14.B.

Many problems still remain to be solved in artificial selection practice, and new ones keep arising as higher levels of sophistication in the breeding industries are reached. For instance, in many classes of plants and animals, high production is obtained from progeny of crosses between lines or strains whose gene pools have been kept discrete. This phenomenon of hybrid vigor, or **heterosis,** requires selection for good combining ability between lines, rather than a high performance of the lines themselves (Box 14.C).

Finally, an example of application of selection principles in the pharmaceutical industry is provided by Figure 14.2. However, selection is not always on the side of the angels, either in agriculture or in medicine. The rising resistance of insects to DDT and other insecticides, of fungi to some fungicides, and of bacteria to antibiotics, is well documented. It appears that the selection for resistance factors in some bacteria may be unusual indeed. In experiments in which bacteria are placed in very high concentrations of a drug, the bacteria simply sit for about 20 hours, unable to multiply. Then suddenly they take off. Electron-microscope examination reveals a large accumulation of resistance factors (Section 6.6), which have apparently replicated autonomously of the bacterial chromosome. Resistance to DDT and other pesticides appears to proceed by normal Darwinian rules, but is nevertheless highly effective when the selective agent is repeatedly and extensively applied. In 1971, in the vicinity of Raleigh, North Carolina, a population of Norway rats resistant to the anticoagulant rodenticide "warfarin" was identified. Such resistance was first noted in northern Europe in 1958, and has since been spreading on that continent.

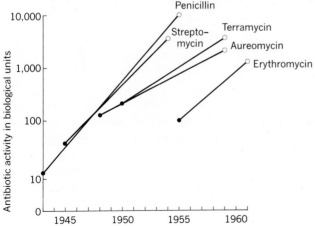

FIGURE 14.2
Increased antibiotic activity obtained by selection from initially low-yielding strains of fungi. The scale is geometric. (Data from S. I. Alikhanian.)

15

POPULATION AND RENEWABLE RESOURCES

The application of genetic principles can increase the productivity of the organisms that provide food and fiber. It has done so in the past, even before genetics was understood as a science. While greater growth of our crops appears to be a solution, growth of human populations and cultures appears instead to be one of our most difficult problems.

15.1 POPULATION EXPLOSION

Most animals have built-in regulatory mechanisms to limit population size. These by no means would be acceptable to humans: release, when there is crowding, of noxious substances, which decreases mating frequency; hormonal inhibition of reproduction induced by the social stresses of crowding; spread of genes causing death or sterility; and so forth. Although a human female normally can produce no more than 10 or 12 offspring, which seem remarkably few when we learn that the tapeworm can lay 120,000 fertile eggs a day, the human population is increasing at a rate that is probably greater than that of any other organism.

Diseases and famine, pestilence and war, infanticide and human sacrifice, unpleasant as it may be to admit it, have regulated population growth hitherto. They no longer do so.

A look at some statistics may help us see the problem, but a word of caution is necessary. All the demographic data to be cited are only coarse approximations. Census figures are notoriously unreliable, and estimates of population size and growth are subject to gross errors. For instance, figures given for the population of China vary at their extremes by 100 percent. The population of Mauretania in Africa is estimated by its government as being threefold the figure given by the United Nations. During the partition of Pakistan and India, the number of migrants from one country to another was computed by dividing the amount of salt issued to them by the average salt requirement. The 1960 census of the United States, in which the data were analyzed by the latest computerized techniques, showed that there were in the United States 1,670 fourteen-year-old widowers, a figure of doubtful validity.

But no matter how much we allow for such inaccuracies, the picture is clear. The tremendous acceleration in the increase of the human population is illustrated by the following figures: it took 4.5 billion years for the first humans to appear, and about 200,000 years for the population to reach one billion people (1830); it took 100 years for it to reach two billion (1930); it took 30 years to reach three billion (1960); at the present rate of increase it took 15 years to reach four billion (1975). About 5 percent of all *Homo sapiens* that ever lived roam Earth today.

We can calculate, as an average, that over the first 200,000 years of human history, the population doubled every 7,000 years. As recently as 1650 the rate of population growth required 200 years to increase the number of people on Earth twofold. Today, the doubling time is about 35 years, varying from 23 years in Brazil to 44 in the United States, and to 76 in Japan. According to these statistics, the 15 million people killed in the battles of World War II were replaced in something like three and a half months. Earth's population today stands at about 4 billion; the annual rate of increase is 2.5 percent or 100 million people. Every day the population rises by 270 thousand; every hour by more than 11,000. Population sizes in millions, both recently and estimated for the year 2000 are:

Year	World	Asia	Europe	USSR	Africa	Northern America	Latin America	Oceania
Mid-1972	3782	2154	469	248	364	231	300	20
UN Estimate, 2000	6494	3777	568	330	818	333	652	35

How are these people to be fed? Today there is barely one acre of cultivated land per person and 60 percent of the people receive less than the average requirement of 2,200 calories a day. By the year 2000, through withdrawal of land from cultivation and population growth, six billion people will have less than half of this amount per person available to them. In the next third of a century, we shall have to triple our food output and production of lumber, and increase our output of energy, iron ore, and aluminum fivefold, just to permit the growing number of human beings to reach a decent living standard.

Technological advances in food production might alleviate the problem of feeding the growing world population. Some authorities think that Earth is potentially capable of supporting 50–100 billion people. Others think we already have more than our resource base can continue to support. At present our food production is growing too slowly. Just to keep pace with the present population growth, without attempting to relieve widespread undernourishment, an increase of 2.5 percent a year in food production is needed. Our recent increases have been on the order of one percent a year. Thus, while North Americans produce more calories than they need, the average food supply available in India and elsewhere is lower than it was before World War II.

Many ways of resolving the situation have been suggested. The lack of fresh water for irrigation is one of the limitations on food production. Increased use of rational and intensive methods of farming, making fresh water from the sea, heating lakes to encourage cloud formation, and fish and game farming are all possible techniques for augmenting food sources now available. New forms of food may be produced. Humans are heterotrophs relying on the photosynthetic activity of plants and protists for energy-binding. This creates a food chain, and much energy is lost in the process of conversion. Although selective breeding of plants and animals has reduced the amount of raw material needed to manufacture foodstuffs, 80–90 percent of the available energy is still dissipated in every link of the chain. Perhaps the longest food chain of all appears in the diet of eskimos. An eskimo must eat five pounds of seal to gain a pound of weight, each pound of seal being derived from five pounds of fish, a pound of which is manufactured from five pounds of shrimp or other invertebrates, each pound of which takes five pounds of algae to produce. In sum, it takes 625 pounds of algae to make one pound of eskimo, with about 99.8 percent of the original energy bound by the algae being lost or recycled in the process.

The recent book *Limits to Growth* increased awareness of the dimensions of the problem, and also provoked considerable controversy. A highly respected economist remarked that he couldn't understand the attention the book was receiving. He felt that while the precise timetables in the book could be challenged, the conclusions were obvious to anyone who understands exponential growth formulas. (Perhaps the most significant event of recent years is that more people—even some in government—are beginning to understand the consequences of exponential growth.)

Barry Commoner has persuasively argued that population growth accounts for only a fraction of our environmental and resource problems, and that a larger effect is due to per-capita increases in demand for and use of our resources (much of it ineffective or even unnecessary use). This is particularly ominous if a goal is raising living standards in the underdeveloped regions of the world, for here population is increasing much faster than where high standards are already achieved. Some recent birth rates, death rates, and rates of natural increase (all per 1,000 per year) are:

Country	Birth rate	Death rate	Percentage increase per year
Algeria	49.1	16.9	3.22
Kenya	47.8	17.5	3.03
Singapore	28.8	15.9	1.80
Trinidad	20.3	6.7	1.36
Japan	18.9	6.9	1.2
Canada	17.5	7.3	1.02

Commoner is probably right in the short run, and the equitable allocation of Earth's scarce resources to its present passengers is a moral dilemma of considerable proportion. But we must also consider the long-run, overwhelming problem of population.

Various kinds of limits to the number of people on Earth have been imagined, including, for instance, eventual scarcity of oxygen or nitrogen, (which are necessary to make and sustain humans) although this is unlikely. Perhaps, the grimmest picture, relieved by a light touch, has been painted by the British physicist J. H. Fremlin.

He has suggested that the absolute limit to the human population on this planet will be determined by terrestrial overheating. People and their activities generate heat that must be dissipated. If the population increased to one quadrillion people, temperatures throughout the world would reach those currently known in equatorial areas. Adequate cooling devices would make it possible to have 60 quadrillion people on earth. The population density would then be 120 million per square kilometer; today it is 18. When the population rose to 10 quintillion, overheating would literally cook people. At the present rate of population growth, this would happen in about 900 years.

Fremlin imagines many difficulties even if the population stabilized at 60 quadrillion, for example, the housing problem. Perhaps technological advances would permit us to build 2,000-story buildings, covering both land and sea, with 1,000 stories of each housing food-production and refrigeration machinery. Of the rest of the space, half would be occupied by wiring, piping, ducting, and elevators, leaving $3\frac{3}{4}$ square meters of living space per person. Food would have to be liquid, clothes would not be worn, cadavers would be immediately processed into food, and each area of a few square kilometers, containing several billion people, would have to be nearly self-sufficient.

Very little movement of people would be tolerated, but still "each individual could choose his friends out of some ten million people, giving adequate social variety," and world-wide television of unexcelled quality would be available. With the present size of the human population, the birth of a Shakespeare is an exceedingly rare event. In the world of 60 quadrillion some ten million Shakespeares ("and rather more Beatles") might be expected to be alive at any given time.

Visionaries who suggest that the problem of overpopulation can be solved by exporting people to other planets can hardly take comfort in the fact that it would take only 50 years at the present rate of population growth to bring Venus, Mercury, Mars, the Moon, and the satellites of Jupiter and Saturn to the same population density as Earth. Colonization of Saturn and Uranus would give only another 200 years' breathing space. At the present speed of travel, to reach the next nearest planet possibly capable of supporting life (see Section 2.7) would take over 325,000 years. But even if travel could be speeded up, the cost would be prohibitive. At the current rate of population increase, the United States would have to spend twenty times its gross national product annually to export the people who are being added to its population every year. And this is not allowing a penny for such prosaic items as food, clothing, or shelter for those remaining. Put in another way, a reduction in the American standard of living by 72 percent from the present level would provide enough capital in a year to export only one day's increment in the world's population. If population were to continue to increase, and technology were able to solve each new problem that arose, we would reach a point where the mass of living humanity would equal the mass of Earth. At that stage, spaceships would have to be constructed of, and fueled by, living human bodies.

Such end points as this, or Fremlin's picture, seem silly, but point out the theoretical possibilities of continued growth. A sillier statement is that growth can continue indefinitely. If we accept that it cannot, then the next logical questions are when and how is it to stop. The main causes of the population explosion are considered in Box 15.A. In brief, they lie in the exponential (Malthusian) nature of population growth and in the increase, owing to advances in sanitation and medicine, in the percentage of human beings who survive to an age at which they can produce offspring. These forces will, in at least the near future, result in population densities well short of the absolute limits proposed by Fremlin, but probably greater than many people would like. (It is interesting, although perhaps pointless, to note that many of the people who are advocating stringent regulation of human reproduction would not be here if their ancestors had taken the advice they are offering.) The widespread malnutrition that is likely to result from increased human density may cause not only stunted bodies, but perhaps stunted minds as well (Section 12.5). Other risks are: pollution; the decimation or even extinction of many of our fellow species; overcrowding with its squalor, ugliness, and perhaps aberrant social behavior; even the collapse of civilization as we know it. These and still other risks must be balanced against the individual, social, and political benefits of large families and large populations.

Many suggestions of how population increases can be controlled—in addition to the very obvious possibility of contraception—have been made. Ireland's population growth has been limited by a traditional prolonged period of celibacy before marriage, a method that seems increasingly unlikely to catch on elsewhere

BOX 15.A SOME CAUSES OF THE POPULATION EXPLOSION

A few data gathered from various sources will illustrate some of the reasons for the current population explosion. The most basic fact is that population size will increase so long as the difference between birth and death rates is greater than zero. Other factors also increase the rate of population growth. One is the tendency to marry younger. In Europe, the average age at which persons enter into marriage contracts is 0.5–2.8 years younger than it was two or three decades ago. This affects both the *rate* at which infants are added to the population, and absolute numbers of births per generation, since younger couples tend to produce larger families than couples who marry later.

Of considerable effect are the increases in life-span that have been achieved during this century, largely owing to the rapid progress in prophylactic and therapeutic methods of controlling infectious disease. At present, life-spans are rapidly increasing in the less developed countries (as they properly should be—but the resulting population problems are thus arising faster there, affording less opportunity to adjust than was available in the developed countries). For instance, it took 130 years to increase life expectancy from 31 to 51 years in Sweden; the same increase was achieved in 8 years in Mauritius. To increase life expectancy from 43 to 63 years in the United States took 80 years; in Taiwan, 20.

Life expectancies in different periods of history have been estimated as follows:

Period	Life-span in years
8000–3000 B.C. (Stone and Bronze ages)	18.0
A.D. 800–1200	31.0
1600–1700	33.5
1800–1900	37.0
Early 1900's	57.4
Mid-20th century	66.5

Only 11.8 percent of dated Paleolithic skeletons represent human beings who lived more than 41 years; today in the United States, 95 percent of the people survive past 40. Some examples of recent changes in life expectancy may be instructive:

Country	Period	
England-Wales	1910–12 53 years	1958 71 years
USSR	1896–97 32 years	1955–56 66 years
Jamaica	1919–21 38 years	1950–52 57 years

Especially dramatic changes have been observed in the underdeveloped countries because of the introduction of sulfa-drugs, antibiotics, DDT, and other means of controlling the spread of infections. At the present time, 60 percent of the world's adults—but 80 percent of the world's children—live in underdeveloped areas. Examples of lowered infant mortality are given by the table below, which shows the percentage of infants surviving a year after birth:

Place	Period	
London, England	1860 70.0	1960 98.0
India	1901 77.0	1962 90.5

One evolutionary effect of the increased probability of surviving until the reproductive age is that, whereas survival *per se* was an important concomitant of Darwinian fitness in the past, today, with the high level of survival, most of the selection in human population is based on differential fertility. As voluntary contraception becomes more widespread and standards of living improve, natural selection could lead to an increase in the proportion of the more prolific parents, which could reduce the population-regulating efficacy of birth control.

in the world. Japan has slowed down its population explosion by encouraging abortion. Legalizing abortion, although it is meeting some religious opposition in the United States, seems a sensible move, but the medical and psychological effects of numerous abortions for the same woman may rule out its value as an important method of population control. Sterilization has been suggested as another method, and the simple vasectomy operation is gaining wide acceptance among males who have decided they have fathered enough children. (The storage of human sperm in cryobanks is a technology available to those who may wish to change their minds—see Section 22.1.)

Even the suggestion that encouragement of homosexuality could keep down population size has been made. Now, so far, no clear-cut genetic basis for homosexuality, or other types of sexual gratification that do not result in conception, has been demonstrated. Should there be one, possibly in the long run this method of population control would be self-defeating: natural selection will tend to weed out genotypes for preferences for sexual behavior not leading to production of offspring. (Studies of the social organization and behavior of some of our mammalian relatives provide grounds for speculation about homosexual behavior. In lion prides, there are commonly two to four *alpha* males—generally brothers—and ten or so adult females—generally sisters unrelated to the brothers. As male cubs reach adulthood, they join wandering bachelor groups, and during adolescence and much or all of their adult lives, their social interactions are almost exclusively with other males. In baboon troops, the adult females normally outnumber the dominant males, and much tactile activity—such as mutual grooming —is between females. In human societies, such social relationships and pleasurable mutual touching might be called homosexual activities or tendencies. In lion and baboon societies, they appear to be normal relationships and activities. Furthermore, the individuals engaging in such behaviors are not excluded from contributing genes to subsequent generations. If the *alpha* males in a lion pride are killed or become weak, they are replaced or ousted by the stronger individuals in a bachelor group, who then typically sire several years' cubs during their tenure in the pride before they in turn are replaced by others. The females in a baboon troop become receptive to heterosexual mating activity during estrus. In both cases, there are advantages to the homosexual relationships, and there is a means of passing on the genes of at least some of the individuals that engage in them.)

Contraception is a method widely in use. Cheap and efficient techniques are available, although not all are wholly satisfactory, either medically or esthetically. However, the more serious problems of ignorance, religious dogma, and sociopolitical considerations still stand in the way of acceptance of family planning. Opposition from organized religious groups appears to be gradually crumbling, and ignorance is being dispelled by education, but other problems still exist. For example, advocacy of limiting the number of children per family in low-status groups sounds hollow indeed when wealthy couples raise huge families (Figure 12.7).

There have been some (for example, male members of certain militant black organizations in the United States) who argue that the campaign for zero population growth, or similar population-limiting schemes, is racist. The main thrust of their argument is that *any* minority will be politically oppressed, and thus a weapon to use in escaping this oppression is a birth rate that will convert the minority to a majority. This strategy clearly has problems in a racially pluralistic society such as the United States, particularly if three or more of our races all decide to try reproductively to achieve a majority. It appears that many female members of these groups are not supporting this political strategy, but recognizing instead the advantages to themselves and their children of smaller families.

There are also whites, who have paternalistically stated that whites (being heavy resource users) should sharply limit their population, but blacks, indians, and others need be less concerned by population size. Projected forward by a time dimension of several generations, this says not only that the present blacks and indians are making only a small average demand on resources, but that their children, and their children's children, will be incapable of (or prevented from) generating resource demands equivalent to those of white children, or white children's children. That position is patently racist. If equity of resource availability is taken seriously, then population is a world concern, and is a problem for people of all races.

In many regions of the world, even with adequate knowledge of and access to birth control measures, the desired number of children per family is still higher than the resource base can reasonably support. It will take time to change these aspirations, but the odds are that they will change. As people's living conditions improve, they realize that it is not necessary to have eight children for three to survive. They then typically have fewer children and higher aspirations for those children they do have.

A 1966–70 study, already significantly out of date, indicated the correspondence between the average number of births and the number desired, broken down by educational level for black and white American women:

Years of education	White women		Black women	
	Average births	Wanted births	Average births	Wanted births
College 4+	2.5	2.4	2.3	2.2
College 1–3	2.8	2.6	2.6	2.3
High school 4	2.8	2.6	3.3	2.8
High school 1–3	3.2	2.8	4.2	3.2
Less	3.5	2.9	5.2	3.1

Note that college graduates of both races want the smallest families, and come closest to having the size they want, with black college-educated women wanting and having fewer children than their white counterparts. These relationships reverse at lower levels of education. A devastating 40 percent of the children born to black women who didn't complete even one year of high school apparently are unwanted. Not all unwanted births become unwanted children, but the costs to the unwanted children, to their sibs and parents, and to society at large. are considerable. More effective access for less-educated women to family planning methods, while not in itself necessarily achieving zero population growth, will be a step in that direction while greatly reducing the human suffering occasioned by unwanted children.

Strategies to encourage zero population growth include intensified education; incentives, including payment for sterilization, for each childless year, for each year the family is held at some specified size, or for bringing new clients to a family planning clinic; community pressure (one of the most effective); an ability to determine the child's sex in advance (Section 8.6); tax and welfare penalties; and shifts in social and economic institutions (such as social security payments replacing retired persons' economic dependence on their children). Of these, the positive (payment) and negative (penalty) incentives are counterproductive, as they penalize the newborn child as well as the parents, while the other strategies generally seem both more effective and more humane. One scenario foresees a number of serious population-related disasters causing widespread deaths, followed by the survivors' reaching a consensus to keep the population size down.

An associated issue is a questioning of the ethic that old people must be kept alive as long as medically possible, even if they are in agony or barely—or not-at-all—conscious. The American pioneer's desire to die with his boots on exemplifies the desire to die with dignity. A concern for the quality of life should not ignore its final chapter.

Among those who attempt to foresee the future, there are some who believe that we face an inescapable dilemma because of the rivalry among nations. If, on the one hand, a given country pursues a vigorous policy of limiting the growth of its population, it will (other things being equal) incur a political disadvantage vis-à-vis a country that does not: on the other hand, lack of restraint of population growth will increase the intensity of the struggle among nations for resources. Garrett Hardin has suggested that it is pertinent to consider the strategies available for use of a lifeboat during a disaster at sea. He suggested that to admit more to the lifeboat than it can hold with a reasonable safety factor invites a catastrophe for all. If those in the lifeboat having conscience altruistically exchange their places with some in the sea, the effect is to eliminate all conscience among the survivors. It is a grim, but realistic, assessment.

Other thinkers, less bleak in their forecasts, hope for an eventual world-wide society—whether a totalitarian state or a free association is envisaged depends on the optimism of the forecaster—without national boundaries and with demographic policies that may make this world a good place to live in. It is possible that in such a society sterilizing chemicals will be added to drinking water, and conceptions will be the result of a premeditated act,* rather than occurring by

---

*As a consequence, one of the best known nursery rhymes may have to be revised:

Mother, may I conceive a child?
Yes, my darling daughter.
Hang your clothes on a hickory limb,
But don't go near the water.

default, as most do now. This could imply the licensing of reproduction, an uninviting prospect, but, in the view of some demographers, an unavoidable one.

One analysis has us balancing the right to have and own children—much as people used to own slaves or harems—with the rights of children to be born into a family that wants them and a society that can support them. Children have had little political power in the past, and perhaps it is time we thought more of them, and less of the things we use them for.

15.2 THE GREEN REVOLUTIONS

Acquisition of the ability to use tools (Section 5.2) was the first great revolution, and was associated with the emergence of the human species. It took place incrementally, probably over a period of several million years. During this time, prehumans and then humans organized themselves in hunting-and-gathering groups. They occupied many of the niches in which food—animals they could catch and edible fruits and other plant parts—was available throughout the year and continuously over many human generations. Humans may not have understood ecology, but they were a balanced part of their ecosystems.

Humans apparently acquired the use of fire, a very special yet generalized tool, about 40,000 years ago. Besides its uses for warmth and defense, it allowed many previously indigestible foods to be added to the human diet. Particularly important among these are such cereal plants as wheat and barley.

The Neolithic Revolution began about 10,000 years ago. It did not begin as a green revolution, as the first organisms domesticated were probably the dog, perhaps for help in hunting, and the goat, probably the first organism to be kept and tended for either food (meat and milk) or clothing (hides and hair). It too was incremental, and lasted over a period of several millennia. The use of pottery became important in the preparation and storage of food. An increasing array of local animals and plants was brought in and domesticated, or sometimes they came on their own. For instance, cattle and pigs were probably crop robbers, and many weedy plants were probably selected for domestication when it was noticed they grew unusually well around the fertile dung heaps near the early camps.

The human species was not only the domesticator of others, but was itself domesticated in the process. For many groups, the wandering ceased, and permanent villages were established. Success was associated with industriousness, patience, forethought, and agronomic skills. The successful grain growers developed a love of home, a feeling for the earth and its crops, and a sense of property not common in present-day hunting-and-gathering groups. Whether genetic selection played a part in this is only speculation; from the speculation follows the question of whether such traits are imposed by a culture, or whether such a culture results from the traits of its members.

The agricultural communities were able to produce enough food to feed all of their agricultural workers and other as well. These people were thus free to work at producing other goods. Copper, bronze, iron, glass, other processed minerals, the wheel, and the plough joined pottery as products of the expanding revolution. These were followed by the less tangible inventions of phonetic alphabets, urban societies, and civilizations. Animals to help with transport and labor, such as the elephant, horse, camel, and donkey, and to help fight crop-destroying pests, such as the mongoose, ferret, and cat, were added to the domesticated group. The complexity of the civilizations increased as humans specialized in the tasks to be performed. The selection pressures on the humans similarly became more complex. At the edges of the fertile lands, some peoples dropped farming and returned to a wandering life, herding reindeer, sheep, cattle, horses, and similar animals. There is some evidence that these herdsmen were more alert and enterprising than the soil-bound cultivators, and the expansion of the neolithic peoples, as well as their exchanges of knowledge and domesticated organisms, is largely due to them, and to the tradesmen plying between established communities. As a result of such exchanges, organisms formerly restricted in their ranges became employed in and adapted to new environments. Warfare, probably never absent, was also important in moving ideas and both human and non-human genes around the civilized world, and extending the current versions of civilization.

Homo sapiens existed for most of its history, about 200,000 years, without significantly domesticating any other organisms. It is curious that domestication then occurred essentially simultaneously in three separate areas of the world. Centered in the Near East, and occurring more diffusely throughout southeastern Europe and Africa, domestication included goats, sheep, pigs, horses, and cattle among the animals, and wheat, barley, sorghum, yams, dates, coffee, watermelon, and cotton among the important plants. Centered in North China, and occurring more diffusely throughout Southeast Asia and the South Pacific, the elephant and buffalo, rice, banana and citrus trees, and many other plants and animals were brought under domestication. And in Central America, and more diffusely throughout South America, the llama, corn, potato, cucumbers, beans, peanut, and other plants and animals formed still a third group of domesticates.

After humans began to record their history, different civilized groups discovered each other. In each area organized communities existed, and rather similar forms of civilization were extant. One theory holds that the events prior to the recorded contacts between these civilizations were indeed independent. The human species, once started and dispersed, was on geographically independent trajectories that led with uncanny timing through domestication to civilization. An alternative theory suggests that there were repeated prehistoric contacts that effected exchange of ideas on domestication, but not exchange of the particular domesticated species. The navigational feats of the Polynesians lend some support to this possibility.

One of the interesting questions facing us today is whether our ancestors were such good economic botanists and zoologists that further searches among the nearly half million wild plant and animal species will yield no important additions to those domesticated species we now possess. To a modern eye and palate,

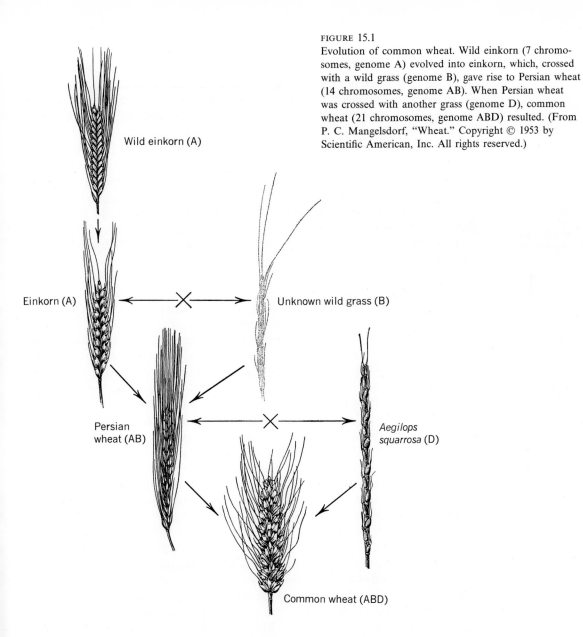

Wild einkorn (A)

Einkorn (A) ✕ Unknown wild grass (B)

Persian wheat (AB) ✕ *Aegilops squarrosa* (D)

Common wheat (ABD)

FIGURE 15.1
Evolution of common wheat. Wild einkorn (7 chromosomes, genome A) evolved into einkorn, which, crossed with a wild grass (genome B), gave rise to Persian wheat (14 chromosomes, genome AB). When Persian wheat was crossed with another grass (genome D), common wheat (21 chromosomes, genome ABD) resulted. (From P. C. Mangelsdorf, "Wheat." Copyright © 1953 by Scientific American, Inc. All rights reserved.)

the ancestors of our present domesticates would seem pretty unpromising. The domestication process, combining luck and skill, used selection, hybridization, and (with plants) polyploidization (Figure 15.1) to produce present varieties. Yet the plants and animals used widely in Europe and the United States have prestige, and are ousting traditional foods in many parts of the world. This will be discussed further in Section 15.3.

With the appreciation of Mendelian and Darwinian principles early in the present century, plant and animal breeding became less of an art and more of a science. Some have called this the modern agricultural revolution, but perhaps it was nothing but a continuation and acceleration of the Neolithic Revolution. Its effects were to release even more people from primary food gathering and production to being available for other useful activities (or as our civilized economy goes increasingly awry, to a condition of being surplus). This revolution was largely effective in the developed countries, which are mostly in temperate climates.

During the past decade, a new "green revolution" aimed primarily at the less developed countries in tropical and subtropical climates has been started. It has lately been getting mixed reviews, ranging from "the solution to widespread starvation" through "buying a little time" to "an ecological and cultural disaster." Much of this green revolution has been based on rice (from research done at a breeding center in the Philippines), wheat, corn, and a new intergeneric hybrid between wheat (*Triticum*) and barley (*Secale*) called triticale (from research done at a breeding center near Mexico City).

It is hard to fault the program itself, and the leader of the wheat program, N. E. Borlaug, received the 1970 Nobel Peace Prize for his and this program's contributions to human welfare (Box 7.B). The research is being done by an interdisciplinary team of scientists (pathologists, soil scientists, nutritionists, geneticists, agronomists, biochemists, economists, etc) who are chosen in part because of their ability and willingness to *do* interdisciplinary research. This is a refreshing change from many other mission-oriented programs, in which the scientists working on an interdisciplinary problem are narrowly trained in their respective disciplines, and either don't or can't effectively communicate on the emergent interdisciplinary properties of the problem.

There is an active program of cycling assistants, fellows, and others from the target countries through the research centers and field stations to give them a chance to learn by doing. The philosophy is "give a man a fish, he eats for a day; teach a man to fish, he eats for a lifetime."

The thrust is to produce broadly adapted, insect- and disease-resistant varieties that are not hybrids (although they may have hybrid origins—see Box 14.C). Attention is given to the quality as well as to the quantity of food produced; for instance, for regions where the essential amino acid lysine is low in present human diets, varieties with a higher lysine content are being developed.

The recent "green revolution" became an international phenomenon that fired the hopes and imaginations of dreamers and planners in many places. From 200 acres of test plots in Asia and Africa in 1965, it spread to 41,000 acres in 1966, 4 million in 1967, and over 50 million in 1971. The Philippines was able to stop importing rice, the Pakistan wheat harvest increased 70 percent in three years, and similar statistics were accumulated around the world. Then came 1972, with typhoon and guerrilla activity in the Philippines, floods and civil war in Pakistan, poor crops in China and the USSR, and drought in the sub-Sahara and India. Most of the new varieties had substantial requirements for fertilizer and water:

the petrochemical shortages cut off the former, the droughts the latter. High-production agriculture depends on a cooperative climate, both politically and physically.

Yet the idea remains good, and eight new research centers are now in being, studying tropical agricultural crops, semi-arid crops, forest crops, and the single-species crop, the potato. It is important to gain information in these areas and on these crops. In particular, tropical ecosystems are quite different from the temperate ecosystems on which much of our present-day agricultural wisdom is based. The year-round warmth of the tropical lowlands is a mixed blessing, bringing with a long growing season the opportunity for both insects and pathogens to rapidly increase their numbers to epidemic proportions on any crop that becomes too common.

A few other special genetic solutions have been suggested for meeting the increasing needs for human food. One of these relates to the recent discovery that some plants use a different—and more efficient—chemical pathway from that used by most plants for the initial incorporation of carbon from atmospheric carbon dioxide into an organic molecule. In particular, compared to most plants, their use of water and carbon dioxide is more efficient at high light intensities. The suggestion is that C-3 food plants be bred with photosynthetically more-efficient C-4 plants to create C-4 crop varieties (the C-3 and C-4 names refer to the location of initial incorporation of carbon during photosynthesis). This will not be an easy task, if possible at all, as there are many differences in the two pathways requiring that many coadapted genes will have to be moved together from a C-4 into a C-3 plant. The worry that the C-4 pathway, which has most frequently been found in plants from tropical or arid climates, might not be able to function except under narrow environmental conditions is partly dispelled by the wide success and high productivity of corn, which is a C-4 plant. Why most plants use the C-3 rather than the more efficient C-4 photosynthetic pathway remains an intriguing and somewhat worrisome evolutionary question.

The breeding of microorganisms or yeasts to convert petroleum into high-protein foods sounded better before the petrochemical crisis became so apparent. But it may still provide a sensible direct use of our dwindling petroleum supplies, in place of their use for the fertilizers and tractor fuels necessary for high-production agriculture.

Some indirect approaches are available, which amount to playing several varieties of genetic dirty tricks on important pest populations. These include flooding target populations with sterile males, or males bearing a high frequency of dominant lethals. The sperm from these genetically defective individuals compete with sperm from normal resident males, and if enough of the defective males can be introduced, the target population is reduced, and in some cases has even gone to extinction. Two types of genetic "time bombs" can be used. One involves introducing abnormal chromosomes that, in heterozygous combinations with the population's chromosomes, form inviable offspring. The other entails introducing conditionally lethal mutants, which can live and reproduce under permissive conditions, but cannot survive under certain recurring restrictive conditions. This is made particularly effective if the conditionally lethal mutants are tightly linked to genes for high fitness under normal conditions, or better yet, to a segregation distortion gene (Section 4.3) that greatly increases their frequency under permissive conditions. Another genetic contribution to solving the problems of crop pests is merely a better understanding of how pests become resistant to pesticides, so that a more rational integrated (i.e., combining several methods) control program can be used.

Returning to the current green revolution, there are some disturbing cultural and economic problems that accompany the solution of the biological ones. High-productivity farming frequently entails many changes in techniques, which ramify through reorganized farm management patterns, perhaps changed ownership patterns, and perhaps greater mechanization. If this revolution leads to the production pattern established in the temperate-climate developed countries, it will mean that more food will be produced by fewer people. If the surplus farm workers cannot be absorbed by other useful employment, economics being what it is, poverty and deprivation for this segment of the population may be intensified in spite of the fact that, somewhere in the system, amounts of food are increasing.

A final thought brings us back to the ubiquitous population problem: Alan Gregg, former vice president of the Rockefeller Foundation (a major supporter of the green revolution program) had some early worries about the overall wisdom of increased food production. Likening the growth and spreading of the human

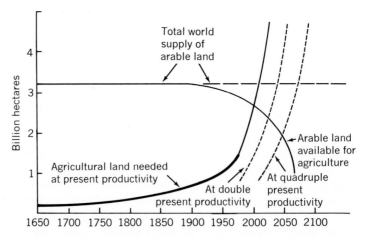

FIGURE 15.2

Total world supply of arable land is about 3.2 billion hectares. About 0.4 hectares per person of arable land are needed at present productivity for growing food. A curve plotted to show how much land will be needed for food production in the future thus reflects the population growth curve. The light line after 1970 shows the projected need for land, assuming that world population continues to grow at its present rate. Arable land available will decrease because arable land is removed for urban-industrial use as population grows. The dashed curves show how much land will be needed if present productivity is doubled or quadrupled. (From D. H. Meadows et al., *The Limits to Growth.* 1972. Universe Books.)

population to a cancer, he remarked that "Cancerous growths demand food; but, as far as I know, they have never been cured by getting it." Some dimensions of the problem are presented in Figure 15.2.

15.3 DIVERSITY AND GENE CONSERVATION

While present-day opinion generally agrees that the change from hunting and gathering to farming served to increase the certainty that food would be available and to create surplus energy, that may not be the correct view for an ecologically balanced hunting-and-gathering group on a typical day (it is probably a sequence of atypical days that keeps population down in such groups). The present-day Kung Bushmen, on the average, work at hunting and gathering twenty hours per week. They have no sign of nutritional diseases, drawing their regular diets from 17 species of meat animals and 23 species of fruits and vegetables, and occasionally adding to this from 37 additional species of animals and 92 additional fruits and vegetables. Forty-six of 466 Bushmen studied were older than 60 years, cared for and respected by a kinship welfare system. Our species and our primate ancestors lived as hunter-gatherers for over 99 percent of the last two million years. While our present numbers are too great to depend on a hunting-and-gathering system, over 90 percent of the human beings that have ever lived were fed by this system. Although agriculture has become a necessity, there is no reason to suppose it was a necessity when it was first developed, although in a Darwinian sense it was advantageous to those groups that mastered it.

A tenet of ecology is that with greater diversity there is greater stability. While diets in our affluent societies are nearly as diverse as that of the Bushmen, and the people and their occupations are more so, many of our urban ecosystems are becoming dreary **monocultures** (i.e., single-species ecosystems) of *H. sapiens* and very little else. The quality of air and water has been degraded, and many members of these monocultures have little contact with forest, or flowing stream, or wildlife, or even domestic animals other than other domesticated humans. In the less affluent human societies, which may be rural, even the diet suffers. Often it is largely based on a single crop, such as rice or corn, and whatever nutrients are lacking or in excess in the crop are also lacking or in excess in the diet. The effect of repeated planting with the same crop species reduces stability of the agricultural enterprise as pests build up, and the soil becomes impoverished from having the same demands made on it year after year. It is probably too late to return to the hunting-and-gathering society of the Bushmen, nor would most of us want to, but perhaps we can learn from them.

Darwin recognized that the most severe form of competition is between genetically identical individuals, as they tend to make exactly the same demands on the environment at the same time. Genetically dissimilar plants growing next to each other, or dissimilar animals occupying the same area, tend to make different demands on the environment, or the same demands at somewhat different times.

They thus make complementary, rather than competitive, demands on the niche. Recent experiments exploring two dimensions of competition, that of neighbors' effects on an individual, and that of the individual's effects on its neighbors, indicate that both have significant genetic components, and will respond to selection. Yields from mixed populations are frequently about 20 or 30 percent greater than yields from the same species, varieties, or genotypes grown in single-species (or single-variety or single-clone) populations. This is most striking among populations sampled from the wild, which have evolved and are existing in a mixed biotic environment. It is much less striking when varieties of domesticates that have been bred as monocultures are mixed.

The problem with diversity in agriculture is that, while a diverse farm may be productive and stable, it is generally not very efficient to manage. Thus, a field that contains an intimate mixture of several crops may require irrigation for one just at the time water should be withheld from another, or require different fertilizers, or have one ripen for harvest at a time when another can be easily damaged. This can be handled in the home garden by giving intensive attention to each plant. But in a large commercial farm, these problems have generally led the farm manager to decide in favor of monoculture, at least field by field, and frequently relative to the entire holding.

Thus, in spite of pleas and suggestions to increase diversity, both by our management of given sites, and by adding to our store of domesticated plants and animals (Figure 15.3), the recent trend is strongly in the opposite direction. Even with the enlightened leadership of the present green revolution, one of its major effects is to replace the small local populations, or "land races," of many locally adapted varieties with the high-productivity varieties coming out of the research centers. Erna Bennett has observed: "it is tragic, indeed, it is astonishing that with the enormous genetic potential available, the most permanent achievement of modern plant breeding with all the equipment of genetics at its disposal appears to be the loss or destruction of much of the world resources of genetic variability."

The lessons have been impressive. While slow to learn, we are learning. A recent lesson in the United States came out of the 1970 corn blight scare. In the prevailing weather of 1970, race T of the southern corn leaf blight proved virulent on the Texas male-sterile cytoplasm (male-sterile cytoplasm is a key to efficient production of hybrid corn). In 1970 the United States had over 46 million acres planted with more than a trillion corn plants that had the Texas male-sterile cytoplasm—more than 80 percent of the nation's total corn acreage. We reached into our reserve of genetic diversity and recovered in 1971 by planting different varieties. In 1945, V-R derivative oats carrying the Pc-2 gene for crown-rust resistance occupied 98 percent of Iowa's 5.3 million acres of oats, and 80 percent of the total U.S. oat plantings. A virulent pathogen appeared, and by 1948, Iowa had shifted 95 percent of its acreage to B-derived oats, resistant to that particular pathogen (but susceptible to others). A century earlier, the late blight hit the Irish potato crop. Millions of Irish starved, and other millions emigrated, many to the United States. At that time, genetics did not exist, and thus there were no geneticists to ride to the rescue with their reserves of genetic diversity to save the day.

Manatee

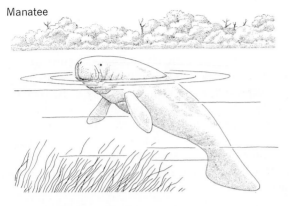

Eland

Capybara

FIGURE 15.3
The manatee, an aquatic mammal, is an example
of an unorthodox source of meat. It can also control
aquatic weeds, which it eats. An adult manatee is
between nine and fifteen feet long. The capybara,
a large rodent that lives in South America, has
also been suggested as a source of meat. Like the
manatee, it feeds on aquatic weeds. An adult is
about four feet long. The eland, a large African
antelope, is an accepted meat animal. Its importance
as a food source is that it is adapted to grazing
on marginal lands that are not suited to agriculture.
(From N. W. Pirie, "Orthodox and Unorthodox
Methods of Meeting World Food Needs." Copyright
© 1967 by Scientific American, Inc. All rights
reserved.)

The potato crop was not rescued for several years, nor were several million Irish.
Many other crop disasters can be laid at the feet of genetic uniformity, and there
are others waiting to happen. For example, practically every coffee tree in Brazil
is a descendent of a single tree, and the coffee plantings cannot be replaced in a
single year the way corn was.

Thus, it appears to be important to preserve and actively to maintain the genetic
diversity of our major crop species. It is also important to preserve a diverse group
of relatives of our important species. Recently it has been recognized that there
is wisdom in conserving the diversity of even our noneconomic, wild species. Thus,
gene conservation has become an important dimension of the general conservation movement and ethic.

We now have a situation outside of any past experience: a confrontation between our species and the world biota, which in the short space of two or three
human generations could imperil a large proportion of the wild species that now
remain. If we stay on our recent course, we will direct the evolution of biota that
are of use to us, and the only ones retaining some evolutionary independence
will be those we are unable to suppress.

Predictions about the future become increasingly uncertain as the pace of change increases. It seems likely that a century or two from now *H. sapiens* will differ more—culturally, technologically, economically, and possibly biologically—from us than we now differ from the early Neolithic revolutionist. Hence, at this point of decision-making, it is our evolutionary responsibility to keep evolutionary options open. This is a modest precept, but is as much as is likely to be socially acceptable at this time. It may grow into an evolutionary ethic if and when we come to regard other species as an essential part of our own existence. Perhaps the greatest difficulty stems from the contradictions of time scales: evolutionary time is compressed into historical time and made subject to decision-making on a sociopolitical time scale. Sir Otto Frankel has suggested the following time scale for measuring human concern:

Subject	Period	Operator	Objective	Time scale
Wildlife	to 8000 BC	hunter-gatherer	the next meal	1 day
Domesticated plants	to AD 1850	"primitive" or "traditional" peasant farmer	the next crop	1 year
	from 1850 from 1900	plant breeder crop evolutionist	the next variety to broaden the genetic base	10 years 100 years
Wildlife		gene conservationist	dynamic wildlife conservation	10,000 years
	today	politician	current public interest	next election

It appears that the ability to synthesize genes may soon be achieved, and the idea of creating a whole new organism is at least plausible. In terms of cost, measured either as money or as energy input, these alternatives would be enormously more expensive than saving genes and species that already exist. As technological and scientific feats, the synthesis of genes and species may be compared with Sputnik and Apollo 8. The massive input of resources into space technology has added enormously to our understanding of Earth, but has added few new physical resources to those already here. Similarly, the synthesis of genes and species may add biological understanding, but will probably not greatly contribute to Earth's biological resources. One problem is that a new gene is not necessarily an adapted gene. And a brand new species, while it can perhaps successfully survive in a protected laboratory environment, is unlikely to fit harmoniously into a community of plants, animals, and microorganisms that have evolved with each other and in Earth's environment for many generations. Its likely fate is to die out quickly when put in contact and competition with native organisms. A much less likely, but frightening, alternative is that it will encounter no effective natural controls, and prove destructive to many of Earth's natives, including possibly us. Without gene and species conservation, risk of such an event is increased.

Can we then simply store organisms, like microfilm, until they may be again interesting or needed? The technology for this option is available, or perhaps soon will be available. For some organisms, storage in spores or seeds is already normal procedure, and technology needs only to extend the period of viability. The development of whole organisms from single cells maintained in cell or tissue culture is an accomplished feat for some species of plants, and may be possible for all species, including animals. There remains one serious objection. Evolution will be halted in such stored organisms. Species that fail to evolve may find themselves in trouble when again asked to exist in an environment that is likely to have changed significantly during the period of storage.

Can we maintain otherwise-extinct species in zoos or botanical gardens? The problem here is that evolution *will* continue. After many generations, these token remnants of the original species will have adapted to, even to the point of becoming dependent on, the environment of the zoo or botanical garden.

The remaining, and apparently best, alternative appears to be the conservation of sufficient natural ecosystems so that a significant proportion of Earth's plant and animal species can continue to exist and evolve, surrounded by and perhaps even in harmony with continuously changing civilization. Of particular interest, in light of its historical importance in Darwin's hypotheses, are the Galapagos Islands. Introduction of animals that became feral caused the near extinction of such birds as the flightless cormorants and penguins, and the giant terrestrial tortoises. In 1968, the government of Ecuador established a national park over the whole archipelago, with the exception of settled areas, and strictly protected the endemic species.

Douglas-fir is a widespread species of major commercial importance. It is native to the northern coast of California, and plantations of this California coastal provenance are among the most successful exotic forest plantings in several countries. The Redwood National Park and several state parks provide protected ecosystems including native douglas-fir of this important provenance. The reserves are surrounded by actively managed, commercially cut douglas-fir forests. Present practice frequently includes aerial reseeding following logging. The seeds are commonly not of California coastal origin (and records show that many exotic-origin douglas-fir seeds were dropped on land that was later included in the Redwood National Park). Even if this practice stops, it will soon be common practice to reforest with genetically selected trees, which will each generation be increasingly unlike the natives. The problem is that the "native" douglas-fir of this important primitive provenance in the park preserves will be repeatedly exposed to clouds of pollen from the vast surrounding commerical forests, made up of both exotic and domesticated douglas-fir. The offspring of the "natives" will be increasingly contaminated by genes from these surrounding forests, and the native gene pool will be compromised and finally lost. Thus, gene samples of such economically important species will only be preservable *ex situ* (i.e., in some other place) if extensive human-caused genetic modification of the native stands occurs in the species ranges.

Wide-ranging species such as lions and elephants are similarly not well conserved in small natural areas in their native ranges. A reasonable compromise is available in such facilities as the San Diego Wild Animal Park. Here the animals are kept in very large enclosures in a semi-wild state, where most successfully reproduce. The problems of a small gene pool and adaptation to the particular environment of the park could be met by having several such parks in rather different environments, and exchanging animals between them every few generations to provide a disruptive gene flow that would counter the forces of drift and selection (Section 4.3).

Finally, what about gene conservation in the most important and difficult species of all? This problem has been faced recently, without adequate solution, every time an uncivilized tribe is encountered in one of the few places not as yet overrun by civilization. The term "Fourth World Peoples" has been used to describe these groups, which are characterized by living in balance with local ecosystems. If civilization collapses without a general poisoning of Earth's ecosystem, these peoples will live on, probably oblivious to the fact that the civilization experiment had even been tried. But unless or until such a collapse occurs, these peoples are defenseless, and will be swamped as the burgeoning populations of the First, Second and Third Worlds require their land and their resources. Should Fourth World cultures and gene pools be protected—if for no other reason than the many values they undoubtedly have to offer us? If so, what are the rights of a member of such a tribe, who might want to try life in one of the Other Worlds? We have no very good policies affecting gene conservation in humans.

16

MUTATION

Having considered the evolutionary force of selection in Chapters 4, 13, and 14, we now turn to the important force of mutation. We defined a mutation as a change in the informational apparatus of the cells. A capacity for mutation is a basic property of life, which permits variation and thus evolution. From the standpoint of the population, it is not an abnormal event.

The variety of such changes is great, and ranges from the substitution of one nucleotide in a DNA molecule to the doubling of the entire chromosomal complement. All genetic changes may be placed in one or another of four categories: (1) intragenic, or point, mutations; (2) changes in groups of genes on a chromosome; (3) changes in whole chromosomes; and (4) changes in a whole chromosome set. These four kinds of mutation can be correlated only imperfectly with the magnitude of the phenotypic effects they produce.

The molecular basis for point mutations, which supply much of the genetic variation significant in evolution, has already been discussed in Box 4.C and Section 6.4.

16.1 CHROMOSOMAL MUTATIONS

Several varieties of intrachromosomal change are possible: deletion, duplication, inversion, translocation. A **deletion** is the loss of part of the chromosome. Deletions including sequences of several genes are usually lethal in the homozygous state. This is to be expected, since absence of gene products essential for the existence of an organism may result from deletions. A deletion in human chromosome 5 is associated with the *cri du chat* syndrome (Section 12.2), a deletion in chromosome 22 with chronic myeloid leukemia, and there is now some evidence that the α-thal-1 allele that causes a form of thalassemia (Section 7.4, Box 7.D) is a deletion of all or most of the α-chain gene. Especially damaging are deletions that include the **centromere,** the part of the chromosome that holds together the sister strands in the dividing nucleus. In Figure 16.1 centromeres are represented by black bars. A chromosome without a centromere is usually lost in the course of cell division.

Duplication of parts of chromosomes is a phenomenon that is important for evolution, because it provides additional possibilities for new mutations to be incorporated into a species. The presence of two genes for the same polypeptide permits one of them to mutate in a new direction without damage to the organism, because the other would still carry on the vital function under its control. The different chains of hemoglobin have very likely evolved from duplicate genes (Box 6.E).

Inversions change the sequence of genes on the chromosome. Consequently, an inversion in a chromosome affects its ability to pair with its (noninverted) homologue during meiosis, and crossing-over may be suppressed in the region where the order of genes on the two chromosomes is reversed. An example of the evolution of inversions has already been shown in Figure 3.6 for the *Drosophila pseudoobscura* group.

Translocations, as shown in Figure 16.1, are exchanges involving two or more chromosomes. Stern, in the experiment described in Box 9.C, used such translocated chromosomes. Translocation heterozygotes produce many incomplete and inviable gametes (Figure 16.2). They have complex chromosomal pairing patterns that can lead to ring-shaped chromosomes and other abnormalities. Among such plants as evening primroses and peonies, translocations have played a significant evolutionary role in the formation of strains and subspecies, much as inversions have in *Drosophila*. Translocations are known to be associated with certain phenotypic defects in humans.

Mutations that are deletions or additions of whole chromosomes may result in **monosomy,** if one of the two chromosomes in a set are lost, or in **trisomy,** if a chromosome is added to a set. At least one case of autosomic monosomy has been reported in humans, the affected child being a mentally retarded four-year-old girl. We have already described monosomy for the X chromosome (Turner's syndrome), as well as additions of one or more X or Y chromosomes (Section 8.5).

In 1867, the British neurologist Langdon Down named a syndrome "mongolism," because he mistakenly thought that the characteristic fold on the eyelid

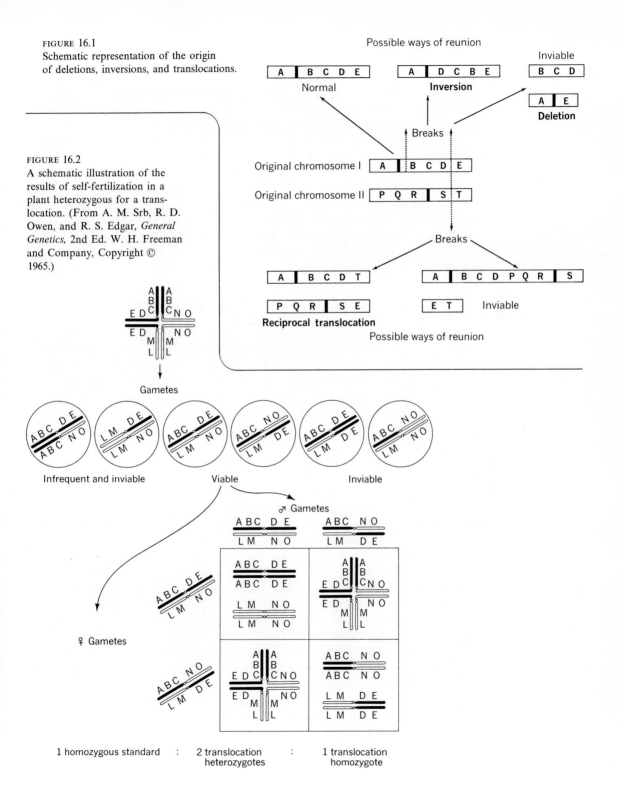

FIGURE 16.1
Schematic representation of the origin
of deletions, inversions, and translocations.

Possible ways of reunion

| A | B C D E | Normal
| A | D C B E | **Inversion**

Inviable
| B C D |
| A | E | **Deletion**

↑ Breaks ↑

Original chromosome I | A | B C D | E
Original chromosome II | P Q R | S | T

Breaks

| A | B C D T |
| P Q R | S E | **Reciprocal translocation**

| A | B C D P Q R | S |
| E T | Inviable

Possible ways of reunion

FIGURE 16.2
A schematic illustration of the
results of self-fertilization in a
plant heterozygous for a translocation. (From A. M. Srb, R. D.
Owen, and R. S. Edgar, *General
Genetics,* 2nd Ed. W. H. Freeman
and Company, Copyright ©
1965.)

Gametes

Infrequent and inviable   Viable   Inviable

♂ Gametes

♀ Gametes

1 homozygous standard : 2 translocation
                             heterozygotes : 1 translocation
                                               homozygote

of affected persons resembled that of Orientals. This unfortunate term has persisted, although now the name "Down's syndrome" is replacing it. The symptoms include retardation of growth and physical maturation, certain characteristics of finger and palm prints, an apparently increased susceptibility to leukemia, and moderate to severe mental retardation. More than 10 percent of institutionalized mentally retarded people have Down's syndrome.

The subsequent understanding of Down's syndrome improved only slowly. Affected infants were born much too frequently for the cause to be a normal mutation, their incidence being 1.0–1.9 per thousand live births (the incidence at conception has been estimated at over 7 per 1,000, the difference reflecting fetal loss due to spontaneous abortion). A strong relationship was noted between incidence and age of parents, and on careful investigation, it was found that the relationship was strictly with age of the mother. The apparent relationship with age of the father reflected only the tendency of older mothers to have older husbands. The incidence of Down's syndrome children born to mothers younger than 30 years was 0.04 percent; with successive five-year increases in maternal age, the incidence increased to 0.11, 0.33, and 1.25 percent, with a sharp increase to 3.15 percent for mothers over 45 years.

There was some tendency noted for Down's syndrome to run in families. In those families in which there were several Down's syndrome births, the mothers were generally younger than mothers who gave birth to a single Down's syndrome child, and in such families certain physical peculiarities of the syndrome were more commonly displayed among relatives than among the normal population. Down's syndrome females would occasionally have children, and about 50 percent of the offspring of these women had Down's syndrome, while the other 50 percent were normal.

A breakthrough in our understanding of the syndrome was made in 1959, when it was discovered that most Down's syndrome patients have 47 chromosomes. There is one too many of the short autosomal chromosomes, and the trisomic group was arbitrarily labeled chromosome 21. (The syndrome is now also frequently called **trisomy 21,** although as we shall see in the next paragraph, that is not an entirely suitable name.) This was the first identification of a chromosomal aberration as the cause of a human disease, and it stimulated a great deal of productive human cytogenetic research.

Shortly thereafter, chromosome 21 translocations to chromosomes 13, 14, 15, 22, and possibly others, were found in Down's syndrome patients who had 46 chromosomes. This explained the recurrence in families, and the muted symptoms in relatives. When the egg (or sperm) from a parent with such a translocation contains both a normal chromosome 21 and the translocation chromosome with part of a 21 included, the resulting zygote is functionally trisomic for the translocated part of the 21 even though it has only 46 chromosomes. A family having a translocated chromosome 21 may repeatedly produce such zygotes, and the reciprocal translocation to 21 may cause sufficient imbalance so that otherwise normal members of the family show mild symptoms of the syndrome.

The common anomaly, however, is the 47-chromosome trisomy 21. This most likely arises due to nondisjunction of this chromosome during meiosis in the egg.

Why it should occur so much more frequently in older women is not yet clear. Perhaps it is a feature of older oocytes. James German has proposed an ingenious explanation, which has been criticized but hardly disproved. He suggests, with support from studies by A. C. Kinsey and colleagues, that the average frequency of marital coitus is negatively correlated with the duration of marriage. In mammals, the meiotic division of the secondary oocyte that produces the egg is suspended, and progresses to completion only after a sperm penetrates it. If no active sperm are present when the oocyte enters the upper portion of the uterine tube, it begins to deteriorate. If an oocyte is penetrated by sperm after deterioration has begun, disruption of the alignment of the chromosomes is likely, leading to unbalanced segregation of the chromosomes. According to German's reasoning, if coitus is less frequent, there is a greater possibility that there will be time for an oocyte to begin to deteriorate before it is penetrated by a sperm.

Other environmental causes, or environmental-biological interactions, are suspected of implication in the increase of nondisjunction with age. These include ionizing radiation received by the mother, the presence of maternal thyroid antibodies, and a mysterious "Australia antigen," which is found in about 30 percent of Down's syndrome patients, in about 9 percent of leukemic individuals, in about 5 percent of patients with viral hepatitis, and not at all in normal controls.

In addition to trisomy 21, several other human trisomies are known. Most of them are for the smaller chromosomes, where the relative imbalance would be smaller. Trisomies of the larger chromosomes have been detected as mosaic sectors of otherwise normal persons. Some symptoms associated with the smaller chromosome trisomies include cleft palate, extra digits, deformed fingers, and congenital heart disease.

Chromosomal abnormalities occur in a surprisingly high frequency, and are found in apparently normal persons enjoying good health as well as in some with gross abnormalities. Nearly one percent (9 per thousand) of newborn infants have diagnosable chromosomal aberrations in their cells. Margary Shaw has summed up the estimated annual number of newborns with chromosomal abnormalities:

Statistic	United States	World[1]
Population size	203 million	3.68 billion
No. annual births	3.6 million	124 million
Sex-chromosome aneuploidy	8,700	298,000
Autosome aneuploidy	3,900	134,000
Structural rearrangements	6,500	224,000
Down's syndrome	3,000	104,000
XYY syndrome	3,700	127,000

[1]Assuming world rates are equivalent to U.S., Canadian, and Scottish rates.

The contribution of chromosomal aberrations to spontaneous abortion is high, with estimates of 25–50 percent of such abortions being due to chromosomal imbalance. For instance, among 423 fetuses aborted before 24 weeks of gestation,

101 had detectable chromosomal anomalies. Among these, 31 percent had a missing sex chromosome, 46 percent were autosomal trisomics, 13 percent were triploids, and the remainder were a mixture of other abnormal chromosome numbers, mosaics, and presumptive translocations.

Chromosomal aberrations that lead to abnormal phenotypes, particularly those sufficiently viable to allow the physically and mentally abnormal children to be liveborn, are of considerable medical and social concern. These children are greatly reduced in fitness, and thus these chromosomal aberrations are evolutionary dead ends. However, adaptive changes in chromosome numbers and arrangement have occurred in the past, as we know from the variation in chromosome numbers and shapes among related species. It appears that chromosomal evolution in the Primates is diminishing the number of small chromosomes with off-center centromeres. At present, 0.1 to 0.2 percent of normal humans have only 45 chromosomes, but essentially a full and balanced complement of genetic material, due to a recent or old fusion of two chromosomes. Perhaps sometime in the future, our species will have 44 or even fewer (but larger) chromosomes, instead of today's normal 46.

Mutations that involve whole sets of chromosomes—that is, those that induce polyploidy, the possession of more than the diploid number of complete sets of chromosomes—are very common in plants. Such mutations are relatively rare in animals, possibly because they produce too much sexual and physiological imbalance and therefore are eliminated. At least three cases of triploidy (three sets of chromosomes) have been described in liveborn humans.

One of the ways triploidy can arise is by failure of reduction of the chromosome number in the egg. For example, in a planarian flatworm, *Dugesia,* with a normal haploid number of 4, a strain was found that laid unreduced diploid eggs, and successive generations with 12, 16, and 20 chromosomes were produced by fertilizing them with normal sperm.

BOX 16.A ESTIMATES OF THE RATES OF SPONTANEOUS MUTATIONS IN DIFFERENT ORGANISMS

Organism	Mutations per locus per million cells or gametes per generation
Viruses	0.001–100
Bacteria (*E. coli*)	0.001–10
Corn	1–100
Drosophila	0.1–10
Mouse[1]	3–11
Human[2]	1–100

[1]Estimates are available from a few loci only; mutation rate has been estimated at about 9 per million for forward and 3 per million for reverse mutations.

[2]Further estimates are given in Box 16.B; these shown here include estimates from cells grown in tissue culture, in which such traits as resistance to various chemicals may be studied.

Induction of polyploidy has proved to be a useful tool in plant breeding. Not only can new polyploid varieties of a given species be produced, but altogether new species can be created. Mating between species almost always produces infertile offspring, because problems of pairing of chromosomes originating from the two different parents make normal gamete production fail in the offspring. If, however, the reduction division is inhibited, viable gametes can result in the production of an *amphidiploid,* containing a full set of chromosomes from each parent. The new form will have no chromosome pairing difficulties. Economically valuable food and ornamental plants have been produced by such methods.

A classical example of the production of an amphidiploid is that reported by the Russian geneticist, G. D. Karpechenko, who found that by doubling the number of chromosomes in a sterile hybrid between a radish (genus *Raphanus*) and a cabbage (*Brassica*), a new form (*Raphanobrassica*) was produced. Unfortunately, the hybrid proved to have no gastronomic merit, and cabbages and radishes still have to be served separately at meals.

16.2 RATES OF POINT MUTATION

In laboratory organisms it is possible to estimate what the mutation rates are under various conditions of observation. Since mutation rates are generally relatively low, the accuracy of estimating them varies with the organism. Viruses and bacteria, that can be grown in millions in a short time, permit the best estimates; mammals, which cannot be raised in such numbers, are difficult. Nevertheless, in experiments with mice, as many as several million gametes have been tested. Since mutation rates in mice are low, say about ten per locus per million gametes (see Box 16.A), even such extensive tests have a large scope for error.

The basic technique of measuring mutation rates in such animals as mice is to prepare stocks with marker genes—for instance, those controlling coat color. These stocks are then inbred by brother to sister matings, and the litters examined for any deviation at the loci in question. Dominant mutations are readily detected in the generation of their occurrence. Recessive ones will segregate out in the first or following generations, depending on litter size.

Such techniques cannot be applied to humans, but reasonably good estimates for dominant mutations can be made. For example, a Danish investigator who studied chondrodystrophic dwarfs—people whose growth is impaired presumably by a deficiency in a pituitary hormone—found that among 94,075 babies born in a certain Copenhagen maternity hospital, there were 10 such dwarfs but only two of them had a dwarf parent. The other eight may be considered to have been caused by newly arisen mutations, giving a ratio of 8 out of about 188,000 genes or 43 per million. Estimates from other data, as shown in Box 16.B, were found to vary around this figure.

Rates for recessive mutations present more of a problem. Because human families are relatively small it is difficult to ascertain if a certain trait has been carried in a heterozygous state or is the result of a newly arisen mutation. Nevertheless, it is sometimes possible to trace a mutation to the individual in which it occurred. The mutation for the blood disease **hemophilia** appeared in one of the parents of Queen Victoria of Great Britain (or else early in her own prenatal life) and may have had some important historical repercussions (as related in Box 16.C). Identification of the mutational event was possible because the diseases

---

BOX 16.B ESTIMATES OF HUMAN MUTATION RATES

The mutation rates given here for dominant and for sex-linked genes are, as is explained in the text, much more accurate than those for recessives. Many human geneticists view estimates for recessives with great skepticism, because of the uncertainties inherent in current methods of determination. They are given here merely to indicate the magnitude of the estimates made.

Trait	Mutations per million gametes per generation	Estimated fitness
Retinoblastoma Dominant; an eye tumor	15–23	0
Juvenile amaurotic idiocy Recessive; blindness, paralysis, mental deficiency, death, onset at about 6 years of age; common in Scandinavia	38	0
Infantile amaurotic idiocy (Tay-Sachs disease) Recessive; symptoms as above, but onset around 2 years of age; common in jews	11	0
Microcephaly Recessive; abnormally small skull	49	~0
Achondroplastic dwarfism Dominant	10–70	0.1
Hemophilia Sex-linked; see Box 16.C	25–32	0.25–0.33
Muscular dystrophy Sex-linked	43–100	0.30
Albinism Recessive	28	<1.0
Aniridia Dominant; absence of iris	5	?
Deaf-mutism Several loci	450	—
Low-grade mental defect Many loci	1500	—
All loci causing death before early adulthood	40,000	0

of members of royal families are on record, and because the trait is sex-linked and therefore is expressed in males.

Attempts have been made to estimate rates for recessive mutations by indirect methods. One method, based on the assumption that the population is in equilibrium, applies the principle expounded in Box 13.C to the computation of m. But the estimates are not reliable, because (1) the postulate of equilibrium is almost always a tenuous one, (2) the value of s is not precisely known, (3) the degree of effect in the heterozygote may vary, and (4) the breeding structure of the population or the extent of departure from random mating is usually not precisely measurable. The figures given in Box 16.B should therefore not be taken as exact.

Due to these problems, and to the fact that samples available for study are generally too small, human mutation rates are not accurately known. However, we know them well enough to speak of orders of magnitude, and these range from an event that occurs once in tens of thousands of cell divisions, or generations, to once in a million or even more (Boxes 16.A and 16.B). Similarly, no very good comparisons of rates between organisms can be made yet. There is a tendency for organisms with longer generation times to have higher per-generation mutation rates than those with short generations, but there is considerable overlap (Box 16.A).

We may also speculate that mutation rate is one of the tools species use to optimize evolution. At least one of the causes of mutation is mutator genes (Section 16.4), and thus there is some basis for genetic control of rate. However, it is likely that the size and chemical constitution of the gene and its chromosome, and the environment of the population are also effective, and that these may be only tenuously connected to a need for stability or changeability in the gene products. Finally, evolution is an irregular process, sometimes requiring rapid adjustment to changed conditions, at other times, merely running to keep in place as commanded by the Red Queen (Section 4.3). Thus, while general levels of mutation rates may be adjusted by general evolutionary responses (setting a balance between frequent enough, but not too frequent), it is likely that the fine tuning of evolutionary change is selection for the mutations themselves, rather than selection for alterations in mutation rate.

16.3 SOMATIC MUTATION

The time at which a somatic mutation occurs in the individual's life is of importance. If a metabolic block appears very soon after fertilization, the phenotypic effects may be similar to those produced by a germinal mutation, even though they are different in not being transmissable. But a somatic mutation resulting in, let us say, an inability to produce liver enzymes in cells already differentiated into muscle tissue might have no consequences whatever. In most plants and some animals, a somatic mutation may become germinal if the organs that descend from the mutated cell produce a flower, or bud off to become a new organism.

There are about 20 trillion cells in the average adult human. Somatic mutation rates are not well known either. Recent techniques for growing human cells in culture, and treating them experimentally like microorganisms, permit relatively precise study of somatic mutation rates in culture conditions. The magnitudes of such rates thus far observed differ little from the values cited in Box 16.B, but extrapolation from cell-culture results to what goes on in a functioning human

---

BOX 16.C HEMOPHILIA

Blood-clotting is a rather complex process. It requires thirteen different identified substances, starting with fibrinogen, a soluble protein in the blood plasma (plasma is blood from which blood cells have been removed; serum is plasma without figrinogen) that is converted into fibrin, the substance that forms the clot. Deficiency of any one of the thirteen substances will result in failure of clotting and consequently in continued bleeding. All of them are probably under genetic control. Nine genes have been definitely identified as controlling one or another of these substances, and at least six more are suspected to have such a function. The best known of these genes is a sex-linked recessive responsible for the deficiency of factor VIII, the antihemophilic globulin that causes the disease spoken of as classical hemophilia.

The disease has been known since ancient times and, even though its genetic basis could not be properly explained until the discovery of sex linkage, some knowledge of the pattern of transmission from mother to son was appreciated in the early years of our era. Thus, in Talmudic law, boys whose two older brothers bled to death from circumcision were excused from the rite, on the basis of what is essentially a sib test. The exemption also extended to sororal nephews of mothers of such children, i.e., to sons of her sister—but not to the fathers' sons by other women. To the contemporaries of those who made them, these provisions were no doubt thought of only as typical examples of the intricacy of Talmudic law; but the provisions indicate some degree of understanding of the basis of inheritance of hemophilia.

The best known pedigree in which the disease appears is that of the descendents of Queen Victoria of Great Britain (1819–1901), an ancestor of many members of European royalty and nobility.

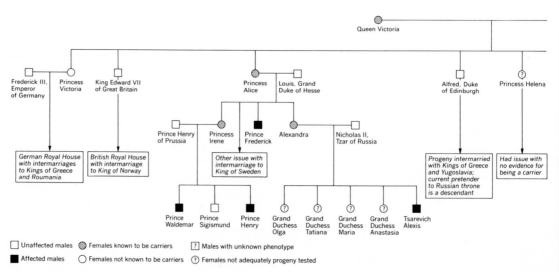

body still involves some elements of speculation. We can say that, as rates are about one per million per locus, each of us has hosted about 40,000 mutations for every gene in our bodies. In many cases, the normal allele will cover for the mutant allele. In many others, that particular gene won't be called on to function in the tissue where the mutation occurred. In other cases, the cell containing the mutation may die as a result of it, to be replaced by a healthy neighboring cell.

As the pedigree shows, the British royal descendents escaped the disease, because King Edward VII, and consequently all his progeny, did not inherit the defective gene. Similarly, no evidence of the disease is found in other royal families descended from Victoria, except for two. In Russia, the only son of the last Tsar was a hemophiliac. Lacking an understanding of the nature of the disease, his mystically inclined mother, the Tsarina Alexandra, became increasingly reliant on a series of faith healers to control her son's affliction. The most notorious of them was the illiterate and dissolute Gregory Rasputin, whose baleful influence at court, financial manipulations, and personal debauchery contributed a great deal to the events leading to the Tsar's abdication and to the subsequent murder of his whole immediate family. Rasputin himself was assassinated on the eve of the Russian revolution by a group of conspirators that included a Grand Duke, another relative of the Tsar by marriage, and a member of the Russian parliament.

Two sons of the last King of Spain, Alfonso XIII, were also hemophiliacs. Although there is no dramatic connection between the disease and the Spanish revolution that sent Alfonso into exile, the gene inherited from Queen Victoria certainly added little joy to the family life of the Bourbons.

The estimated mutation rate for hemophilia is given in Box 16.B. Some females are known to suffer from the disease, but they are extremely few, since they could only be produced as a result of an unlikely marriage between a hemophiliac and a woman heterozygous for the trait, or a mutation in similarly infrequent unions.

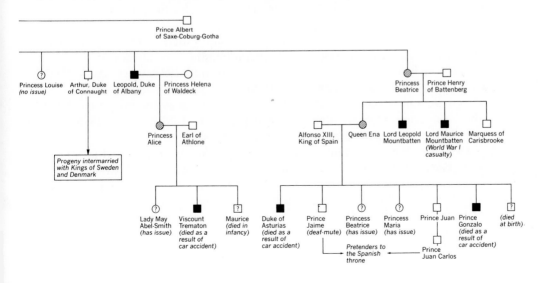

Prince Albert
of Saxe-Coburg-Gotha

Princess Louise
(no issue)

Arthur, Duke
of Connaught

Leopold, Duke
of Albany

Princess Helena
of Waldeck

Princess
Beatrice

Prince Henry
of Battenberg

Progeny intermarried
with Kings of Sweden
and Denmark

Princess
Alice

Earl of
Athlone

Alfonso XIII,
King of Spain

Queen Ena

Lord Leopold
Mountbatten

Lord Maurice
Mountbatten
(World War I
casualty)

Marquess of
Carisbrooke

Lady May
Abel-Smith
(has issue)

Viscount
Trematon
(died as a
result of
car accident)

Maurice
(died in
infancy)

Duke of
Asturias
(died as a
result of
car accident)

Prince
Jaime
(deaf-mute)

Princess
Beatrice
(has issue)

Princess
Maria
(has issue)

Prince Juan

Prince
Gonzalo
(died as a
result of
car accident)

(died
at birth)

Pretenders to
the Spanish
throne

Prince
Juan Carlos

And in rare cases, the cell may do its job better than it might otherwise have done.

The possible role of mutations in the causation of cancer is of considerable importance. At present, no simple answer is available, and it seems likely that we will find cancer to have multiple and complex causes. When cancer cells are grown in culture, their cellular descendents have cancerous properties. We have already mentioned (Section 16.2) that leukemia incidence is high in Down's syndrome patients, and that a deletion in chromosome 22 is associated with chronic myeloid leukemia. Among persons having the disease *xeroderma pigmentosum*, which is caused by homozygosity for a recessive autosomal allele, the incidence of skin cancer is very high, and is associated with inability to repair mutations induced by ultraviolet light. Heightened incidence of various types of cancer is associated with a number of other known heritable conditions. Further, evidence that viruses induce many cancers is increasing. It is not clear whether the virus mediates a transduction, or the cancerous growth is simply directed by the virus DNA. (Neither of these virus-associated mechanisms is strictly a mutational event.)

One theory of aging considers the accumulation of somatic mutations to reduce the efficiency of the body's cells, and thus age it. This theory, however, fails to explain the rejuvenation that occurs at meiosis, and thus at best cannot be a general explanation of aging (Section 7.1).

16.4 CAUSES OF MUTATION

The causes of mutation may be organized as genetic factors, physical factors, and chemical factors, with chance playing some role in each of these.

There are so-called *mutator genes* that increase mutation rates in their carriers. Some of them affect only particular bases. In general, it is possible that part of the genetic information in cells contains instructions for the production of a certain proportion of errors in self-reproduction.

The physical factors affecting mutation include radiation (X-ray, ultraviolet, and various high-energy particles), temperature, and possibly sound.

The search for mutagens was not successful until 1927, when **H. J. Muller** (Box 22.A), by devising appropriate techniques, was able to demonstrate that X-rays were mutagenic. He later received a Nobel prize for this discovery. At about the same time, another American geneticist, **L. J. Stadler** (1896–1954), showed the mutagenic properties of X-rays on barley.

There are three important mutagenic sources that are considered to be parts of the **background radiation:** the natural radiation from (1) **cosmic rays,** (2) radioactive elements in Earth's crust, and (3) radioactive elements ingested in food. (In addition, visible light appears to have a mutagenic effect on some bacteria.) The effect of cosmic rays varies with altitude. Between sea level and 15,000 feet there may be more than a five-fold difference in amount of cosmic rays. No direct evidence, however, is available about the actual differences in mutation rates at various altitudes. It is true that the number of neonatal deaths increases with altitude (in the United States, for instance, the death rate per 1,000 births varies from about 18 in the coastal plain at 300 feet to more than 21 in the Rocky Mountains at 5,000–6,000 feet) but full examination of the data suggests that the reduced available oxygen, rather than the direct or indirect mutagenic effects of radiation, is responsible for the rise in death rate.

The amount of radiation from Earth's crust differs with location. If we take the average amount of radiation in France and Japan to be 1, there are areas in the world where it exceeds 15 (Espirito Santo in Brazil) and even 85 (Kerala province of India).

Geographic and geologic peculiarities in a strip of land in Kerala province have caused an accumulation of sands containing the heavy radioactive element thorium. This fact was discovered by German coconut buyers, who found the natives attempting to cheat them by filling sacks with sand instead of the nuts. The comical side of the story is that thorium is much more valuable than copra or coconut oil. In the pre-atomic age, it was used in the manufacture of gas mantles. The Germans attempted to capitalize on the situation by building a thorium-extracting plant. But the plant was completed just before World War I and was promptly confiscated by the British. A study was made of the wild rat population on that radioactive strip of land. Over a period of 300 generations of isolation these rats would have received about eight times as much radiation as rats in neighboring areas. No differences in phenotypic variability were found between the two groups. This does not exclude the possibility that some genetic effects may be present, but if there are, they were not detectable by the survey method of study employed.

Human exposure to natural radiation by ingestion of food is a natural consequence of the incorporation by plants of radioactive elements from both the soil and the air; slowly decaying isotopes are eventually ingested either directly or after passing to other links in the food chain.

Our present knowledge of radiation-induced mutations is based on bacteria, mice, drosophila, and other organisms in which there is no doubt that radiation does indeed cause mutation. Our knowledge of radiation-induced mutation in humans, however, is not so firmly based. Radiation clearly increases the incidence of some cancers, such as leukemia. (Many, but by no means all, mutagens are carcinogens, and vice versa.) But like the rat studies, studies of humans from areas of high radioactivity have not turned up unusual variability. The massive human exposure to radiation at Hiroshima and Nagasaki provides our best available evidence on the effects of radiation, and the populations that received the radiation have been well studied. The numbers of people studied were large enough for detection of changes in rates of infant mortality, malformed babies, and still-births to about twice their normal rates or greater. But few significant changes were found. *Homo sapiens*, it appears, is not an unusually radiosensitive species. This hardly means the effects of radiation on human mutations are negligible, but it seems now that they are not disasterous.

It is known that mutation rates, like the rates of other chemical processes, are affected by temperature. Most mammals have a temperature-regulating device that keeps the body warmer than the scrotum. Swedish investigators found that wearing tight trousers can raise scrotal temperatures from 30.7° to 34°C. This, on the basis of crude estimates of known mutation rates and the temperature effect, could increase the incidence of mutation by 85 percent, that is, nearly double it. It is an amusing possibility that perhaps half of what is considered to be spontaneous mutation originating in males is in reality contributed by current fashions in clothing, and that the Scottish kilt may have more merits than its wearers suspect.

Cells, be they bacterial or human, are highly reactive systems, incorporating, creating, and modifying a great many chemicals. Many of these chemicals are meant to react with the genetic material, and others do so by mistake. Many of them are thus mutagenic, including some that are essential for the normal functioning of the organism (for instance, certain vitamins). Some minimum rate of mutations is therefore guaranteed by these chemical mutagens alone.

Artificial chemical point mutagenesis was discovered in Scotland and the USSR about the time of World War II. Chemically induced polyploidy had been discovered earlier. The list of substances capable of inducing mutations is growing steadily. It includes both naturally occurring and manufactured chemicals, such as mustard gas, hydrogen peroxide, formaldehyde, carbolic acid, calcium chloride, various alkaloids, and a great many others. Their effects have been tested on bacteria, fungi, various plants, drosophila and mice. They may produce point mutations, deletions, chromosome breakages, nondisjunctions, and other chromosomal abnormalities. It is difficult to tell what effect chemical mutagens have on human mutation rates. While there is no doubt that we are constantly exposed to substances that have mutagenic activity, and ingest a great amount of them, it is not certain how many of them reach the gonads, or, for that matter, somatic cells, without having been broken down.

Mutagens may act in several different ways. They may cause nicks in the DNA molecule (radiation and many chemicals cause such nicks). They may interfere with DNA repair systems, causing many more mutations to be fixed in the DNA than would otherwise be allowed (alkylating agents seem to interact here). They may react with DNA components, causing unnatural pairing at replication, and resulting in base substitutions (several chemicals and perhaps some physical factors may act in this way). They may react directly with the DNA, again causing mispairing (base analogues and many antibiotics and anticancer drugs are in this category). Some molecules may intercalate—that is, insert themselves into the stack of DNA base pairs, making a frameshift mutation (Section 6.4, Figure 6.3). Several small molecules, such as acridine dyes and nitrosofluorene, act in this way.

Before proceeding with the discussion of increased mutagenesis, we may conclude this chapter by noting that in addition to mutagens, there is also a class of substances that exhibit *antimutagenic* activity. These substances tend to depress spontaneous mutation rates, although without having any effect on the rate of mutation induced by radiation.

17

INCREASED MUTAGENESIS

This chapter is concerned with the possible problems arising from the increased exposure of all forms of life, and in particular human life, to additional radio-activity and a great variety of new and untested chemicals not previously present in our natural environment. These are being delivered directly and indirectly as a result of military action, medical practice, on-the-job exposure, agricultural practice, food additives, other legal and illegal commodities, and general releases into our common environment.

17.1 RADIATION

A brief review of the elementary physics of radiation phenomena appears in Box 17.A. In addition, several terms employed in measuring radiation intensity must be introduced. Their precise technical meanings need not be remembered,

BOX 17.A RADIATION

Radiation is a process by which energy travels through space. *Electromagnetic* radiation is basically a self-propagating electric and magnetic disturbance that affects the internal structure of matter. *Corpuscular* radiation consists of streams of atomic and subatomic particles that have the capacity to transfer their kinetic energy to whatever they strike.

It will be recalled that atoms consist of a nucleus containing various numbers of electrically uncharged *neutrons* and positively charged *protons*, around which negatively charged *electrons* orbit because of the attraction between the oppositely charged particles. If atom *A* comes close to atom *B*, its nuclear attraction may pull one or more of *B*'s electrons into its own orbit. *A* will then become negatively charged because it has more electrons than protons, and *B*, having more protons than electrons, becomes positive. These atoms are then electrified *ions,* and, because of opposite charges, form ion pairs. Radiation that produces ion pairs is called **ionizing radiation.** Although some nonionizing radiation (such as ultraviolet or visible light) is mutagenic, it is ionizing radiation that concerns us more.

As the illustration of the electromagnetic spectrum shows, ionizing radiation is produced by the extremely short waves of X-rays and gamma rays.

The properties of the different kinds of radiation are shown in the following table:

Radiation	Source	Penetration
Electromagnetic		
X-rays	X-ray tubes	Moderate
γ-rays ⎫		Moderate
Corpuscular ⎰	Radioactive	
β-rays[1] ⎱	substances	Low
α-rays[2] ⎰		Very low
protons[3] ⎫	Nuclear	No
neutrons[4] ⎭	fission	High
Mixed		
Cosmic rays	Of extra-terrestiral origin	High

[1]High speed negatively charged electrons.
[2]Positively charged bare helium nuclei.
[3]Positively charged hydrogen nuclei.
[4]Electrically neutral particles.

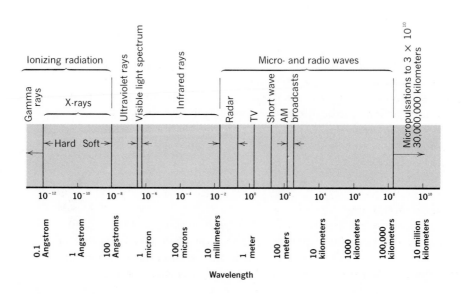

but since they are used in the discussion, the terms are introduced here. The *roentgen* (r), named after the discover of X-rays, is a unit of exposure used for X-rays and gamma rays. One r is equivalent to the amount of radiation that produces 2×10^9 ion pairs per cubic centimeter of air. The *rad,* a unit of absorption, equals 100 ergs of energy per gram of irradiated matter. The *rem,* or roentgen-equivalent-for-man, expresses the biological effect of 1 rad on a human. It is computed by multiplying rads by the empirically determined relative biological effectiveness of the particular kind of radiation measured. Because the effects of radiation differ with the way in which the dose is administered, the rem may not be a very useful measure, but dosages are still expressed in this form in much of the literature. Although the roentgen, the rad, and the rem are not quantitatively identical, we shall use all three, as appropriate.

Most of the figures cited for humans are expressed in terms of exposure per generation—that is, about 25 years, or the average period of exposure of the gonads before the gametes that form a zygote are produced.

In Section 16.4 we discussed the sources of natural radiation and the variation in exposure with altitude and geographic location. We can now restate the situation in quantitative terms. Estimates of the radiation exposure per generation from Earth's crust range from 0.7 rads in France, through an average of 2.5 (from 70 to 107 microrads = about 0.1 r per year) in the United States, to 15 in Espirito Santo and up to 84 in Kerala. At sea level a person may receive 1.1 rems from cosmic rays and 3 or 4 rems from Earth's radiation. At 15,000 feet the comparable figures may be 6, and again, 3 or 4. The totals are increased somewhat because of ingestion of radioactive substances. Hence, the full dosage received per generation is ordinarily about 4 to 10 rems.

To quantify the various aspects of mutagenesis in humans necessarily requires a great deal of guesswork. In the early years of the atomic age much extrapolation from drosophila data was used, but more recent and extensive experiments with mice, particularly in the Oak Ridge laboratory, throw considerable doubt on the validity of drawing definitive conclusions at this time. It is not so much that the genetic material of drosophila differs from that of mouse or human but rather that many different factors contribute to the degree of mutagenic effect observed under different conditions. For example, many drosophila experiments were based on irradiating mature sperm, which now turns out to differ in its response to radiation from sperm in earlier stages of gametogenesis. Male responses differ from female responses, and further differences in response are also caused by temperature and level of oxygen tension. New evidence supports the existence of mechanisms that repair damage caused by radiation, and indicates that different chemicals vary in their effect on the rate of mutation under radiation. In short, while we have found out a great deal about rates of artificially induced mutation under specific conditions and in specific organisms, we are a long way from generalizing what has been happening in humans and is happening today. Box 17.B gives a sampling of some of the mouse experiments, the results of which rest on somewhat more secure grounds. It would be dangerous, however, to generalize about human genetics on the basis of what is known of seven selected loci of mice.

17.2 RADIATION HAZARDS

Many science fiction stories begin with, feature, or end with an atomic cataclysm, in which life is either wiped out or monstrous mutants are the typical survivors. Even Ian McHarg used this device in his philosophical view of our experiment with culture (Section 5.3). Such scenarios are all too possible, and we share the concern of many who write them. This section, however, deals with a scenario that stops short of cataclysm, but instead presents some poorly calculated risks associated with the use of radioactivity.

Before addressing ourselves to specific sources of radioactivity, we present the following table, drawn from several recent estimates:

Source of radiation	Average human dose per year (mrem)
Natural	
Cosmic radiation	30–60
Earth radiation	50–60
Ingested natural radiation	20–30
Total natural	100–150
Additional	
Medical & dental	30–75
(Diagnostic)	(35–72)
(Therapeutic)	(1–5)
Occupational	.2–.8
Nuclear test fallout	<1–1
Nuclear power industry	<1
Miscellaneous (TV, air travel, etc.)	<1–2
Total additional	32–80

It seems clear that the radiation exposure from fallout, from nuclear power developments, and from occupational exposure (treated as part of the overall population average) is now small compared to that from natural radiation. Most of the additional radiation comes from medical and dental technology. In the United States, this medical technology is being improved both in terms of sensitivity of the practitioners and of their instruments. In 1964, medical diagnostic exposure was estimated at 55 mrem per year; by 1970 this had been reduced to about 36 mrem. The exposure from therapeutic radiation is much less, and that from dental radiation still less.

It has been estimated that the average dosages per generation from occupational exposure now run about 5 mrem, with considerable variation among occupations. (In one radiation research laboratory, the ten highest recorded levels of exposure ran about 5 r per year.) Better equipment and greater attention to safety will tend to lower these rates for individuals, but increased nuclear technology will work in the opposite direction on the population level.

One response to our energy demands is to increase nuclear power production. There are two critical considerations. One is safety of the reactor itself, in that an explosion (probably not the nuclear variety) or other reactor failure could suddenly and massively contaminate an area in the vicinity of the reactor. The second is the problem of leakage of radioactive products, both from the plant itself, and from the containers used to store the radioactive wastes. There are three major types of reactors: the currently used fission reactor; the breeder reactor; and the fusion reactor. At least two fundamental breakthroughs will have to be made

BOX 17.B RADIATION EXPERIMENTS ON MICE

To compare mutation rates under different techniques of radiation, W. L. Russell and his associates at Oak Ridge prepared a strain of mice homozygous for seven recessive marker genes. Irradiated dominant homozygotes were then mated to this strain. The occurrence of a mutation at any one of the seven loci was signalled by the appearance of a recessive homozygote in the offspring.

Roentgens administered in a dose	Doses	Interval between irradiations	Mutations observed per locus per 10^5 gametes
1,000 r	1	0	6.7–10.3
500 r	2	2 hours	11.5
200 r	5	1 week	19.5
200 r	5	1 day	26.6
500 r	2	1 day	49.9

Justice cannot be done here to the variety and scope of the series of experiments undertaken. In some of them more than a half million young mice were checked for various effects of irradiation. Only three of the experiments are summarized here to illustrate different aspects of the problems studied, and no attempt is made to generalize.

One experiment was designed to investigate the effect of the interval between the time at which females were irradiated with fission neutrons and the time of conception. In the 90,000 offspring born within the first seven weeks after irradiation, 59 mutations were observed. In the 120,000 offspring born after seven weeks, none were found. Similar results were obtained with X-rays and gamma rays. Whether the difference between the two periods is to be ascribed to differential sensitivity of the gametes at different stages of formation, to repair mechanisms, or to some other cause, cannot be told as yet.

In another experiment designed to study the effect of fractionating a given dose of irradiation, male mice were irradiated at the rate of 90 r per minute until each had received a total of 1,000 r:

Again, definite interpretation is not yet possible. One hypothesis might be that the heavy single dose either kills the cells or damages them to the extent that many never form sperm.

The third experiment, also on males, dealt with differences in intensity of radiation, and as the table shows, demonstrated that acute doses in the range investigated are more harmful than chronic ones.

Source of radiation	Rate in r per minute	Dose	Mutations observed per locus per 10^5 gametes
γ-rays from	0.001	300	4.3
cesium 137	0.001	600	5.9
X-rays	9	600	8.1
	90	300	8.7
	90	600	13.3

The important point about these experiments is that they show that many factors affect the degree of mutagenic activity of ionizing radiation, making exact quantitative statements about hazards to mice—or humans—complicated and difficult.

before the fusion reactor can become even theoretically operational; if it is successfully developed, it should be relatively safe, producing little radioactive waste. Several reactor accidents and waste leakages have already occurred with present fission reactors in the United States, Yugoslavia, and the Soviet Union, and a technician in New York died following 3,000 r total body radiation.

The breeder reactor will bear the most watching. The British breeder reactor at Harwell has already experienced some disquieting difficulties. The breeder design places liquid sodium and water in near proximity, and nuclear bombardment of the jacket that separates them causes significant instability in joints, gaskets, and so on. An accident or leak resulting in mixture of these two liquids would result in a violent reaction, which could release radioactivity from the core. In addition, the radioactive waste products of the breeder present vexing problems of very long-term storage. This one has much potential for genetic mischief.

A decade ago, radioactive **fallout** from the testing of nuclear weapons was a major topic of debate. The explosion of nuclear bombs produces a variety of unstable radioactive isotopes of many elements, and these decay at different rates by emission of beta particles. Strontium 90 has a half-life of 28 years (that is, every 28 years its radioactivity diminishes by a half), and cesium 137 a half-life of 30 years. It is such slowly decaying isotopes that present a long-term danger. They remain radioactive for a long time after they fall from the atmosphere onto the soil, and can be taken up by plants and pass through them into animal products, such as milk.

The **local fallout** (that in the immediate vicinity of the explosion) consists of the heavier elements. The **tropospheric fallout** (that in the lower part of the atmosphere) can be carried for many miles beyond the vicinity of the explosion. The **stratospheric fallout** (that in the upper part of the atmosphere) is produced by bombs larger than one megaton. It is made up of the relatively light particles, including strontium 90, and it eventually circulates throughout the world.

While there is no doubt that continued "unclean" testing or actual nuclear warfare can have nothing but dire genetic results, the accumulated fallout effects have so far been rather small compared to those from the use of X-rays. It has been estimated that the average increase in exposure from fallout before the 1963 test ban treaty amounted to no more than 0.05 r at sea level and a little more than 0.10 r at the altitude of Denver. This represents less than one percent of the background radiation, or what each one of us would receive in the course of three and a half months from natural sources. In effect, the increase is equivalent to that which would obtain should the average marriage age, or interval between generations, be increased by 3 or 4 months.

In retrospect, it appears that many of those warning of fallout dangers were in fact using a genetic argument as a political tool to reduce nuclear weapons research, and thus perhaps the probability of nuclear war. (The relative disinterest in the mutational effects of medical diagnostic radiation was instructive in this regard.) The basic issue is not one of genetics: it is political, and above all, ethical. Whether the development of nuclear weapons—or deterrents if you prefer euphemisms—at some costs in human life and suffering will prevent greater disasters than not developing them is not a question that geneticists are considered competent to answer. Yet any citizen—geneticist, bartender, or college freshman—can harbor a suspicion that the statesmen of Washington, Moscow, Peking, London, and Paris are not competent to answer it, even though their deliberations are based on many factors that may well be more important than the genetic hazards of fallout. (For the record, we should make explicit our personal bias. Although we are not militant pacifists who oppose war under any and all conditions, we are antimilitarist and consider wars and armament races to be immoral.)

Finally, although as noted in the tabulation at the beginning of this section, the miscellaneous category, including television, averages 2 mrem or less per person per year for the population, the viewing habits of some children are cause for concern. Television emits ionizing radiation from essentially a point source, which means that the dosage is inversely proportional to the square of the distance from the source. Persons watching TV from across the room receive only a negligible amount of radiation. However, many children spend hours each day sitting directly in front of the set.By doing this, they could be receiving in the neighborhood of 40 mrem per year, putting themselves at risk for both germinal mutations before their reproductive years, and somatic mutations at a time when their bodies are still actively growing.

Having considered the various amounts of exposure to radiation, let us examine next what their actual effects on individuals and populations may be. In general, the danger from excessive radiation lies in the development of cancers, tumors, and leukemia. It has been estimated that every additional 30 r of exposure will double the number of cases of leukemia in a population. Prenatal and juvenile exposure frequently results in growth abnormalities that are probably physiological rather than genetic. Microcephaly (unusually small head) and several eye defects are common among children who received large doses *in utero* when their mothers were being treated with pelvic radiation, and retarded mental development is usually associated with the microcephaly.

There is some tolerance to exposure of limited areas of the body, which can even take doses of up to 1,000 r. In cancer therapy, where the attempt is made to kill the diseased cells without affecting the surrounding normal ones, fractionated doses of 1,500 r are administered from such sources as cobalt 60. But the tolerance to total body radiation and to radiation of the gonads is much less. A dose of 50 r can cause temporary sterility and even 10 r may effect pathological changes in white blood cells.

All of this, of course, calls for enforcing various precautionary measures, such as shielding all but the organs being investigated or treated by irradiation. Many states have outlawed the use of X-ray machines for fitting shoes. In California the practice of scanning visitors to penal institutions with X-rays has been abolished under pressure from biophysicists and geneticists. The technology of airline security scanners is changing so rapidly that an overall evaluation of this is, at present, difficult.

Estimates of genetic damage are highly speculative, but for whatever they are worth, some figures may be given. The following figures were derived by James F. Crow, who stresses their indicative nature rather than their precision. They express the possible effects of exposing a human population of 200 million to 10 r:

Form of damage	Number of cases in the first generation	In subsequent generations
Chondrodystrophic dwarfism	1,000	200
Gross mental or physical defects	80,000	720,000
Infant and childhood deaths	160,000	3,840,000
Embryonic and neonatal deaths	400,000	7,600,000

These figures apply to exposure in a single generation. If later generations were exposed to additional doses, new and higher levels of equilibria for the mutant alleles would be established. Figure 17.1 diagrams two situations of increase in mutation rate for detrimental recessives under assumptions of a temporary and a permanent increase.

While large genetic changes as a result of exposure to nuclear-bomb radiation have not been detected (Section 16.4), some examples of the somatic conse-

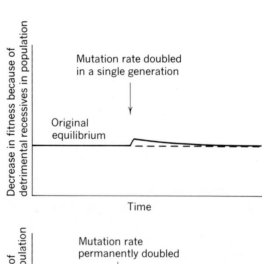

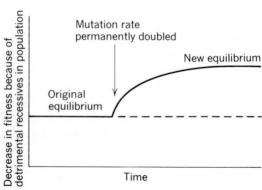

FIGURE 17.1

The upper graph shows the effects of doubling the mutation rate of detrimental recessives for a single generation, followed by a return to the previous rate and slow elimination of the extra detrimental genes over many generations. The lower graph shows the effects if the doubling of mutation rate is permanent.

quences of exposure may be found among data from Hiroshima and Nagasaki. Of 183 children exposed *in utero,* 33 had microcephaly, and 15 were mentally retarded, a rate of mental retardation five times that of children born at about the same time whose mothers were farther from the bomb. Each year the death rate among those survivors who were closest to ground zero has been higher than that of unexposed control groups, and about half of the excess deaths are attributed to leukemia. This higher leukemia incidence among persons closer to ground zero has continued up to the present time among the Hiroshima survivors, but recently the yearly leukemia incidence among Nagasaki survivors has decreased almost to the level of control groups, having peaked about 7 years after the bombing. Some explanation for the difference may lie in the fact that the Nagasaki bomb released fewer neutrons than did the Hiroshima bomb—and neutron yield is a significant measure of a bomb's potential to do biological damage—but the reduction in leukemia deaths among the Nagasaki survivors has been somewhat offset by a recent rise among them in other types of cancer, compared to incidences among controls.

Chromosomal abnormalities such as those illustrated in Figure 17.2 were found in cells of 35 percent of persons younger than 30 years and 59 percent of persons older than 30 exposed to estimated radiation in excess of 200 rads, while none were observed in the controls. The chromosomal abnormalities were in the form of rings, fragments, translocations, or two centromeres on the same chromosomes. Similarly, in a group of 43 Marshall Islanders accidentally exposed to fallout when the winds shifted after detonation of a nuclear device in Bikini, 23 were revealed to have chromosomal aberrations when their white blood cells were examined in culture ten years later. The one redeeming fact is that so far no connection has been established between these abnormalities and the state of health of their carriers.

To conclude this section let us examine the results of an interesting experiment done at Chalk River, Ontario—the Canadian equivalent of Oak Ridge. Rats were exposed to cumulative doses of 2,400–7,200 r over a period of 2–12 generations. They were subsequently tested for maze-running ability and on the average exhibited a decline, roughly equivalent to a drop of 5.35 points on the human IQ scale. Mentally retarded rats in this experiment may not have reduced fitness, as the Japanese microcephalic children probably do. Furthermore, reasoning by analogy and extrapolation from rat IQ's to human IQ's hardly permits one to draw any valid conclusions. Yet it is still worth noting the possibility that behavioral traits may be also subject to genetic deterioration due to increases in mutation rate.

17.3 PRACTICAL USES OF RADIATION

The nuclear age, with all its risks and perils, is bringing *Homo sapiens* to a degree of control over nature undreamed of earlier. The risk-benefit equation is particularly hard to balance where there are genetic risks, for those who receive the

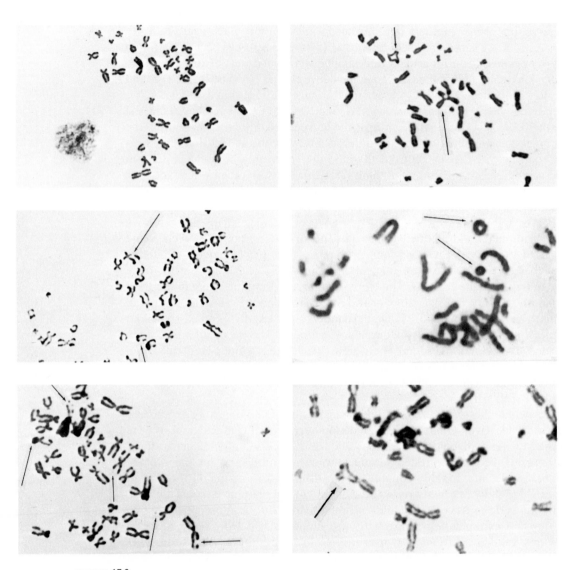

FIGURE 17.2
The left column of photomicrographs shows examples of abnormalities arising from irradiation of human chromosomes in tissue culture. The top photomicrograph shows a control. Note the breaks and fragments in the middle and bottom pictures. The right column of photomicrographs show the consequences of breaks in several chromosomes with subsequent abnormal reunions of broken ends. (Photomicrographs courtesy of Theodore T. Puck.)

benefits and those who run the risks are often not the same people: frequently those who will be at risk are several generations in the future. In assessing risk for either radiation-caused biological harm, or that caused by use of chemicals (Section 17.4) or other environmental changes, the following dimensions will be useful: (a) risk relative to natural background causes; (b) risk estimates for specific genetic conditions; (c) risk relative to current prevalence of serious disabilities; and (d) risk in terms of overall ill health. Risk (d) should probably be applied in terms of both people and general ecosystems.

Most of the ways we use atomic energy (power production, various engineering feats, medicine, and public health) fall completely beyond the scope of this book. However, a number of applications of radiation techniques in the pharmaceutical industry, in agriculture, and in pest control, are of practical genetic significance.

As shown in Figure 14.4, selection has been highly successful in increasing the yield of antibiotics. Most of the genetic variation utilized to obtain selection progress was induced by ultraviolet and X-ray irradiation. In horticulture and plant breeding, X-rays and gamma rays have been similarly employed to produce useful single-gene mutations and to increase the genetic variation of polygenically determined traits, which permitted previously established plateaus of performance to be surpassed. Among the ornamental plants, new varieties of chrysanthemums, tulips, carnations, and snapdragons have been produced and marketed. It might be noted that U.S. law affords patent protection to such novelties, which is, of course, a great encouragement for breeders to employ artificial mutagenesis in the development of new varieties.

Examples of agriculturally important radiation-induced mutant forms abound. In Sweden, new forms of barley have been developed that are better adapted to the conditions where this cereal crop is grown, and peas have been developed that give a higher yield. In both Sweden and Germany, as well as elsewhere, mildew-resistant barley has been produced. In the United States, peanut yields have been increased and rust-resistant oats have been obtained from irradiated and selected seed. In the USSR, polyploid beets produced by chemical mutagenesis have been reported to yield 15 percent more sugar than the foundation stocks. In Japan, new variation in many quantitative traits of rice was induced by irradiation, while in Australia the same was done with subterranean clover.

So far, it is not possible to predict the direction that the phenotypic expression of a mutation produced by one or another agent will take in higher organisms. Whether mutants with desirable characteristics will appear is a matter of chance and volume of the material exposed. But the ultimate aim of research workers in this field is to find means for directed mutagenesis. This may not be too remote a possibility; in microorganisms mutagens have been found that affect only specific bases.

In animals, with the exception of silkworms, artificial mutagenesis has so far received little practical application. Although the results of some pilot experiments on drosophila indicated that radiation-induced variability can lead to gains in the characters which are selected for, the only experiment carried out with a domestic animal gave negative results. The experiment was an attempt to produce polygenic variation in a flock of chickens selected for egg production. One or more translocations were obtained, and variation in some quantitative traits was increased, but selection did not result in added gains in economically desirable traits. That no such gains were produced can be attributed to the scale of the attempt, which was quite small compared to the scale possible in plant and drosophila experiments.

At least two applications of radiation to genetics have been devised to manipulate the sex of the silkworm. A biologist in the Soviet Union and another in Japan independently found that 100 percent male offspring can be obtained by destroying the nucleus of the egg and permitting it to be fertilized by two Z-carrying sperm (it may be recalled that male silkworms are homogametic). This procedure is economically useful, because male cocoons yield more silk than those of females. The other application, devised for the same purpose, is apparently now in commercial use in Japan. Irradiation was used to translocate a small piece of an autosome to the Z-chromosome. The section of the autosome carried with it a recessive allele w, the effect of which is to keep the white outer membrane of the silkworm egg from darkening, which it would normally do after several days of development. Matings between $Z^w Z^w$ males and $Z^W W$ females yield $Z^W Z^w$ male offspring whose eggs turn dark, and $Z^w W$ daughters whose eggs remain white. An electric-eye sexing machine can then pick out the white eggs and electrocute them, thus ensuring that only male cocoons are produced. The first edition (1968) of this book contains the following paragraph:

> Sexual manipulation is also involved in the use of irradiation in pest control, which was originally developed by E. F. Knipling, a scientist in the U.S. Department of Agriculture. The screwworm fly is a livestock pest that causes millions of dollars in damage. The female lays a mass of 200–300 eggs in open wounds or abrasions of livestock, including the navels of newborn animals. When the larvae hatch they feed on the host, often to the point of killing it, especially because their presence attracts more egg-laying flies. The controlling technique Knipling worked out consisted in massively irradiating male flies in preadult stages to induce many dominant lethal mutations and render them sterile. The males were then released in great numbers in the affected areas, where they entered into sexual competition with normal fertile males. A large proportion of females that copulated with irradiated males then produced eggs that did not hatch. Repetition of the procedure for several generations eliminated the screwworm fly population completely in previously badly infested areas of Texas, New Mexico, Arizona, and California.

In 1973, the following data were included in an article entitled: "The Screwworm Strikes Back."

Cases of Screwworm in the United States

Year	1962	1969	1970	1971	1972
Cases	50,000	219	153	473	92,192

The first number given is for 1962—the year eradication was begun in the Southwest. The mechanism of the screwworm comeback is not understood yet, but it is likely associated with selection for behavior that somehow distinguishes native fertile males from introduced sterile ones. We dislike the damage the screwworm does to both the health of the livestock and the economy in general, but as evolutionary biologists we can't help favoring the despised screwworm with a small smile, during its continuing fight against our science and technology.

17.4 CHEMICAL MUTAGENESIS

As we learn more about the chemical basis of the gene, its replication, and repair, the use of chemicals to direct specific mutations or classes of mutations looks increasingly promising. But we are also coming to appreciate that contact with mutagenic chemicals may prove to be genetically more harmful to humans and other species than the radiation we had earlier come to fear.

Currently, less than one-tenth of one percent of all drugs, pesticides, food additives, and industrial chemicals have been investigated for mutagenicity, and virtually no chemical has had the kind of systematic testing that is regarded as adequate. Most of these chemicals are completely new in the evolutionary experience of Earth's organisms. Currently, we know little about which chemicals can penetrate to human germ-cell lines, and not much more about what happens when they do penetrate.

It has been strongly argued that we should begin monitoring human populations to detect a rise (or change) in the mutation rate—simply to see if we are experiencing a major effect due to something. Such monitoring would not pinpoint the causes, but only the fact of a change. Obtaining even such crude information would require a major effort. To detect, with reasonable confidence of being correct, a rise from the present mutation rate in the neighborhood of 50 percent would require a study population of 6–10 million people (producing 120,000 to 200,000 births per year).

An alternative approach is the screening of suspect (or all) chemicals. This also is an enormous task, and in 1969, the Environmental Mutation Society established the Environmental Mutagen Information Center to coordinate such screening. The choice of organisms to use for such screening is a crucial problem. Luckily, unlike in toxicology, the target of the suspect chemical is known. It is DNA, which is available in a wide variety of test organisms. However, there are reactions that take place within organisms which convert mutagenic compounds into nonmutagenic compounds, and vice versa. For instance, cyclamate is not mutagenic, but can be converted to cyclohexylamine, which is mutagenic. If humans perform a biochemical conversion of this sort, the appropriate test organism must also do it. Mammals, which are expensive, time consuming, and laborious to work with, are more likely to share such reactions with humans than are the much faster-reproducing and cheaper bacteria. There are also differences, from organism to organism, in cellular uptake and detoxification mechanisms. Even *within* species of mammals, there are sex and strain differences in sensitivity to mutagens. For instance, such differences have been detected in mice for the mutagenicity of EMS (ethyl methanesulfonate, a highly mutagenic substance).

One clever compromise between the efficiency of microorganisms and greater similarity to humans of mammals is the **host-mediated assay.** The suspect chemical is administered to a relatively few mammals, the more different kinds, the better. Microorganisms are then introduced into the treated animals, where they will encounter the metabolized products of the suspect chemical, as well as the original if it is still present, in and among the cells of the animal. These very large numbers of microorganisms can then be recovered and efficiently screened for mutations, which will increase in frequency if the suspect chemical or its derivatives in the mammal are mutagenic to the DNA of the microorganisms.

In setting priorities, several things must be considered. Clearly, it is important to detect common chemicals with powerful mutagenic properties. Detecting rare chemicals with powerful mutagenic properties is less urgent. A big problem is the group of widely used substances that may have weak mutagenic activities, as these effects are very difficult to demonstrate, yet because of their ubiquity, they could cause an enormous number of additional mutations. Even if they are demonstrated, legislatures and people may be willing to accept the risk of continued use of the chemicals if the probability of mutation per single exposure is low.

Caffeine may be used as an example. It was first suggested as a major mutagen in humans over ten years ago. Since then, it has been shown to break human and plant chromosomes in tissue culture, and apparently to increase the mutation rate in bacteria. But early results from feeding mice caffeine were conflicting. It now appears that caffeine is *not* a mutagen in the classical sense, but that it interferes with one or more DNA repair mechanisms. This may explain the conflicting results. If other mutagens were operating during the experiment, caffeine might appear mutagenic by allowing mutations that would normally be repaired to instead be incorporated in the gene pool. We don't know if caffeine acts this way in humans. Until we do, what is a reasonable response? One that has been suggested is that pre-reproductive males remove their pants after coffee breaks, or perhaps better, that ice packs for the testicles be served with the coffee, thus reducing the scrotal temperature and thereby reducing the number of new mutations whose repair may be blocked by the caffeine (Section 16.4).

Another argument for screening of chemicals is that there is a good relationship between teratogens (which cause developmental abnormalities), carcinogens, and mutagens. At least many geneticists would like to think this is so, as there is a lot of money available to support cancer research. However, the causal relationship may not be all that good, but some biologically reactive compounds may do more than one thing.

Drugs in particular should come under special scrutiny, since by their nature they are biologically active chemicals. Caffeine is a case in point. Another that has received much recent attention is LSD (lysergic acid diethylamide); like those on caffeine, the results are conflicting—they are also much more controversial. Chromosome damage is the main genetic symptom associated with LSD use, although it is clear that a number of developmental and psychological problems are also associated with its use. There are two major problems with the evidence. As in many studies of mutagens, the chemical was administered to the test animals in much higher doses than would reasonably be taken by humans. And many chemicals (for instance aspirin) when administered in high concentrations disrupt normal cell functioning, and may cause such damage as chromosome breaks. At more reasonable lower concentrations (where effects are hard to demonstrate), in general excess chromosome damage or increased mutation rates have not been demonstrated. The second problem, where studies of human users were done, is the difficulty of separating the observed chromosome damage that may be due to LSD use from damage due to other causes. Since LSD is an illegal drug, its users frequently get impure preparations. More important, it is difficult to find a chronic user who has not also used one or a variety of other potentially harmful chemicals. Of those studies using pure LSD, where cells were monitored before and after LSD use, there is little evidence for increased chromosome damage. Nevertheless, the argument that LSD causes genetic damage has been widely used in the anti-LSD campaign.

Pesticides (some are better called biocides), for similar reasons, deserve special attention. A mixture of 2-4-D, 2-4-5-T, and a few other things (the "agent orange" of the Vietnam war) has drawn much publicity, based on its alleged mutagenic and teratogenic effects on American livestock and Vietnamese embryos. There is some evidence that both 2-4-D and 2-4-5-T may cause chromosomal abnormalities, and dioxin, one of the impurities in "agent orange," may be a teratogen. So may many other widely used pesticides that have not received such attention. The outcry against "agent orange" had more to do with its use in Vietnam than its unusual danger to the biosphere.

Thus, in this section, we have cited genetic arguments used by people interested in opposing the Vietnam war, or in opposing LSD use. In Section 17.2 we noted that genetic arguments have been used in opposition to nuclear-weapons development, and in Section 12.5, as a means of oppressing a minority race. In each of these examples, the arguments are used during early stages of the research, before the facts are clear and in reasonable perspective. In some cases, the researchers are actively involved in the propagandizing, in other cases, other people are using the findings of uninvolved researchers. It might even be argued that the research is stimulated and thereby helped by the publicity. But nevertheless, whether they are used for causes we support or oppose, we are disturbed when research results are prematurely used in the political arena.

17.5 MUTAGENESIS AND HUMAN VALUES

It would be a reasonably simple matter to assess the damage caused by increased mutation rates if a Platonic viewpoint could be adopted. That is to say, if there were an *ideal* genotype, homozygous at every locus, any deviation from it could be considered detrimental, and the harm done to our species by mutagenic agents would be directly proportional to the number of mutations they induced. But we know how far from reality oversimplification of this sort is.

It is possible to speak of a norm with reference to phenylketonuria, amaurotic idiocy, deafmutism, and many other such defects; taken all together the incidence of these is about one percent of live-born infants. It is also possible to speak of norms for such gross chromosomal aberrations as trisomy 21 or the *cri du chat* syndrome found in another one percent of children born alive. But the bases of many other developmental malformations and disorders are so little understood that value judgments and definitions of norms are impossible. Even more significant in this connection is the existence of vast arrays of polymorphs, which express themselves not only biochemically but in every single kind of human attribute, including native abilities, temperament, and social behavior.

We have seen how abnormal hemoglobins may be of advantage to individual carriers in particular environmental situations. High reproductive fitness, the essential criterion of natural selection, might by definition always be thought to have a positive value. But even from the purely biological point of view, as Haldane has pointed out, this is true only for rare and scattered species competing with other species. As soon as the population becomes dense, intraspecific competition becomes inevitable and resources become overtaxed. Indeed, many species have become extinct because of selection in a direction advantageous to the individual but disastrous for the species. One possible example is the extinct Irish elk that developed the largest known antlers, and, perhaps, as Haldane says "literally sank under the weight of its own armaments." When we consider not only biology, but also social values, judgment as to what is optimal or ideal with reference to fitness or some other attribute becomes highly beclouded.

This is even truer in a world changing as rapidly as ours is, where the properties we admire today may become completely irrelevant to the world of tomorrow. Within a single lifetime, people have been known to change their idols from Lenin, Marx, and Sun-Yat-Sen to Lincoln, Schweitzer, and Einstein, and others have, no doubt, changed in the opposite direction. Even well-meaning and vigorous opponents of the typological approach sometimes fall into a trap in discussing this problem. A single quotation from a great biologist, humanist, and a champion of human diversity will suffice: "It is too easy to let our imagination strive for something with a body as beautiful as a Greek god, healthy and resistant to cold and heat, to alcohol and to infections, with the brain of an Einstein and the ethical sensitivity of a Schweitzer."

It may be questioned whether Bantu tribesmen or Chinese intellectuals share the standard of classical Greek beauty. And to some people the ethics of Schweitzer in his attitude to native Africans are a little difficult to distinguish from those of segregationists in the United States or Africa.

In any case, to return to the more direct issue under consideration, because we live in a world of heterozygotes and of unique genotypes, any attempt to assess which of the numerous multiple alleles at our numerous loci (alleles that produce practically an infinity of combinations) are superior, would be a puerile undertaking. A general nonspecific evaluation of the effects of increasing the clearly

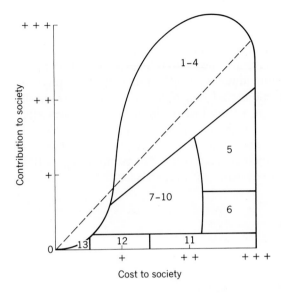

FIGURE 17.3
Contributions and costs to society of
different phenotypic classes. (After
S. Wright.)

undesirable mutational reserves of the human gene pool is, nonetheless, worth-while, if only to clarify our thinking on this matter. Such an appraisal has been undertaken by S. Wright, and the rest of this chapter is largely based on his discussion.

Figure 17.3 schematizes the relationship between the contributions and costs to society of different classes of phenotypes as visualized by Wright. Group 1 includes the bulk of the population, whose costs and contributions balance at a modest level. Group 2, whose costs and contributions balance at a relatively high level, includes professionals of average competence with a higher than average education and standard of living. Group 3 is a small one composed of individuals making extraordinary contributions at modest cost. Group 4 comprises people whose cost to society in terms of education is high, but whose contribution is even higher. Because these four groups include a great variety of types, who as a whole return to society at least as much as they cost, they are combined in Figure 17.3. Although the mutations that place individuals in these classes may be personally injurious, they do not damage society.

The remaining groups consist of individuals whose costs outweigh their contributions. Group 5 includes individuals who have the capacity to contribute to society but do not do so because of such reasons as inherited wealth. The genetic component of the phenotypic variation of this class is probably exceedingly low, and hence will not be affected by increased mutation rates.

Group 6 is the category which is exceedingly costly to society because of its antisocial activities. To use Wright's labels for its members, this group comprises charlatans, political demagogues, criminals, and others who prey upon society. Whether increasing mutation rates would place more individuals into this group is difficult to assess. Such behavioral traits as egotism, aggressiveness, and impulsiveness are as likely to have a genetic component as those considered in

Section 12.4, but practically nothing is known about their mode of inheritance, so that little can be said about them here.

Group 7 includes individuals subnormal in health and physical constitution. Group 8 is composed of those who have low mentalities but are able to take care of themselves. Group 9 comprises individuals who suffer relatively early physical breakdown from accidents or infections and degenerative diseases. Group 10 includes the cases of mental breakdown from one of the major psychoses after maturity. Groups 7–10 are combined in Figure 17.3 because the appraisal of mutation effects that can be made is about the same for all. The critical issue is the type of genetic basis underlying their characteristics. Oversimplifying matters considerably, we can divide all mutant genes into four classes:

A. Polymorphic due to
  1. heterozygous advantage,
  2. frequency dependency.
B. Mutation-selection balanced and
  1. completely recessive,
  2. with adverse effects in the heterozygote.

A fundamental difference exists between mutants of types A and B in the effects produced by changing their mutation rates. As has already been noted (Figure 17.1), increasing the mutation rate of type B increases the incidence of the defects. In contrast, increasing the mutation rate of type A would have only a small effect on incidence. Thus, increasing the mutation rate of the Hb^s gene would not increase sickle-cell anemia proportionately.

The answer to the question of how much damage increased mutation rates will have on groups 7–10 depends then on which kind of mutation is the more important in determining the genetic components of phenotypic variation. We know that both types A and B exist, but their relative importance still engenders somewhat heated debate among geneticists. Of particular importance in this connection are the investigations of the American geneticist B. Wallace, who found type A-1 mutations arising in drosophila after irradiation. Although the interpretation of his data has been questioned, subsequent work on drosophila and on flour beetles has lent it considerable support. Data on humans are even more confusing.

Some evidence suggests that schizophrenia, the chief contributor to Group 10, and one of the socially most costly diseases, may also be of the type A. Perhaps, as Wright says, "the burden of overt schizophrenia is the price society pays for benefits conferred by persons of slightly schizoid type among the heterozygotes or the 90 percent homozygotes that do not break down." On the other hand, Huntington's chorea, a progressive degeneration of the nervous system that begins in middle age and is due to a dominant gene (see Section 19.3) is definitely of type B.

Group 11 includes individuals who suffer complete physical or mental incapacity throughout a normal life-span and hence are extremely costly to society; Group 12 is less costly because its members die early. There is a considerable genetic component in these two groups, as exemplified by PKU. Most geneticists hold that the majority of genes responsible for the conditions placing individuals in these two groups are of type B. If this view is correct, then social and personal costs would be adversely affected by increases in mutation rate, but it is difficult to determine whether a heterozygous advantage in terms of some desirable human abilities (not in fitness) exists in carriers of genes for some of the defects placing their carriers in these groups.

Group 13 comprises fetal and embryonic deaths, which are due not only to such nongenetic factors as maternal viral infections, but also to various chromosomal abnormalities. Although an increase in these would add to personal grief, the cost to society is negligible. It may thus be seen that the question raised regarding the social effects of increasing mutation rates has no easy solution.

18

HUMAN DIVERSITY

Most people have very strong ideas about race, or what constitutes different races, although all surely do not agree on the same meaning. These are concepts that—in one form or another—are firmly impressed upon the young of most twentieth-century cultures. Just as young people seem to learn to hunt and fish more easily than they learn many other tasks, the young seem to learn these concepts easily. Some serious intellectuals have suggested that—on statistical and biological grounds—race is not a valid concept for the human species. But *something* is widely recognized under the general term "race." Whatever it is, it is real, and included in the concepts and misconceptions surrounding it are the sources of some of the major problems facing the human species today.

18.1 RACE

As you've probably noticed, we're having trouble with the names of races in this book. That's partly because some of our information sources are splitters, and some are lumpers, and among both groups, the lines between their different races are drawn in different places and in different ways. Further, fashions in names change. The first edition of this book employed the term "American Negro." Since then, there has been much pressure to replace the Spanish word for black with the English one. Besides being fashionable, this name change is useful, as it emphasizes that this American race is indeed significantly different from the several African groups and races that form part of its ancestry. We've also adopted the convention of beginning words that we are using as names of races with a lowercase letter, and capitalizing only names that refer to religions, tribes, nations, or nation-groups. Thus european and african genes may be found in European or African individuals, and both european and african genes are found in American blacks. The "black" as applied to American blacks is fairly specific, and does not include members of African or Australian races who are black, nor other people of color. Nor is it an exclusive term, as some of its advocates would have it, to be reserved only for certain persons deemed worthy to be members of an elite without regard to skin color, race, or origin.

The emotion-laden debate on human equality and human diversity that has gone on for centuries is still continuing with great intensity, despite the clarification of issues that has resulted from nontypological evolutionary thinking and genetics. The description of the problem by Dobzhansky can hardly be improved upon:

> It is often alleged, even by some reputable scientists, that biology has demonstrated that people are born unequal. This is sheer confusion—biology has proven nothing of the sort. Indeed, every person is biologically, genetically, and therefore irrevocably *different* from every other. However, genetic diversity is not tantamount to inequality. And vice versa—equality of opportunity, or of status, or economic equality are not predicated either on genetic identity or on genetic diversity. Monozygotic twins, though genetically similar, may engage in different occupations and achieve unequal socioeconomic status. Human equality or inequality are not biological phenomena but sociological designs; genetic diversity is a biological reality. Equality before the law, political equality, or equality of opportunity, stem not from genes but from religions, ethical, or philosophical wisdom or unwisdom.

The debate has been especially virulent when dealing with differences among groups described as races (see Box 18.A). The dictionary definitions of the term "race" are vague, and have a remarkable tendency to produce misunderstanding and confusion. For this reason, the word has been largely avoided in previous chapters of this book, but it is appropriate at this point to give it a genetic definition. According to Dobzhansky: **Races** *are populations which differ in the incidences of alleles of some genes* (see Box 18.B). This is not the common definition that anthropologists have used in the past. But gradually, as the influence of population genetics has extended to anthropology, the meanings assigned to "race" by geneticists and anthropologists have been converging. For example, a recent anthropological proposal has been to define races as groups between which gene flow has been restricted (W. S. Laughlin). Both of these definitions are, first

BOX 18.A SOME HISTORICAL NOTES ON RACE

The history of the various concepts about race is exceedingly old and rich with ugly detail. No attempt to give a systematic outline is made here, and only a selection of illustrative points is presented. Much of the material in this box comes from the historian of the subject, T. Gossett.

Xenophobia and race prejudice can be traced in India perhaps to five millenia before our day; in China, to the pre-Christian era; and in the Western world, to the Biblical tradition of the Hebrews as the Chosen People. According to Talmudic legend, the three sons of Noah—Shem, Ham, and Japhet—gave rise respectively to the Semitic people, the dark-skinned races, and the other gentiles. The Hamites were considered inferior because of the curse laid on their founder for being disrespectful to his drunken father, according to the book of Genesis, and for a variety of other crimes related in supplementary legends.

In Greek mythology, the origin of differences between black and white populations is laid to Phaëthon, the son of Helios. Unable to control the sun chariot, which his father permitted him to drive once, he drove too close to some parts of the earth, burning the people there black, and too far away from other parts, causing the people of those regions to pale from the cold. Most of the Greek philosophers attributed the differences between populations to climate.

It was in 1684 that the French physician Francois Bernier proposed in an article in a Paris journal that humanity is divided by appearance into four groups: Europeans, Far Easterners, "The Black," and Lapps. The great German philosopher and mathematician Leibniz in 1737 objected to the notion, likening the differences between the races to those between plants or between animals, and attributed them to the effects of climate, as the Greeks did. Nevertheless, Bernier's proposal was soon established in the mainstream of science and philosophy. Although Linnaeus (whose classification was consistent with his notions of special creation) thought that all humanity was a single species, he recognized four subspecies, *H. s. europaeus, H. s. asiaticus, H. s. afer,* and *H. s. americanus.* The debate on causes of differences and the scale of superiority of one race over another had some religious overtones in the eighteenth and nineteenth centuries; but it

was particularly the rise of colonialism and of Social Darwinism that caused the spread of racism in Europe, and in America it was the need for justifying the enslavement of blacks and the genocide of native Americans that led to the doctrine of white supremacy.

In Europe the cult of Aryanism, arising from Sanskrit legends about blond conquerors of darker-skinned inhabitants of India and Persia, spread and established itself, eventually culminating in the atrocities of Hitler. Even Galton believed in Nordic superiority (just as, in America, Jefferson and, for most of his life, Lincoln, held the assumption of Negro inferiority).

The brutalities of Anglo-Saxon imperialism, and those of Southern racism during the post-reconstruction period are too well known to be described here, but the following table is instructive in terms of statistics (available only since the 1880's):

| Years | *Lynching victims in the U.S.* | |
	White	*Black*
1882–1888	595	440
1889	76	94
1892	69	162
1906–1915	61	620
1918–1927	39	416

It is Gossett's thesis that the support of the intellectuals made racism in America possible. He credits the great American anthropologist of German-Jewish origin, **Franz Boas** (1858–1942), with first turning the tide against racial prejudice in this country. Although up until only a few years ago quotas for jews were maintained at such institutions of learning as Columbia University, Princeton University, and New York University, the battle among intellectuals appears to be won. There are still pockets of resistance, as demonstrated by acrid exchanges that appear in the pages of such journals as *Science.* But the general changes in attitudes expressed by both biologists and intellectuals in other fields within a relatively short period of time are indeed remarkable. Whether the nonintellectual population of both the South and the North is prepared to follow intellectual leadership in these matters is, unfortunately, a moot point.

of all, nontypological: no phenotypic prototype based on skin pigmentation, skull shape, hair form, or shape and size of body is assigned to a race. Second, both definitions imply the possibility of temporal changes in races. For example, the American black race is less than 350 years old. It arose from a mixture of genes from the African Forest Negro and Bantu, with contributions from the northwestern european, the alpine, the mediterranean, and other gene pools.

These definitions are, of course, not sufficiently precise to permit universal agreement in delimiting one race from another. There are lumpers and splitters among those who use the term, just as there are among those who use the word "species" (Box 18.C). But philosophically speaking, this is the way it must be if the classification is not to be static, but rather as dynamic as the gene pools themselves.

In any event, the genetic definition does not permit the confounding of criteria of race that has become so common in the layman's world. Most importantly, it puts to an end the notion of racial purity, or the existence of "pure-races," that has dominated the mentality of particularly narrow-minded racists and others for so long. Another confusion that the definition circumvents is the supposed relationship between biological race and language, which provided one of the bases for Hitler's notions of Aryanism. One of the chief proponents of the theory of Aryanism (see Box 18.A), the nineteenth-century philologist Friedrich Max Müller, who originally believed that race and language were related, spent the last years of his life deploring the belief: "I have declared again and again that if I say Aryans, I mean neither blood nor bones nor hair nor skull, I mean those who speak an Aryan language." Of the many examples that dispute the relationship, we need cite but one—that of Iceland. Although the language of Icelanders is Scandinavian, the blood-group and allele-frequency profiles make them biologically akin to Celts, whose language is quite different.

Parenthetically, it may be noted that there are claims for some relationships between allele frequencies and linguisitic preferences. Thus, in Europe, it has been alleged (although most linguists are not convinced) that the English *th* sound is found in the languages of populations in which the frequency of the O allele (of the ABO blood-group system) exceeds 0.65: this sound is found in the languages of Iceland, Britain, and Spain (but not Portugal); and in those of Greece and Italy, where it was present in the language of its ancient inhabitants, the Etruscans. In the languages of populations having an O allele frequency between 0.60 and 0.65, as in those of the Scandinavian peninsula and Germany, the sound used to exist but no longer does, and in the languages of populations having a frequency under 0.60 the phoneme *th* is absent.

Confusion also exists in regard to race and religion. Jews are often spoken of as a single race. In fact, however, their populations are derived from several gene pools. We have seen in Section 13.1 how varied the frequencies of the G6PD allele are among Israelis of different geographical origin. In general, allele frequencies in jewish populations tend to be similar to those among gentiles immediately surrounding them. Thus Yemenite jews are low in A and B and high in O, as are Yemenite arabs; and Cochin jews are reasonably high in A and B, as are Cochin hindus.

One notable exception to this generalization is the Roman ghetto. Between 1554 and 1870 it was virtually sealed off from the rest of Italy and little gene flow into or out of it occurred. As a result of this isolation, the blood-group frequencies of present-day Roman jews differ from those of the other populations of Italy. For instance, the percentage of the *B* allele among them is 27 percent but in no other populations on the Italian mainland is it above 11 percent.

Skin color is used by many as a handy guide to race. (The adaptive significance of skin color will be discussed further in Section 18.3.) However, odor may lay claim to being more diagnostic than color, with the members of european and african races possessing characteristic body smells to a much greater degree than

---

BOX 18.B DIFFERENCES IN ALLELE FREQUENCIES BETWEEN POPULATIONS

We give below some examples of differences in allele and phenotype frequencies, gathered from several sources, but primarily from the compilation by W. C. Boyd. The first important point to realize is that similarities between populations with respect to a single locus can be very misleading and that many loci are needed to diagnose relationship. For instance, if we look at the MN blood groups in Navajos and eskimos, we find:

| Allele | Frequency among | |
	Navajos	Eskimos
M	.917	.913
N	.083	.087

But, if we look at the ABO locus, the resemblance in frequencies in the two populations is no longer present:

A	.013	.333
B	0	.027
O	.987	.642

When the same comparisons are made between the English and the Australian aborigines, the data are as shown in the table at the top of the next column. Clearly, despite the similarity of frequencies at one locus, in neither example does the population belong to the same gene pool.

| Allele | Frequency among | |
	English	Aborigines
A	.250	.216
B	.050	.023
O	.692	.766
M	.524	.176
N	.476	.824

Second, the figures should be understood to represent particular samples of data; in some of these samples, the number of people tested was small and the data are thus subject to sampling error. In other words, different samples taken from the same population can vary.

Five different phenotypic traits have been selected to show differences in representative samples of different populations.

1. Diego blood group. This is a blood-group type controlled by a locus independent of that of the other blood groups. (Chapter 20 discusses these.) The dominant allele at this locus is Di^+. The use of Diego allele frequencies is a good example of the immunoanthropological approach (see Section 18.2). The frequency of Di^+ is low among the peoples of eastern Asia; the eskimos do not have it; it is rare in North American indians but common in some South American

do members of oriental races. It should be apparent by now that no one or two characteristics sufficiently define an ecotype, or satisfy Dobzhansky's genetic definition of race.

The question of how to classify individual people frequently confounded classical anthropologists, who depended on a checklist of morphological characteristics. Sometimes, it came down to a legal or cultural decision, which interestingly, the genetic view of race tends to support. As an example, the child of an african and a northwest european (Box 18.C) would be classed as a black in the United States. In Haiti, the same child would probably be included in Haiti's white minority race. In both cases, the decision is defensible within our genetic

tribes. Since American indians are generally believed to have come to this continent across Kamchatka and Alaska, this pattern of distribution suggests the operation of the founder effect, with the bearers of the Di^+ allele arriving in America in a migration preceding that of the eskimos.

2. High excretion of a naturally occurring amino acid, β-aminoisobutyric acid, or BAIB. There are constant differences between individuals in the amount of this substance excreted in the urine, and a high rate appears to depend on a single recessive gene.

3. Haptoglobin type. Haptoglobin is the hemoglobin-binding protein of plasma. Four phenotypes for the kind of haptoglobin present in humans have been identified. They apparently represent four different genotypes.

4. Ability to taste *phenylthiocarbamide*, PTC. This is a substance that to some people tastes exceedingly bitter (although other reactions have also been reported), while to others it has no taste at all. Nontasters are recessive homozygotes. The polymorphism at this locus seems to have some adaptive significance. PTC is related chemically to goitrogenic substances, and among patients with nodular goiter (a disease of the thyroid) the proportion of nontasters is higher than expected by chance. Similarly, PTC tasters have fewer caries of primary teeth than nontasters.

5. Fingerprint pattern. This is a highly heritable trait (see Box 11.C), although its mode of inheritance is not very well understood. It would be a misnomer to refer to the figures given in the last two columns of the table as allele frequencies (percentages not stated in these columns are for loops).

Population	Frequency of		Proportion of high BAIB secretors	Haptoglobin types				Proportion of PTC tasters	Fingerprint patterns	
	Di^+	Di^-		0	11	21	22		Percent arches	Percent whorls
Australian aborigines	0	1.00		0	.12	.68	.20	.27		
Chinese	.05	.95						.93	.03	.50
English	0	1.00	.09	.03	.10	.55	.32	.69	.07	.25
Eskimos	0	1.00	.23					.59		
North American indians	.02	.98	.59					.97	.05	.50
South American indians	.86	.14								
Japanese	.07	.93						.91		
American blacks			.29	.04	.26	.48	.21	.84		

BOX 18.C A GENETIC CLASSIFICATION OF HUMAN RACES

Anthropological classification of *Homo sapiens* into races on the bases of morphology and pigmentation is being replaced by a classification based on immunological and biochemical genetic differences (see Section 18.2 for some discussion of the new anthropology). A number of schemes, varying largely in the degree of subdivision proposed, have been devised. An example of one such scheme, the proposal of the Boston biochemist and immunologist W. C. Boyd, is given here, but without the details of the manner in which the blood-group allele frequencies used as the basis of classification differ among the different races. It should be clearly apparent that Boyd is a lumper: he proposes subdivision of all humanity into 13 races, whereas the anthropologist S. M. Garn, while recognizing nine major geographical races, lists 32 local races within them.

A. EUROPEAN GROUPS.

1. *Early europeans.* This is a hypothetical group, now largely extinct, represented by the Basques (see Figure 18.1) and possibly by the Berbers of North Africa.

2. *The lapps.* A small population but one sufficiently distinct from others to merit a separate position.

3. *Northwest europeans.* The ABO frequencies of this group are also shown in Figure 18.1.

4. *Eastern and central europeans.*

5. *Mediterraneans.* This race includes southern Europeans, the inhabitants of the Middle East, and many of the people of North Africa.

B. AFRICAN GROUPS.

6. *Africans.* This group includes the populations of black Africa, although there are many local subgroups that differ from each other.

7. *Asians.* In Asia, too, there are many local geographical subraces.

8. *Indo-dravidians.* The various sub- or *microgeographical* races dwelling on the Indian subcontinent are included in this group.

C. AMERICAN GROUP.

9. *American indians.* A very heterogeneous race of many isolates, including the eskimos.

D. PACIFIC GROUPS.

10. *Indonesians.*

11. *Melanesians.*

12. *Polynesians.*

13. *Australian aborigines.*

All of these geographically distinguishable races have distinct blood-group gene profiles.

---

definition of race. The child becomes part of a recognizable population, and there are generally some breeding barriers between this population and others. The fact that there are some unusual or "atypical" members of a population doesn't bother the definition a bit, as long as these unusual members are really biologically integrated in the population. If a population gets big enough, and different enough, we will probably start identifying it as a separate race, as has happened with American blacks (a mixture of european and african races), or the neo-hawaiians (a mixture of european, polynesian and oriental races). It is quite possible, by choice or by chance, that two full-sibs or even identical twins could find themselves classified as belonging to different races. Conversely, the definition allows a single active person to be a functioning member of more than one racial group, or population, which could include contributing genes to each.

It is interesting to review the attitudes of geneticists toward mating between races, and such a review has recently been provided by W. Provine, who teaches the History of Science at Cornell. He noted that early in the century, most geneticists (with a few notable exceptions) solemnly warned against matings between persons of widely different races, largely because of the expectation that "disharmonic" combinations would appear among the F_1 or (more likely) among the F_2. By the late 1920's and 1930's, most geneticists cautiously adopted an agnostic position on interracial matings. In response to a little biology and a lot of Hitler, most geneticists then swung to the position that interracial mating was not biologically dangerous, and might even be beneficial, in that new heterotic combinations might be produced. During this entire period, there were few or no controlled experiments, but there were plenty of "natural experiments" available for geneticists of all three persuasions to observe. Provine didn't find it surprising that little research was done on the effects of interracial matings, but did find it surprising that geneticists of all three eras were willing to make such positive statements about crosses between human races based on so little reliable evidence. It seems clear today that the offspring of interracial matings are both viable and fertile, and that the great variability that has been observed in the health and performance of such offspring is heavily influenced by social, economic, and cultural factors. We can draw on controlled experiments from plants and animals for support of the detrimental, neutral, or beneficial view, with a fourth voice— pleading uneasily for preservation of interesting gene pools—being this decade's new addition.

Enough has probably been said by now to clarify what the geneticists mean by "race." As a summary, excerpts from a statement prepared under the auspices of the United Nations Educational, Scientific, and Cultural Organization are given in Box 18.D.

18.2 BIOCHEMICAL ANTHROPOLOGY AND IMMUNOANTHROPOLOGY

There is no need here to discuss in detail the origin of human races. Essentially the same evolutionary forces—mutation, selection, migration and drift—that produce speciation (described in Section 4.3) act in race formation, with one major difference in their operation. In speciation, the process goes to completion (see the evolutionary tree in Figure 5.1). But among human populations, the relative shortness of the periods of partial isolation and repeated gene interchange prevented establishment of effective barriers to interbreeding. All human races are capable of crossing to produce fertile offspring. Because of this major difference, the process of race formation is more clearly diagrammed by a reticulum, or network, than by a tree such as that used to represent speciation. Some races are dissolved in the course of history by interbreeding and disappear or lose their identity, as did many during the successive waves of migrations to Europe from the East during the time when western civilization was beginning. Other races are of recent origin, as are the American black race, the South African "colored" population (which has genes from Bushman, Hottentot, Bantu, european, malaysian, and indian pools), and the neo-hawaiian complex of european, polynesian, chinese, japanese, and pilipino genes.

The historical reconstruction of race formation and dissolution is being greatly aided by the discoveries of increasing numbers of gene markers, such as serum

---

BOX 18.D UNESCO STATEMENT ON RACE

In July 1952, UNESCO issued a statement prepared by an international group of geneticists and anthropologists on the nature of race and race differences. In minor ways, it is now somewhat out of date, partly because of the great deal of new information on allele differences between populations obtained since then, and partly because of newly acquired understanding of the feedback between biological and cultural evolution. But the principles embodied in the statement are still valid. Listed below are verbatim extracts from the eight major points that were made.

1. Scientists are generally agreed that all men living today belong to a single species, *Homo sapiens,* and are derived from a common stock, even though there is some dispute as to when and how human groups diverged from this common stock.

2. Some of the physical differences between human groups are due to differences in hereditary constitution and some to differences in the environments in which they have been brought up.

3. National, religious, geographical, linguistic, and cultural groups do not necessarily coincide with racial groups.

4. Broadly speaking, individuals belonging to major groups of mankind are distinguishable by virtue of their physical characters, but individual members, or small groups, belonging to different races within the same major group are usually not so distinguishable.

5. Studies within a single race have shown that both innate capacity and environmental opportunity determine the results of tests of intelligence and temperament, though their relative importance is disputed.

6. The scientific material available to us at present does not justify the conclusion that inherited genetic differences are a major factor in producing the differences between the cultures and cultural achievement of different peoples or groups.

7. There is no evidence for the existence of the so-called "pure" races. In regard to race mixture, the evidence points to the fact that human hybridization has been going on for an indefinite but considerable time.

8. We wish to emphasize that equality of opportunity and equality in law in no way depend, as ethical principles, upon the assertion that human beings are in fact equal in endowment.

With the exception of point 6, we cannot take issue with the assertions made in the extracts above. But the example of the interaction between culture and the gene-pool content given in Section 7.4 does throw some question on the validity of the sixth point. A better formulation would emphasize the fact that genetically determined cultural differences do not warrant popularly held notions of racial superiority or inferiority; this point is in fact made in the complete statement, but in another context.

A similar report more specifically addressing itself to the status of the American black was issued in 1963 by a committee of the American Association for the Advancement of Science.

proteins and blood-group alleles. Instead of relying on various anthropometric measurements of traits, whose polygenic inheritance caused dissimilar genotypes to have similar phenotypes and thus obscured the biological relationships between populations, anthropologists now can use traits for which the allele frequencies can be computed from phenotypic frequencies. For instance, the study of the distribution of the ABO blood-group alleles in Western Europe depicted in Figure 18.1 has contributed to the solutions of some archaeological and linguistic puzzles. Thus, for a long time there has been unexplained evidence of a link between western Britain and the Mediterranean that dates back to Neolithic times. The similarity in frequency distributions of A and O between the two (compare the top lefthand and bottom righthand corners of the first two maps in Figure 18.1) suggests that populations of Mediterranean origin did inhabit Ireland and Scotland at one time, leaving their imprint in the present gene pool.

Similarly, the suggestion that the Hungarian gypsies originated in India—as they themselves have claimed, and as the linguistic connection between their language and Sanskrit has indicated—has been supported (although too much reliance should not be placed on a single locus) by the comparative distribution of the A and B blood-group phenotypes summarized in the following table:

| | Blood-group percentages | | | |
Population	O	A	B	AB
Hungarians	31	38	19	12
Hungarian gypsies	34	21	39	6
Indians	31	19	41	9

Even more exciting possibilities for historical research lie in the development of techniques of blood-typing material from the dead. Some blood-group substances, which are present in the blood cells, are found and are identifiable in hair, dried muscle and saliva, and most important for ancient and poorly preserved remains, in bone. This makes it possible to investigate the allele frequencies, not only of contemporary populations, but also of their ancestors and of extinct populations. Thus, when the techniques of *paleoserology* have been perfected, it may be possible to solve the identity of the Etruscans, the pre-Roman inhabitants of Italy whose origin has baffled historians for so long.

Another example bears on recent investigations of the royal Egyptian mummies that were thought to be the remains of Tutankhamen and Akhenaten (Amenophis IV). The investigations indicate that the "mummy of Akhenaten" is more likely the mummy of Smenkhkare, successor to Akhenaten as Pharaoh of Egypt in the eighteenth dynasty, and brother of Tutankhamen. Both mummies contain blood group A_2 (a subgroup of A) and both are MN. While this hardly proves that the two mummies are the remains of sibs, the blood-group evidence is consistent with that hypothesis.

A recent problem, which we hope will be corrected, is an understandable

FIGURE 18.1
Distribution of alleles for the ABO blood-group locus. Maps showing isogen lines in Western Europe (*isogens* are lines placed on a map to delimit areas of the same allele frequency, as an isobar does for barometric pressure). On the left, *A* has a frequency of <25 percent in the dark shaded areas and >25 percent in the light ones. In the middle *O* has a frequency of >70 percent in the dark and <70 percent in the light areas. On the right, the frequency of *B* is <5 percent in the dark areas and >5 percent in the light areas. (Adapted from G. deBeer.)

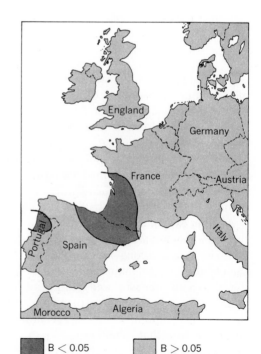

reaction to the previous racist practice of maintaining blood in blood banks segregated by race. The response was to prohibit any note of racial origin in the blood-bank records, thus moving from injustice to irrationality. (At least it seems irrational to some of the geneticists and anthropologists who would like to use this rich source of information.) Data from large samples, which can be efficiently gathered from blood-bank analyses of the blood groups and serum proteins during routine blood typing, can be computer analyzed to construct groupings such as the one shown in Box 18.E. As more and more biochemical polymorphisms are discovered and genetically understood, such statistical reconstruction of present-day similarities and differences between existing populations will be increasingly valid.

18.3 ADAPTATION

As races form, the differences that develop in their characteristics, other than those established by drift or recent immigration, are due to natural selection. Some possible examples of adaptive values of blood-group and vision alleles have already been alluded to, but racial differences exist in many other traits. Our information on most of these traits is very incomplete.

We know something about the various environmental conditions under which different races of humans developed and live, and physiological investigations about the nature of adaptation to these conditions are possible. Much is already known about adaptations of animals to climate. For instance, we are aware of the frequent occurrence of smaller body size and longer extremities in warmer regions; both characteristics are clearly related to conservation and dissipation of heat. Whether peoples who invented clothing and housing as artificial homeostatic devices would be subject to selection for body size and proportions in the same way as other mammals is by no means clear. Yet ratios of stature to weight appear to increase in human populations as one proceeds from cold to warm climates (Figure 18.2). For example, the Tusi, the ruling class of cattle-grazers of the Runid people in East Africa, live next to the Pygmy rain-forest hunters who are shorter on the average by slightly more than a foot; but height-weight ratios in both groups are considerably higher than in any European group.

Without clothing, and with only small fires for warmth, humans *can* withstand air temperatures below freezing. It is of interest to note that among 716 cases of frostbite suffered by U.S. soldiers during the Korean War, blacks were six times as prone to this injury as whites; blacks also tended to have the more severe cases. (However, some have suggested that during this period of military desegregation, newly integrated black soldiers were given an unfair share of sentry duty in cold weather.) The Korean data correlate well with ice-bath experiments in which eskimos and American indians had the least heat loss, whites an intermediate amount, and blacks the most heat loss. (It is obvious that human survival in the Arctic or in any areas that have sustained cold weather had to await the development of structures and clothing.)

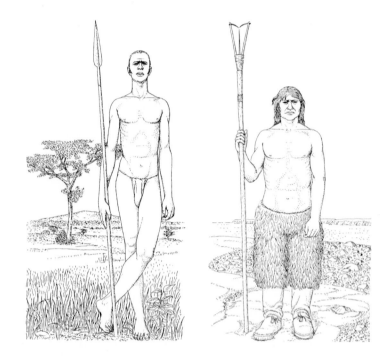

FIGURE 18.2
Human adaptation to climate is typified by nilotic negro of the Sudan (*left*) and arctic eskimo (*right*). Greater body surface of the negro facilitates dissipation of unneeded body heat; proportionately greater bulk of the eskimo conserves body heat. (From W. W. Howells, "The Distribution of Man." Copyright © 1960 by Scientific American, Inc. All rights reserved.)

BOX 18.E HUMAN PHYLOGENIES

An early attempt by L. L. Cavalli-Sforza and A. W. F. Edwards to apply the techniques of *cladistics* to human data is shown below. This particular cladogram is based on allele-frequency data from only 5 blood-group loci. Long horizontal lines indicate many differences, and short horizontal lines indicate few differences. Thus, Ghanians and Bantus have few differences, and are more similar to each other than to the next-most similar group, the Ethiopians. The Ethiopians probably diverged from the Bantus and Ghanians before they diverged from each other, and the Ethiopians have changed little at these loci since that divergence. It is the horizontal, and not the vertical, dimension that provides information in a cladogram, as the computer arbitrarily lists population order at a divergence point. Thus, by pivoting the top and bottom halves of the clado-gram, the Australian aborigines and Ghanians could as easily be listed next to each other in the middle instead of listed first and last, but the sum of horizontal lines between their point of separation would be the same.

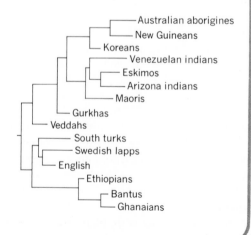

Toleration of extreme heat and altitude, however, is increased only by appropriate physiological and morphological adjustments. Culture does not help much. The shape of nose, for instance, is correlated with different climatic factors. Nasal breadth is highly correlated with mean annual rainfall and mean hottest-month temperature, while nasal height is mostly correlated with mean coldest-month humidity.

It seems clear that tropical negroes do not do well in extreme cold and may not flourish at high altitude, but are well equipped to tolerate heat, particularly the moist variety. Mongoloids seem best equipped to tolerate cold, and it is of interest that the high-altitude areas of the Himalayas and Andes are occupied by sub-races of the mongoloid race. Mongoloids also thrive in the tropics of both hemispheres, and as a whole seem to be the closest to an "all-purpose" race (although that is quite different from saying any one mongoloid would do well in all these conditions). Whites, who largely occupy the more fertile temperate lands, do well in moderate cold and desert heat, but have a history of difficulty in the moist tropics.

Which particular foods are eaten, except in a few societies, is largely a recent product of culture (Section 15.2). The adaptations of local populations to particular foods would be mostly the result of only a few millennia of evolution. The example of lactose intolerance is instructive. U.S. agencies slowly became aware of the problem during some of their foreign-aid programs, when powdered milk was used by the recipients to whitewash sheds, or dumped out amid muttered rumors of CIA-sponsored mass poisoning. Cow's milk contains 4–5 percent lactose (human milk is over 7 percent), and the dried or condensed milk sent to food-short people may contain from 15–38 percent lactose. The enzyme lactase is needed to break lactose down to its component sugars, glucose and galactose. Nearly all mammals, including humans, produce an adequate level of this enzyme at birth. But typically, during development the levels of this enzyme drop, perhaps as a useful mechanism to prevent adults (who can eat other foods) from competing with infants for food. Most adult humans in the world are lactose intolerant. Percentages of intolerance in various groups thus far tested are:

U.S. caucasian 2–19	U.S. blacks 70–77	U.S. orientals 95–100
Finnish 18	Ibos 99	Thais 99
Swiss 12	Bantu 90	Eskimos 88
Swedish 4	Australian aborigines 85	American indians 58–67

One of the exceptions to the above pattern is provided by a comparison of the Yoruba in western Nigeria, the Ibo in the east, and the Fulani in the north. The Yoruba and Ibo were hunter-gatherers who became farmers. They settled in regions infested with tsetse fly, and thus never acquired cattle. The Fulani have been pastoral people for thousands of years, and many are still nomadic, herding their cattle from one grazing ground to another. The Yorubas and Ibos are each about 99 percent lactose intolerant, while only 22 percent of the Fulani are intolerant.

Many cultures use milk for food, but in modified form. Thus, cheese and butter each have less than one percent lactose, and many fermented milk products (kemiss, kefir, matzoni, yogurt, nono) have less than 2.5 percent lactose, allowing people with low lactase activity to eat them with little discomfort. Most American blacks' African ancestors were Yoruba, or Ibo, or similar West Africans with little lactose tolerance, and thus the increased tolerance of American blacks is probably due to northern european genes, plus possibly some adaptation (physiological and genetic) to the diet of this culture.

In the United States, we recently had an advertising campaign that proclaimed

EVERY BODY NEEDS MILK

This was then changed to the milder

MILK HAS SOMETHING FOR EVERY BODY

For tens of millions of Americans who are past infancy, and for most of the people of the world, drinking milk produces the symptoms of any fermentative diarrhea: a bloated feeling, flatulence, belching, cramps, and a watery explosive diarrhea.

Many other morphological and physiological characteristics, such as the number of sweat glands, the amount of facial hair (relative beardlessness might be an advantage to the eskimo because of icicles that would form on the beard), and the amount of body hair have been speculated about, but few firm conclusions have been drawn. Perhaps the characteristic that has received most attention is skin pigmentation.

Vitamin D (which mediates the absorption of calcium from the intestine and deposition of minerals in growing bone) is produced from a precursor activated by sunlight penetrating the human skin (in particular, by the ultraviolet portion of sunlight). Vitamin D is not present in most (unsupplemented) diets, occurring in significant amounts only in the liver oils of a few fish. Too little vitamin D results in the bowlegs, knock-knees, and twisted spines associated with rickets in young children, and in the softening of bones associated with other difficulties in older people. Too much can be toxic or lead to calcification of soft tissues.

Human skin has two adaptive mechanisms for regulating the penetration of ultraviolet light, the production of either black melanin granules or yellow keratin disks. Tropical peoples are generally dark-skinned, and people from high latitudes light-skinned (eskimos, who are dark-skinned, eat a lot of fish). The dark skin color produced by melanin or keratin may have other functions (although reflecting heat is not one—light colors would normally reflect heat better), but a principal function appears to be allowing some ultraviolet light to penetrate, but screening out enough so that overdosing with internally produced vitamin D is prevented. At higher latitudes, many mothers customarily put their infants out of doors for "fresh air and sunshine" even in the middle of winter. In almost all races, children are lighter-skinned than adults, which is sensible, since children need more calcium for bone growth than do adults. Seasonal responses are taken care of by suntanning, and it is significant that both the keratinization and melanization components of suntan are initiated by the same wavelengths that synthesize vitamin D. Sudden massive doses of vitamin D are minimized by the alarm system called sunburn, which keeps untanned individuals suddenly encountering abundant sunlight from overexposing. It is easy to see how these mechanisms would have evolved, as humans migrated to different climates. Individuals whose vitamin D synthesis gave them healthy bones would have a selective advantage in many ways over those having abnormal bones because of either too much or too little calcium.

18.4 THE AMERICAN BLACK

Of problems in the biology and sociology of race, those of the American black are among the most serious. Because the widespread social effects of these problems contribute so greatly to the socioeconomic and moral crises in the United States today, it seems important that all segments of our population be informed. Yet, strange as it seems, many recent textbooks on human genetics, written by well-informed and probably well-meaning biologists, don't have a single entry on the American negro or American black in their indexes.

The American black racial group is a recent one. The following compilation, made from shipping manifests of the British slave trade and the port of Charleston, South Carolina (the main U.S. port of entry), indicates its multipopulational origin—the pooling of peoples from many African races:

Ports of Origin	Regions raided	Peoples	Percentage of slave population
Senegal, Gambia	Senegal, Gambia	mainly from interior	13
Sierra Leone	Sierra Leone, Portuguese & French Guinea	coastal and interior	6
Windward Coast	Liberia, Ivory Coast	various from area	11
Gold Coast	Ghana	about $\frac{3}{4}$ southern, $\frac{1}{4}$ northern	16
Bight of Benin	Togo, Dahomey, west Nigeria	various from area	4
Bight of Biafra	east Nigeria, Cameroons	about $\frac{3}{4}$ Ibo, rest from area	23
Angola	Gabon, Congo, Angola	many peoples, from coast to far inland	24
Mozambique & Madagascar	Mozambique & Madagascar	from area	2
Region unknown			<1

Importation of slaves began in 1619, and essentially halted in 1808 (this did not halt legal and some illegal immigration from Africa and several intermediate stops). There were matings of slaves with caucasians and American indians, but few records were kept, and thus an estimation of indian, caucasion, and african intermixture cannot be based on historical records. The degree of intermixture is better based on calculations of allele frequencies of blood groups and other genetic markers.

Although these computations are subject to considerable uncertainty due to variability in the contributing populations, possible selection operating to change allele frequency, and considerable heterogeneity within the broadly defined American black population, the following generalizations are emerging. About 20 percent of the autosomal genes in the current American black population are of european origin, and the American indian contribution is negligible. Estimates of percentages of european genes in northern black populations are typically much higher than those for southern black populations, ranging from near 30 percent in the north to as low as 3.6 percent among the "Gullah" of South Carolina. Some peculiar isolates have developed, such as the "Wesorts" of Maryland and the "Alabama Cajuns," both starting from mixtures of American indian, african, and european genes. These groups, although of hybrid origin, are now highly inbred, with many of the deformities, disabilities, and diseases typical of inbred populations. For years, they could not marry whites, and would not marry blacks, so they have married each other, cousin to cousin, generation after generation. One would hope that in time, with legal and social constraints eased, such isolates would be biologically integrated into the main human population of the United States. But it is not inevitable that this will come to pass: In Japan, there is a people called the *buraku-min* who have suffered economic and social discrimination for fifteen generations, although they are defined as a group neither by having a particular color of skin nor by any other major distinguishing physical features; since 1871, the discrimination has been illegal, but custom and social position have maintained the oppression.

We have already mentioned other new hybrid races in process of formation. One of interest is in Brazil, drawing on the same three major races as the American black. The Brazilian population is estimated to contain about 30 percent african, 11 percent American indian, and 59 percent european genes. Not only does this hybrid race have different proportions of contributing major races from those of American blacks, but the contributing subraces of American indians were certainly different, and those of the african and european races were probably significantly different. The contribution of african genes to the hybrid North American white race is also of interest, and is estimated at present as being in the neighborhood of one percent.

If random mating were to be instituted for the whole of the population of the United States, we might expect a distribution of skin color as shown in Figure 18.3, with the equilibrium average pigmentation after a thousand years of random mating being just a shade darker than that of today's average white.

In fact, however, random mating does not occur and is not immediately likely to occur across color barriers. Among the blacks themselves, there is positive assortment that is partly related to considerations of social status (see Chapter 21 for a more general discussion of human mating systems). Despite recent

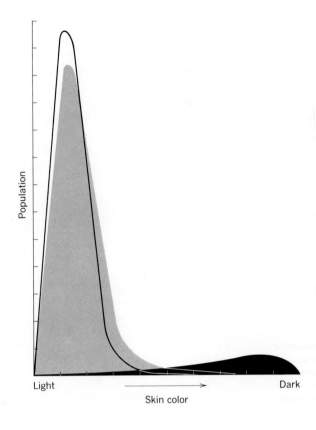

FIGURE 18.3
The area under the curve farthest to the left
represents the distribution of skin color
among present-day whites; the black area,
the distribution of skin color among present-
day blacks; the grey area, the equilibrium
distribution expected after a millenium of
random mating. (From Curt Stern, *Principles
of Human Genetics,* 2nd Ed. W. H. Freeman
and Company. Copyright © 1960.)

publicity on individual racial intermarriages, the incidence of marriage between whites and blacks is still low, and seems to be less than half the rate of black-white matings per generation that prevailed prior to the Civil War and the freeing of the slaves. Much of this early mating was probably between white men and black women, which leads to the probability that the frequency of european-origin Y chromosomes is higher than that of european autosomes in the black gene pool, and that of european-origin X chromosomes would be correspondingly lower. Our data are not yet good enough to verify this.

The pattern of future matings between Americans is, of course, not predictable. The effects of expanding social consciousness, and of education about the biological nature of races, about the absence of pure races, and about the lack of foundation for the belief that interracial crosses have harmful effects may lead toward randomness. But the rise of the blacks as a power group may have a restraining effect on the black attitude toward interracial mating. There exist other intangibles that are related to the blacks' fear of loss of identity among lighter-skinned persons, and to the social and psychological difficulties experienced by offspring of interracial marriage. In any case, the biological issues are necessarily long-range because of the 25- to 30-year interval between generations. All of these considerations may be modified if it is indeed eventually possible for genetic engineering to change the genetic instructions for skin color. What is urgent, however, is that the biological truths about races be introduced into the American mores so that they may influence our views and actions on racial equality and social justice.

That valid judgments about the biological significance of differences in tests of mental abilities are difficult has already been stressed in preceeding sections. Various proposals have been advanced by probably well-intentioned people that suggest how meaningful investigation of genetic rather than phenotypic differences in intelligence and achievement might be carried out. Some of these individuals suffer from the obstinate inability to see the methodological difficulties and inherent biases of their schemes. Some anthropologists and others even

---

BOX 18.F POPULATIONS ANALYZED BY LEWONTIN

The races and populations listed below are presented for your own careful analysis. You probably won't agree with Lewontin in all particulars. The list indicates the diversity of human populations that exist, and were available for this analysis. It is by no means an exhaustive list.

CAUCASIANS

Arabs, Armenians, Austrians, Basques, Belgians, Bulgarians, Czechs, Danes, Dutch, Egyptians, English, Estonians, Finns, French, Georgians, Germans, Greeks, Gypsies, Hungarians, Icelanders, Indians (Hindi speaking), Italians, Irani, Norwegians, Oriental Jews, Pakistani (Urdu-speakers), Poles, Portugese, Russians, Spaniards, Swedes, Swiss, Syrians, Tristan da Cunhans, Welsh

BLACK AFRICANS

Abyssinians (Amharas), Bantu, Barundi, Batutsi, Bushmen, Congolese, Ewe, Fulani, Gambians, Ghanaians, Hobe, Hottentot, Hututu, Ibo, Iraqi, Kenyans, Kikuyu, Liberians, Luo, Madagascans, Mozambiquans, Msutu, Nigerians, Pygmies, Sengalese, Shona, Somalis, Sudanese, Tanganyikans, Tutsi, Ugandans, U.S. Blacks, "West Africans," Xosa, Zulu

MONGOLOIDS

Ainu, Bhutanese, Bogobos, Bruneians, Buriats, Chinese, Dyaks, Pilipinos, Ghashgai, Indonesians, Japanese, Javanese, Kirghiz, Koreans, Lapps, Malayans, Senoy, Siamese, Taiwanese, Tatars, Thais, Turks

SOUTH ASIAN ABORIGINES

Andamanese, Badagas, Chenchu, Irula, Marathas, Naiars, Oraons, Onge, Tamils, Todas

AMERINDS

Alacaluf, Aleuts, Apache, Atacameños, "Athabascans," Ayamara, Bororo, Blackfeet, Bloods, "Brazilian Indians," Chippewa, Caingang, Choco, Coushatta, Cuna, Diegueños, Eskimo, Flathead, Huasteco, Huichol, Ica, Kwakiutl, Labradors, Lacandon, Mapuche, Maya, "Mexican Indians," Navaho, Nez Percé, Paez, Pehuenches, Pueblo, Quechua, Seminole, Shoshone, Toba, Utes, "Venezuelan Indians," Xavante, Yanomama

OCEANIANS

Admiralty Islanders, Caroline Islanders, Easter Islanders, Ellice Islanders, Fijians, Gilbertese, Guamians, Hawaiians, Kapingas, Maori, Marshallese, Melanauans, "Melanesians," "Micronesians," New Britons, New Caledonians, New Hebrideans, Palauans, Papuans, "Polynesians," Saipanese, Samoans, Solomon Islanders, Tongans, Trukese, Yapese

AUSTRALIAN ABORIGINES

FROM R. C. LEWONTIN IN TH. DOBZHANSKY ET AL. (EDS.) EVOLUTIONARY BIOLOGY, VOL. 6, APPLETON, CENTURY, CROFTS—BY PERMISSION OF THE PLENUM PUBLISHING CORPORATION.

opine that such studies are irrelevant or too vulnerable to misinterpretation and too fraught with political danger to be undertaken. This may or may not be true, but it is likely that generations of discrimination have made direct comparisons of mental traits between blacks and whites confounded by the psychological components of race.

A look at recent life expectancy figures is relevant here. In the United States in 1900, for a white male the life expectancy at birth was 48 years; for a black male it was 32 years. In 1940 and in 1958, the respective figures were 62 years and 67 years for whites; and 52 years and 61 years for blacks. To assume that the differences in the environments that made life expectancy in 1900 33 percent lower for blacks had no effect on intellectual performance in life, or on a psychological test, would seem to be an utter folly. Furthermore, the dramatic changes in half a century must be attributed largely to environmental causes, since not enough time has elapsed to permit operation of genetic causes of the prolongation of the life-span.

The first section of this chapter included a quotation from the geneticist Dobzhansky, and it may be appropriate to conclude this section with one from the anthropologist S. L. Washburn, who is also the source of the data given in the preceding paragraph. In his presidential address to the American Anthropological Association, he said,

> I am sometimes surprised to hear it stated that if Negroes were given an equal opportunity, their IQ would be the same as the whites'. If one looks at the degree of social discrimination against Negroes and their lack of education, and also takes into account the tremendous amount of overlapping between the observed IQ's of both, one can make an equally good case that, given a comparable chance to that of the whites, their IQ's would test out ahead. Of course, it would be absolutely unimportant in a democratic society if this were to be true, because the vast majority of individuals of both groups would be of comparable intelligence, whatever the mean of these intelligence tests would show.

Whether one agrees or disagrees with the first part of the statement, the last sentence summarizes the moral of the preceding discussion.

18.5 THE RELATIVE IMPORTANCE OF RACE

As experimental data have come in from a great variety of species, a certain generalization has become first possible, and then compelling: there is an enormous amount of genetic variability maintained *within* populations. R. C. Lewontin has recently (1972) completed an analysis of the distribution of variability between major human races, between subraces within the major races, and within the subraces (Box 18.F). This analysis was based on available data for nine polymorphic blood groups and eight polymorphic serum proteins and red-blood-cell enzymes.

Lewontin agonized over the racial classifications he should use, and the relative weights he should use. Should the huge population of Japan be given the same weight as the Yanomama tribes of the Orinoco? He decided that it should, and in addition, to include roughly equal numbers of African, European, Oceanian, Asian, and native American people in the analyses. The effect of the first decision is to increase the amount of apparent variability among populations within major races, and the effect of the second is to increase the amount of apparent variability between the major races.

The results were a surprise to almost everyone. The proportion of total variability between major races varied between 0.2 percent (for the Lewis blood group) and 25.9 percent (for the Duffy blood group), with an average of 6.3 percent for the fourteen data sets that could be analyzed (three lipoproteins could not be analyzed at this level). The proportion of total variability among subraces, or populations, within the major races varied between 2.1 percent (for adenylate kinase) and 21.4 percent (for the Lutheran blood group), with an average of 8.3 percent. The genetic variability within populations varied from 63.6 percent of the total variability (for the Duffy blood group) to 99.7 percent (for the Xm lipoprotein), with an average of 85.4 percent.

Thus, on the average, only 15 percent of human genetic diversity in these variable systems is accounted for by differences between large human groups. Furthermore, over half of that is differences between populations and subraces, with only about 6 percent of the variation occurring between major races. Why is that so surprising to most of us? Lewontin suggests, and we agree, that the racial classifications which have been erected are based on a biased sample of characters (nose, lip, and eye shapes, skin color, hair) that are useful in distinguishing groups, as well as individuals within those groups. The characters Lewontin studied measure essentially genetic, rather than phenotypic, variability, and are a much less biased group of characters relative to any subconscious preselection for patterns of distribution.

Thus, while it appears from this analysis that racial variability is significant, and even important (Lewontin concludes otherwise, but we feel that 6 percent, or 15 percent, of anything of importance is worth careful study, be it taxes or human variation), it is also clear that most genetic variability is within the races and populations of those races. This emphasizes once again that the individual should be the unit of interest and judgment, and even individuals within a single family can differ from each other enormously.

Lewontin's analyses were essentially free of environmental components of variation. Indeed, we know little about these. It may be that their distribution between and within races is quite different from the genetic components. If we again call on our experience with other species, the environmentally caused variability associated with data collected from different locations or under different conditions is frequently much greater than the variability associated with environmental variation within a replication of an experiment. But if racial judgments and classifications are to be made at all, they should be based on the genetic, and not the environmental, differences between the groups. We think our attention will most profitably be turned to variability among people within races.

19

FORMAL
HUMAN GENETICS

Humans are a difficult but fascinating species to study genetically. Many of the studies draw their data from medical records, and thus many genetically known traits are of medical importance. A new field, genetic counseling, is being developed to put some of this knowledge to use.

19.1 HISTORICAL

Although genetics as a science is considered to have been born with the rediscovery of Mendel's laws in 1900, speculations about human inheritance trace back to the very beginnings of our civilization. Plato's proposal for a eugenic program, and the apparent understanding of the mode of inheritance of hemophilia by the writers of the Talmud 1,500–1,600 years before the disease was formally described as hereditary, have already been mentioned. In 1752, the French astronomer and philosopher Maupertuis, one of Mendel's great predecessors, published a four-generation pedigree showing that a number of the members of a certain family were born with extra digits and a computation demonstrating that such a collection of developmental errors among relatives was not likely to have been a chance event. Familial incidence of color blindness was described in 1777. In 1814, Joseph Adams, a British physician and apothecary, published a book in which he not only recognized diseases with dominant and recessive bases of inheritance, but also acknowledged the existence of genotype-environment interaction. He also discussed the increased incidence of defects in marriages between relatives, which today is known to be due to the increased probability of the union of gamytes carrying the same deleterious recessive allele.

Galton's pre-Mendelian contributions to the study of human inheritance have already been discussed (Box 10.B). In 1901, **Karl Landsteiner** (1868–1944) discovered the ABO blood groups, although it took another quarter century before the multiple-allelic basis of their inheritance was properly understood. Landsteiner, with collaborators, also discovered the MN blood-group locus in 1927, and the Rhesus factor (see Chapter 20) in 1940.

In 1901, the British physician **Archibald E. Garrod** (1858–1936; Figure 19.1), described eleven cases of alkaptonuria (Box 6.F), and noted that at least three of the patients were offspring of consaguineous matings. A year later, taking a clue from Bateson, he identified this **inborn error of metabolism** as being due to a single recessive gene. He extended the concept of hereditary enzymatic deficiency to other diseases in his book *"The Inborn Errors of Metabolism"* published in 1908. Garrod thus is the founder not only of modern human genetics, but also of biochemical genetics, which, however, did not begin its rapid development until after his death.

In the half-century following Garrod's initial discovery, many contributions to medical and formal human genetics were made that were largely based on pedigree analysis and population genetics. But only in the last two decades did the great advances that we have been discussing come about. They are due to a

FIGURE 19.1
Sir Archibald Garrod.
(Courtesy of the Royal Society.)

number of technical breakthroughs in several disciplines, and many of them were based on research seemingly not directly related to human welfare. Mission-oriented research, that is, research directed towards the solution of some specific applied problem (for example, prevention and cure of cancer), cannot be efficient or for that matter, successful, unless it feeds on basic research. The jibes that often appear in the press—frequently supplied by economy-minded congressmen —about the worth of some of the basic research for which Federal support is sought, are often based on misinformation and lack of understanding of the nature of scientific endeavor. To have ridiculed Mendel's studies on garden peas as being useless would have been all too easy, but without Mendelism there would be no medical genetics.

Among the recent breakthroughs of most significance to human genetics were those of molecular and biochemical genetics. The development of cytogenetic techniques of handling human chromosomes is another. The designing of computer techniques and information retrieval machinery formed another frontier of advance in studies of demographic and population genetics. Immunogenetics and studies on graft and organ transplant tolerance are still other areas of current significance. And to repeat one mentioned earlier, the discovery of methods of maintaining human cells in culture is making a great impact on our understanding of biological processes.

19.2 EXEMPLARY

A recently published catalogue of human genes that are definitely known or reasonably postulated describes 943 dominant, 783 recessive, and 150 sex-linked phenotypes. If polygenically determined traits are added to these, the totality of characters investigated genetically becomes an even more formidable number. A comprehensive survey of human genetics is, of course, not within the province of this book, and only a sampling to illustrate specific points can be presented here. Molecular and biochemical examples of the operation and effects of human genes have already been given; inherited morphological, physiological, and behavioral traits have also been described. Chapter 20 contains more information about such characters as blood groups. The purpose of this section is merely to extend the variety of examples for a broader appreciation of the fact that genetic components are ubiquitous, and that they effect the full range of human attributes.

Our understanding of the genetics and biochemistry of metabolic variation is increasing: every year many more electrophoretic variants of human enzymes become known. Some of the enzymes are found in red blood cells, others in the serum, placenta, liver, and other tissues. Each is responsible for mediating some specific metabolic reaction, such as is exemplified by the several blocks in the metabolic pathway of phenylalanine (Box 6.F). Cognate variants of these enzymes have been identified in many other species. For example, different esterases (enzymes that accelerate chemical reactions of the ester class of substances) have been discovered in rabbits, mice, birds, newts, drosophila, pines, and corn.

Many hereditary metabolic diseases associated with mental deficiency have now been described, and for some of them, the specific enzyme that is deficient has been identified. In addition to those that have already been discussed, we may mention the maple syrup urine disease, which is controlled by an autosomal recessive gene and so named because the urine of affected individuals smells like maple syrup. Both physical and mental retardation accompany the disease. Apparently a block in the metabolism of some essential amino acid causes excretion of the odoriferous substance.

Less harmful are two other genes determining breakdown of certain ingested substances. It has been postulated that a single dominant allele controls the excretion of methanethiol after ingestion of asparagus. Its frequency in an English population is estimated at 0.23. The strong odor characteristic of the urine of carriers after they have eaten asparagus is detectable even if they eat as little as three or four stalks. Persons who are homozygous for the recessive allele at this locus can eat as much as a pound of asparagus without displaying the trait. The other gene, which also appears to be harmless, specifies the breakdown of the red pigment of beets. The enzyme betamin, produced by the dominant allele, breaks down the pigment betacyanin, but a recessive allele (which has a frequency of 0.31 in an English sample tested) does not, causing the urine of recessive homozygotes to be red after they eat moderate amounts of beets. (However, people with iron deficiency may also excrete the pigment from beets, even though they have the dominant allele.)

Differences in ability to taste, which are likely to have a hereditary basis, are found not only with respect to phenylthiocarbamide (PTC), referred to in Box 18.C, but also for a variety of other substances. In a Swedish experiment, a number of subjects were asked to describe the taste of various chemical compounds. The following table shows a remarkable variation among people in the taste of the same substance, or at least in the way they describe it. The numbers indicate how many persons reported the taste sensations described by the column headings:

| | Predominant taste quality | | | | |
Compound	Salty	Sour	Bitter	Sweet	Other
Salt	249	2	11	18	29
Hydrochloric acid	7	243	10	8	24
Quinine	2	3	226	48	47
Strychnine	2	0	180	43	55
Sugar	0	3	21	251	19

The American biochemist R. J. Williams has investigated taste sensitivity for a number of substances and found that each subject has a different "taste profile." As Figure 19.2 shows, however, identical twins were very much alike.

Differences of this type make each of us biochemically unique, with different dietary preferences, different dietary needs, and different excretion of substances,

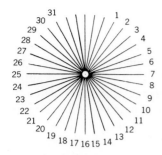

FIGURE 19.2
Biochemical individuality as expressed by taste sensitivity
to 31 different substances. The length of lines above **A**
represents average sensitivity of all persons tested. In **B-E,**
relative lengths of the lines indicate how sensitivities of four
persons—among whom **D** and **E** are identical twins—deviate
from these averages. (From R. J. Williams, "Introduction,
General Discussion, Tentative Conclusions." *Biochemical
Institute Studies IV,* University of Texas Press.)

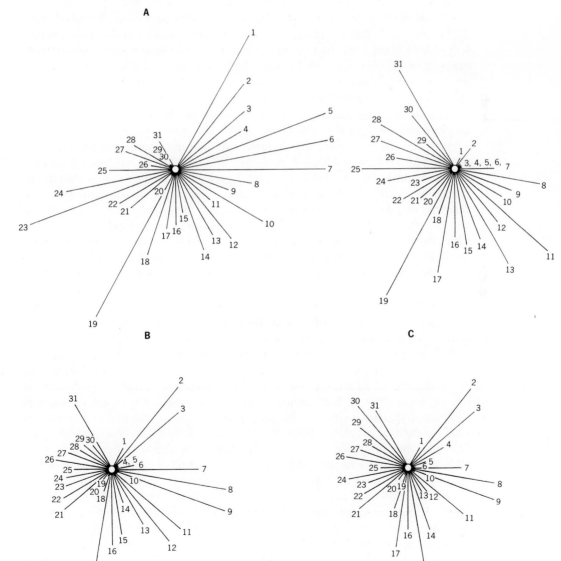

even if we are not aware of them. Not only does our sense of smell vary, but our body smells are different, and apparently under genetic control. Dogs, having a more sensitive perception of odors than humans, have been reported to be confused when trying to identify one or another of identical twins (Box 11.C). Once more, such differences point to the absurdity of the typological approach with ideals or archetypes in categorizing human beings. If the average for some variable trait is taken as the norm, then every human being would be abnormal when only a few characters are considered together. Even if we allow the variability for each character to include 95 percent of the population (that is, only 5 percent would be considered abnormal), and if we then consider a combination of 15 traits, less than one percent of the population would be normal for all of them.

The genetics of hearing has not yet received the attention it deserves, particularly in view of the importance of music in most human cultures. But interestingly, ear wax has been studied. It appears that there are two main types, wet wax produced by a dominant allele W, and dry wax produced by the homozygous recessive ww. It is remarkable that the entire range of allele frequencies is found in various human populations, the extremes for the W allele being 0.02 among northern chinese, and 0.93 among American blacks.

For the majority of inherited characters studied in humans, the primary gene products affected are not known. It is often difficult even to imagine the kinds of differences in biochemical developmental pathways—starting with a mutational change in a polypeptide chain—that produce phenotypic variation in such characters as the ability to fold or roll one's tongue, the normal pattern of clasping one's own hand (right or left thumb uppermost), tone deafness, placement of voice with respect to normal speaking pitch, and any number of other characteristics for which a hereditary basis has been postulated. As a matter of fact, none of the traits above has been unequivocally proved to depend on simply expressed single gene differences. Nevertheless, from all we know, they depend on differences in biochemical pathways, and it is the task of developmental genetics to investigate them. In organisms like drosophila, the enzymatic basis of such developmental traits as failure of head emergence is now being studied.

With respect to well-established human mutations that express themselves in morphological or physiological changes, a great variety of abnormalities may be mentioned in addition to those we have already considered. They include, among many others, short digits, absence of canine teeth or lateral incisors, cataracts, hypersensitivity to cold, progressive deafness, absence of fingerprints, malformation of nipples, a form of muscular dystrophy, webbed fingers, white forelocks, and prognathous jaw (the Hapsburg lip, illustrated in Figure 19.3), all inherited as dominants, although some also have recessive forms. Other recessive genes known to determine abnormalities include those for a type of anemia, hairlessness, congenital paralysis of cranial nerves, glaucoma, hydrocephaly, and hereditary goiter accompanied by deafness. Among sex-linked deleterious genes that have not been so far mentioned are those for congenital cataracts, absence of central incisors, progressive and congenital deafness, and scaly skin (for others, see Section 8.4).

FIGURE 19.3

The expression of the dominant prognathous jaw, the *Hapsburg lip*, over four centuries.
A: Emperor Maximilian I (1459–1519). **B:** Emperor Charles V (1500–1558), grandson of **A.**
C: Archduke Charles (1771–1847). **D:** Archduke Albrecht (1817–1895), son of **C.** (From
Strohmayer, *Nova Acta Leopoldina*, 5, 1937.)

19.3 FOUNDER EFFECTS

Founder effects, a special type of genetic drift (Section 4.3), may be expected to have influenced the genetic composition of any population begun by just a few breeding individuals (about ten pairs or fewer). These effects may also be important in continuously large populations, if one or a few individuals contribute very large proportions of the offspring that become the succeeding generations. The royal families of Europe (which may be thought of as a deme) had, for a number of generations after Queen Victoria, a high frequency of the gene for hemophilia (Box 16.C). However, with the passing of so many of the royal houses and the rapidly changing pattern of mating, with an increasing number of royalty-commoner marriages, this group is no longer even a partial isolate.

Another well-documented example of a founder effect—in an isolate that has similarly disappeared—is for the dominant gene causing night blindness. Over a period of 11 generations, some 135 descendents (out of several hundred) of a Provençal butcher, who settled in a small village near Montpellier, were discovered to suffer from the defect. These findings were published in 1907, but when a follow-up study was attempted some forty years later, his descendents were found to have dispersed from the village, destroying the partial isolate of several centuries' duration. No doubt, the defective allele is now scattered throughout various parts of France.

Another dominant gene that has been thoroughly investigated is that for *porphyria variegata*, the failure to metabolize porphyrin, a derivative of a respiratory pigment connected with hemoglobin. This defect, diagnosed by the presence of porphyrin in the stool, causes formation of brown skin patches and sensitivity to barbiturates, produces abdominal pains, neurotic symptoms and can lead to paralysis and death. In South Africa, the ancestry of some 8,000 carriers of the gene can be traced to a couple married in 1688. In that year, eight girls from a Rotterdam orphanage were sent to South Africa to provide wives for Dutch farmers. One of them, Ariaantje Jacobs, married Gerrit Jansz, himself a recent emigrant from Holland, and four of their eight children figure in the pedigrees of the currently affected persons. Incidentally, a defective gene is not the only cause of porphyria. A disastrous outbreak of it among Turkish children, who suffered liver damage as a result, came from their eating seeds treated with a fungicide.

A less spectacular, but historically more significant, case occurred in the late 1700's. King George III, the British monarch at the time of the American Revolution, suffered a series of attacks beginning in 1788. The physical symptoms of his disease were recorded but largely ignored, and both attending physicians and politicians concentrated on his neurotic mental behavior. For four months, parliament gave its entire attention to the king's illness and the constitutional issues it raised. Following another attack in 1810, the Regency Act of 1811 was passed and he was replaced as the active ruler by the Prince of Wales. Meanwhile, on the medical side, the attention to his affliction led to the establishment of psychiatry (then called the "mad business") as a serious branch of medicine.

Recent review of the royal records indicates that George suffered from a classic case of porphyria, and that it appears not only in his ancestors and descendents, but in two other royal houses descended from Mary, Queen of Scots, who was afflicted with similar symptoms.

Further examples of the spread of a gene from a single ancestor are provided by *Huntington's chorea*. This insidious disease produces a progressive degeneration of the nervous system and leads to a complete breakdown of physical and mental powers. It is inherited through a dominant gene, which may possibly have an overdominant effect on fitness, and has a variable age of onset. The average age is 40–45, which means that people do not know if they are afflicted until after their reproductive periods are largely over. Since there is as yet no way of identifying carriers before the symptoms of involuntary jerking movements begin, children of diseased individuals are doomed to a life of anxiety in the knowledge that they have a 50 percent probability of being afflicted too.

In the Australian state of Tasmania, in a population of about 350,000, there are 120 recorded cases of Huntington's chorea. Every one of the 120 persons has as a common ancestor a Huguenot woman who left England in 1848 to settle in the Antipodes. In the United States, there are 7,000 known sufferers scattered from New England to Hawaii. Pedigree tracing is difficult in the United States, but more than one thousand cases derive from a woman who lived in Suffolk at the beginning of the seventeenth century and whose three sons (all by different fathers) immigrated to America. And it is likely that she had other descendents or affected relatives, because the incidence of the disease in Suffolk today is twelve times that for England as a whole. Furthermore, another large New England kinship carrying the disease can be traced back to a shipyard owner from a town only ten miles away from the Suffolk woman's residence.

As another example of founder effects in humans, the religious community of the Old Order Amish in Lancaster County, Pennsylvania, may be given. A recessive type of dwarfism, with homozygotes having extra digits and other physical deformities, is a rare disease known to have affected some 100 persons in the whole world since 1860, when it was first medically described and recorded. More than 55 of these (24 in recent times) were members of the Amish group. Thirty-seven Amish families have affected members at present, and a line of ancestry for each of the 37 has been traced to Mr. and Mrs. Samuel King, who arrived in the United States in 1744. Of the 8,000 present-day Amish in this community, about 13 percent are heterozygous for this gene.

Tay-Sachs disease is a cerebral degenerative disorder, inherited as a recessive, and manifested as a failure to produce the enzyme β-D-N-acetylhexosaminidase A. It is a heartbreaking disease, with typically beautiful, normal-appearing children deteriorating both physically and mentally in spite of all treatment and dying when they are 2–4 years old. It has unusually high incidence in Ashkenazic jews. It has recently been suggested that this could be due to a founder effect, operative 6–7 centuries ago before or during the eastern European migrations of Jews fleeing persecution during the Crusades. A counterproposal notes that there may be some heterozygote advantage for carriers of the gene, and further notes some evidence of resistance to pulmonary tuberculosis among probable carriers. As we've indicated earlier, two such hypotheses are by no means mutually exclusive. It is quite possible that, a long time ago, a family carrying the gene successfully avoided some disaster that overtook many of their fellow Jews. Their descendents, many of them carriers, were scattered in the course of normal or chaotic events, and (probably by chance) were also unusually successful in surviving a series of decimations—some political, some religious, some biotic (such as plague). The probability that this particular sequence of events would occur can be shown to be very low, but the probability that something like it did occur for one or a few of the rare recessive genes in a population is quite reasonable (Section 3.6). In retrospect, Tay-Sachs may have been one of the diseases thus increased. Once large numbers of the gene were present in many eastern European jewish populations, then it is possible that occasional selection, such as for resistance to pulmonary tuberculosis, maintained or even further increased the frequency of this allele.

A large family living in and near Colton, South Dakota, has a high frequency of spinal cerebellar degeneration. The first symptoms typically appear during young adulthood, progressing during about a fifteen-year period through impairment of balance and coordination, then of speech and breathing, and usually ending in death due to pneumonia. One member of the family traced the disease back to the great-great-grandfather of the present generation, G. J. Vandenberg, some of whose children came to the United States from Holland. Vandenberg had eight children, four of whom were later stricken. One of these affected offspring had nine children, and with particular bad luck, seven of them got the fatal allele. In late 1970, a family reunion helped by the National Genetics Foundation attracted ninety-five people of Vandenberg ancestry from five states and from Argentina. Doctors present found possible signs of the disease in twelve children, and definite symptoms in eight adults. Doctors advised relatives to delay having families until they could be certain they had escaped receiving the allele. Responses varied. One 22-year-old man whose father has the disease was unwilling to forgo fatherhood himself, asking, "Aren't I entitled to the same kind of life as anyone else?" But a woman with the disease, already the mother of a son, declared that he will be her only child (he has a 50 percent chance of being free of the allele) and a 21-year-old nephew of an afflicted person underwent a vasectomy to be sure his copy of the allele, if he has one, stops with him.

19.4 DIAGNOSIS AND COUNSELING

The significance of genetics in medicine and public health should by now be clear. Yet it is only recently that medical schools have begun to incorporate genetics into their curricula and research programs. New diseases and syndromes are being added practically daily to the list of those known to be genetically determined. Yet many professionals in the medical field are as innocent of genetic knowledge as the uniformed laity, and few hospitals that are not affiliated with universities have either a genetically trained physician or a consulting geneticist on their staffs.

A welcome development has been the establishment of genetic units for diagnostic purposes and of counseling centers for doctors, hospitals, adoption and welfare agencies, and individuals. A variety of newly acquired tools are being used for these purposes. One of these, a catalogue of human genes, presents a great advance for ease of reference to what is known about hereditary diseases. This is supported by computerized diagnosis, in which symptoms are entered into a computer that then retrieves from its memory those relevant syndromes that may have genetic bases, what the genetic bases are, and what therapy may be possible. (In the general practice of medicine, other clinical, laboratory, and therapeutic information is also being computerized.)

Until recently, genetic counseling for any particular couple was usually not begun until after they had borne a child afflicted with some genetic disease or defect. Thus, counseling came into play only after some fraction of the possible genetic damage had already been done. The calculation of **recurrence risks** remains an important dimension of genetic counseling; a recurrence risk is the probability that a second child, born to a couple that has one child with a genetic abnormality, will have that same abnormality (Box 19.A). But the ability to identify carriers of recessive genes, or late-acting or incompletely penetrant non-recessive genes, plus the development of **amniocentesis** (Box 19.B), which allows prenatal diagnoses of chromosomal and metabolic diseases, has transformed genetic counseling from a relatively passive to an active aspect of medical practice.

Besides being skilled in the art of sensitive humane communication, the genetic counselor needs three technical tools. First is accurate diagnosis, as incorrect diagnosis may lead to some very bad advice, and many of the possible diseases have similar symptoms. Second is a carefully constructed family history, which must of course be studied in the light of a good knowledge of genetics. Third is a

---

BOX 19.B AMNIOCENTESIS

Amniocentesis, the removal of fluid from the amniotic cavity surrounding a developing fetus, became useful and important in the early 1960's for the detection of unborn infants who ran the risk of Rh incompatibility with their mothers (Section 20.2). It soon became clear that much useful information could be obtained by examining cells and other constituents of the amniotic fluid. As the illustration shows, a sample of fluid surrounding the fetus is taken by inserting a sterile needle into the amniotic cavity and withdrawing a small amount of fluid. The fluid, derived mostly from fetal urine and secretions, contains fetal cells (not drawn to scale). Care must be taken not to puncture the placenta or the fetus. The sample is centrifuged to separate cells and fluid. A variety of tests can be made as suggested by the information on the righthand side of the illustration.

The most obvious question concerns the safety of the procedure. By about the sixteenth week of gestation, enough amniotic fluid has accumulated for a sample to be easily obtained by needle puncture. The fetus is small and therefore not likely to be injured, and sufficient time is still available to grow cells recovered from the fluid, perform diagnostic procedures, and proceed with a therapeutic abortion if that is indicated. Sometime in the future, it may be possible to institute corrective procedures and thus prevent many of the detectable diseases; but today this is possible for only a very few of the diseases that can be detected by amniocentesis. In the hands of experienced doctors, the amniocentesis procedure does not present much danger of serious injury to either the mother or fetus. The fetal parts can be manipulated away from the site of the needle entry, and the placenta can be located by ultrasonic methods.

A more vexing problem concerns possible developmental damage to the fetus resulting from the removal of fluid or accompanying change in pressure. A newly organized Amniocentesis Registry will collect and evaluate long-term effects of the procedure on both fetus and mother.

Genetic counseling combined with amniocentesis gives even to persons who are at high risk of producing defective offspring the possibility of choosing not to have defective children without requiring the decision to avoid the possibility of parenthood altogether, as the following example illustrates. A young woman came to a genetic counselor with the information that three of her sisters had Down's syndrome. It was determined that she was a carrier of the translocated chromosome 21, and thus that there was roughly a 30 percent chance, each time she conceived, that the child would have Down's syndrome. She later returned, three months pregnant, and underwent amniocentesis. The fetus had a Down's syndrome karyotype,

knowledge of the development of the disorder, so that remedial action (if possible) may be taken, and the family can be prepared and thus better equipped to cope.

One new diagnostic tool, dermatoglyphics, or study of fingerprint and palm and sole patterns, is being developed both for hereditary and other types of diseases. Diagnosis of various chromosomal abnormalities, which are also recognizable cytogenetically, is aided by the fact that they are accompanied by characteristic dermatoglyphic features. Over 40 types of chromosomal aberrations have such perceivable attributes. Similarly, at least ten disorders arising on a single-gene basis, including PKU, Huntington's chorea, and the Amish type of

and a therapeutic abortion was performed. When next pregnant, the woman again underwent amniocentesis, and this time the test revealed a normal karyotype. She gave birth to a healthy daughter.

Some have voiced the warning that in a society in which abortion is becoming acceptable, amniocentesis may be capriciously used. At the moment it is not clear how we will arrive at standards to decide what the fitness or desirability of a given fetus is, or whose standards they might be.

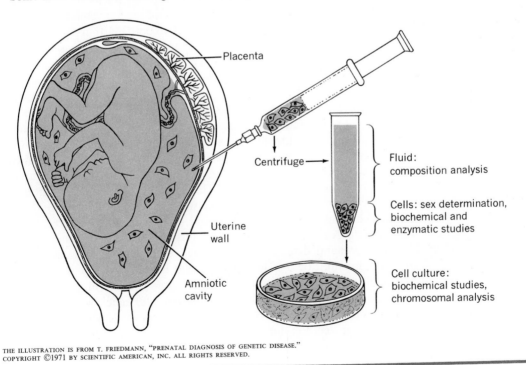

Placenta

Centrifuge

Fluid: composition analysis

Cells: sex determination, biochemical and enzymatic studies

Uterine wall

Amniotic cavity

Cell culture: biochemical studies, chromosomal analysis

dwarfism, are apparently associated with recognizable departures from normal dermatoglyphs, as are a dozen or more diseases of unknown genetic bases or due to toxic or traumatic factors.

There are many reasons for seeking genetic counseling. One of the most common, which brings in parents or other relatives, is a person who has been identified as having some genetic defect or disease. Often the afflicted persons themselves come in. The advice generally sought concerns the probability, if a child is born, of its having a certain disease, or suggestions for cure or treatment of the disease. Frequently, a question about possible guilt for a defect is consciously or subconsciously asked.

Less often, healthy people with no specific disease or defect in mind come in to find out if they might have a genetic problem that could affect the health of the children they might someday have. It is hoped that this class of inquiry will increase as genetic counseling becomes capable of dealing with greater numbers of both people and kinds of hereditary defect. In 1972, heterozygous carriers for over 60 different genetic diseases could be detected, and the number is growing yearly as more diseases are understood, and efficient techniques for identifying heterozygotes for the responsible alleles are developed. We've already mentioned the use of hair follicles to detect Lesch-Nyhan female carriers (Box 12.A) and of G1PUT levels in red blood cells for carriers of galactosemia (Section 12.2).

In 1971, in a campaign supported by local synagogues and Jewish women's organizations, nearly 8,000 Askenazic Jews of child-bearing age in the Baltimore-Washington area were tested for heterozygosity for Tay-Sachs disease. As expected, about one in 30 was found to be a carrier, and for four couples that were in the midst of pregnancy, it was determined that both partners were heterozygotes. When two heterozygotes—knowing that they are carriers—conceive, they may elect amniocentesis to determine the genotype of the embryo. A therapeutic abortion can prevent the birth of a child whose future would at present be hopeless if the embryo is found to be homozygous recessive. On the average, three out of four couples who have amniocentesis done under these circumstances will learn that the embryo is either heterozygous or homozygous dominant; these three couples can more confidently prepare for the birth of their child, knowing that it won the roll of the genetic dice against Tay-Sachs disease. (The total cost of carrier detection, prenatal diagnosis, and pregnancy termination for all under-30 U.S. jews would be about $6,000,000. The hospital costs, ignoring emotional cost, of the Tay-Sachs children these people will produce in the absence of such screening will be about $35,000,000.) A similar, even more massive screening has been done among American blacks for carriers of sickle-cell anemia (Section 7.4). However, prenatal detection of conceived embryos is much less certain, partly because the adult type of hemoglobin that is the site of the defect makes up only a small fraction of the embryo's hemoglobin. Furthermore, the disease is not so severe as Tay-Sachs, and there is some hope for relief or even cure, and thus abortion of discovered homozygous embryos is not as clearly indicated. Prevention through the decision not to have children by a couple both identified as carriers, however, is still a worthwhile activity where the head can rule the heart in matters of love.

Another type of counsel available is for barren couples who want to have a child, especially those with histories of unsuccessful pregnancies. Genetic reasons for such pregnancies can be diagnosed by blood typing if there are incompatibilities (see Chapter 20) or by cytological examination. A University of Michigan genetic counselor reports the case of a Roman Catholic mother, one of whose children had the phenotypic symptoms of trisomy 21. The woman wanted to encourage one of her other children to become a priest. She went to the counselor because she thought that if any of her sons were a carrier of the abnormality (in this instance, a translocation), he should be the one to enter this celibate profession and, thus, avoid passing on the defect to the next generation. Although disclaiming responsibility for the decision of which child would be urged to become a priest, the counselor was able to determine which sons were carriers and which were not by looking at their chromosomes.

People who are related and are contemplating marriage, or having children, also sometimes come in for genetic counseling. Chapter 21 will expand upon the advice given. Briefly, it's a bad risk for the people and their children, although a great source of data for human geneticists. For instance, the frequency of first-cousin marriages in the general population is less than one percent, but 33 percent of the alkaptonuriacs, 17 percent of the albinos, and 11 percent of the totally color-blind people are children of such marriages.

Recently, genetic counseling has begun to be used in psychiatry. Genetic knowledge of four major areas of mental disease now seems sufficiently advanced so that this can be done. These are: presenile dementias (including, for instance, Huntington's chorea); mental retardation; schizophrenia; and the affective psychoses (for instance the psychoses of certain manic-depressives and unipolar depressives). Thus far, clinical psychiatrists do little of it, partly because psychiatry has been slow to accept evidence that genetic factors play an important part in many psychiatric disorders. Patients and their families often get misinformation that is painful, or a barrier to effective life planning, or both. Some of this misinformation continues to come from professionals. For instance, to the parents bearing the wretched burden of having been told in "therapy" that their bad attitudes and home environment *caused* the schizophrenia of their children, our present understanding of its genetics can bring immense relief (Section 12.2).

A type of question frequently presented to the psychiatric genetic counselor comes from the spouse of a newly diagnosed schizophrenic, wanting to know the risk for their children. Since the children are already born, estimation of recurrence risk is not the central issue. At present, there is no advice to give such a parent except the most difficult: take no special measures, and provide what would be a good environment for the growth of any child. (Perhaps it makes some sense to stress a warning against the children's use of psychoactive drugs, or trying "mind-expanding" experiences, such as poorly supervised sensitivity training groups.)

The following two remedial actions were both undertaken by the patients themselves, who had come to understand their diseases and prognoses, and only later came to the attention of counselors. A wife with Huntington's chorea who wanted children found a satisfying substitute through giving temporary foster care. She chose to take foster children temporarily instead of adopting children in view of the prospect of her disablement and death at an early age. The second case was that of a man advised that his risk of developing a degenerative neurologic disease was 50-50. He took a well-paying and very dangerous job, with a generous insurance program covering his family. The job: handling labile explosives. An early symptom of the disease: hand incoordination. He understood.

Diagnosis of zygosity in twins is also part of genetic counseling. Finally, disputes about parentage can sometimes be settled by genetic investigations. Such medicolegal genetics forms part of the broader field of immunogenetics, which is the subject of the next chapter. Meanwhile, it is only to be hoped that the use of genetic counseling increases and becomes a commonplace social service. Freedom of action on the advice is a matter entirely apart, but people with genetic problems should seek advice, not only out of personal concern, but surely also as a social obligation.

To meet this need, genetic counseling units have opened, or existing ones have been expanded, at many medical centers and teaching hospitals within the past few years. As of 1973, nearly 200 such centers existed worldwide. Most accept patients and families only when referred by medical professionals in private practice or in public clinics. Some areas far from major centers are served by "satellite" centers, where specialists from the centers visit on a regular or appointment basis.

A recent dimension added to genetic counseling is in recognition of the fact that families often need psychological as well as medical and scientific help. Thus family counseling has been incorporated in genetic counseling. The problem of dealing with guilt, particularly on the part of parents of an afflicted child, is particularly important. One purpose of genetic counseling in such a situation is to help the parents and other members of the family deal with the afflicted child and with each other, compassionately and intelligently. This is done by helping them understand things as they are, and why they are as they are.

20
IMMUNOGENETICS

Antigens are part of most proteins, and of many other large molecules. Antibodies are plasma proteins, and are generally produced in large quantities in response to the presence of a particular antigen. This is the basis of reaction and immunity to disease. Antibodies produced in response to the antigens present on invading bacteria and viruses fight off a current infection and prevent later reinfection. In this circumstance, antibody production is good, and has high evolutionary survival value. It is less good when the antigen is present on pollen, as an allergic response called hay fever may result. It can be bad when the antigen is in human blood, for it can negate the beneficial effects of a transfusion. The same can be said of a human tissue transplant, for the antibodies attack it and eventually cause the transplant to fail. In the worst circumstance of all, the body becomes sensitized to its own tissue, and begins making antibodies against itself. Such an autoimmune reaction results in a disease that is usually fatal.

20.1 HUMAN BLOOD GROUPS

In Section 18.2 some examples involving inherited differences in red-blood-cell antigens were given, and in Section 14.1 attention was drawn to the possible selective value of the ABO system. The many polymorphisms found in human blood types are of tremendous usefulness in human genetics and in anthropology. Although some of the systems contain complex multiple allelic relationships, basically the inheritance of blood type is straightforward. Since blood types are determined by fully penetrant codominant genes, they provide ready material for the study of individuals and populations.

Box 20.A lists information about many currently known antigens, under control of various loci. New blood-group types continue to be found. The evolutionary antiquity of human blood types has been studied by testing human antigens and antibodies in apes and monkeys. Many of the primates are polymorphic for two or more of the human ABO-system alleles, and the reactions in the chimpanzee, orangutan, and gibbon are particularly like those in humans. The one human blood antigen known to be sex-linked, X_g, appears to be sex-linked in the gibbons also. In addition to the antigens they share with us, apes also have specific simian blood-group systems.

In our discussion here we shall emphasize two of the systems, ABO in this section, and the Rhesus (Rh) in the next. The ABO system comprises six antigens: H, present in varying amounts in all humans (note, however, the rare "Bombay phenotype" in Box 20.A); four subvarieties of A, which we shall group together under A; and B. Antibodies to A and B are constitutive substances, that is, they are present in the serum whether exposure to their antigens has occurred or not, and in this sense these particular antibodies are unusual. The following table, in which *A, B,* and *O* are used to designate the different alleles, summarizes the genetic basis of ABO inheritance, the results of agglutination (in tests or in the body), and compatible types for transfusion:

Blood group	Genotype	Red blood cells have antigens A, B:	Serum has antibodies to antigens A, B:	Clumping will occur if introduced blood is:	Blood groups acceptable in transfusion
O	*OO*	neither	anti-A and anti-B	A, B, AB	O
A	*AA, AO*	A	anti-B	B, AB	A, O
B	*BB, BO*	B	anti-A	A, AB	B, O
AB	*A B*	A, B	neither	none of the four clump	all four

Because the red blood cells of a type-O person have neither A nor B antigens, this person is sometimes called a "universal donor." Similarly, because of the absence of antibodies to A and B antigens, an AB type is called a "universal

Group	Antigens in red blood cells	Antibody present in serum	Reaction to serum (listed to left) of red blood cells from group			
			O	A	B	AB
O	—	Anti-A Anti-B				
A	A	Anti-B				
B	B	Anti-A				
AB	AB	—				

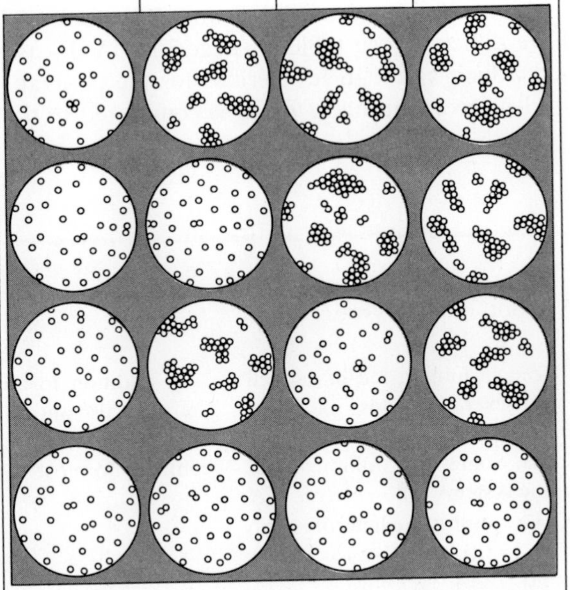

FIGURE 20.1 The clumping reaction of red blood cells exposed to antibodies. (From Curt Stern, *Principles of Human Genetics,* 2nd Ed. W. H. Freeman and Company. Copyright © 1960.)

recipient." Each of these labels is something of a misnomer, as neither of these genotypes protects against antigen-antibody reactions due to other blood-group systems, or against difficulties due to other factors (such as hepatitis). Note that two genotypes are listed for group A and two for B: in each case, the heterozygote and homozygote are not phenotypically distinguishable.

Figure 20.1 shows how blood is typed. A sample of blood whose type is to be determined is mixed with serum whose antibody content is known. The mixture is then examined under a microscope, where evidence of clumping is readily discernible.

In addition to problems raised in blood transfusion by the polymorphism at the ABO locus, there is a further medical aspect to this system. If a fetus whose mother is of *OO* genotype has either the *A* or *B* allele from the father, A or B antibodies from the mother's serum can diffuse through the placenta and cause severe damage to the fetus's blood cells. The disease, called **erythroblastosis foetalis,** is *hemolytic* (blood-destroying), and may cause abortion. A Japanese study found that from such matings as between an A-type father and an O-type mother, there was a significant shortage of live A offspring compared to live O offspring. Presumably, the missing A phenotypes were lost in unsuccessful pregnancies. Such losses have been estimated to occur in 8–35 percent of the conceptions in such matings. In the Sioux indians of South Dakota, among whom the frequency of the O type is 91 percent, O–A or O–B combinations occur in less than 10 percent of the marriages, whereas among the Bangkok Siamese, 37 percent of whom are O, nearly a quarter of the marriages (0.37×0.63) would be O-A, O-B, or O-AB. (Other maternal-fetal combinations—namely, type-B mothers with type-A or -AB offspring, and A mothers with B or AB offspring—also can have difficulty; see table on page 352). A disease more serious for the afflicted individual, although less serious for the population, is erythroblastosis induced by Rh incompatibility, which is considered in the next section.

BOX 20.A BLOOD-GROUP SYSTEMS

There are about 80 known human red-blood-cell antigens, grouped into several systems. Each system is controlled by a different locus, with two or more alleles per locus. Some information about the major polymorphic systems follows (from V.A. McKusick and R. C. Lewontin); the heterozygosity estimates listed in the last column are for an English population.

System	Year of discovery	Number of antigens known	Estimate of average heterozygosity[1]
ABO	1900	6	0.51
MNS	1927	18	0.70
P	1927	3	0.50
Rhesus[2]	1940	17	0.66
Lutheran	1945	2	0.08
Kell-Cellano	1946	5	0.12
Lewis	1946	2	0.30
Duffy	1950	2	0.52
Kidd	1951	2	0.50
Diego	1955	1	0
Auberger	1961	1	0.49
X_g[3]	1962	1	0.46
Dombrock	1965	1	0.46
Stoltzfus	1969	1	?

[1]Based on the assumption of random mating.
[2]See Section 20.2.
[3]Sex-linked.

In addition to antigens of the major systems, there are also antigens that either have been found only in single families (private systems) or are common to most humans (public systems):

Private systems		Public systems
Levay	Romunde	I
Jobbins	Chr	Vel
Becker	Swann	Yt
Ven	Good	Gerbich
Cavaliere	Bi	Lan
Berrens	Tr	Sm
Wright	Webb	
Batty		

The genetic relationships of these systems to the polymorphic systems are not known. Other genes may also be considered as belonging to the category of blood-group determinants. One, for instance, is the rare recessive "Bombay phenotype" for the absence of the H antigen (of the ABO system), which otherwise is present in all people.

White blood cells also have antigenic properties, but their genetics is only now beginning to be worked out. In the broadest sense of the word, the polymorphic serum protein types (haptoglobins, transferrins, and gamma globulins, see Box 3.D and Section 20.3) may also be described as blood-group systems.

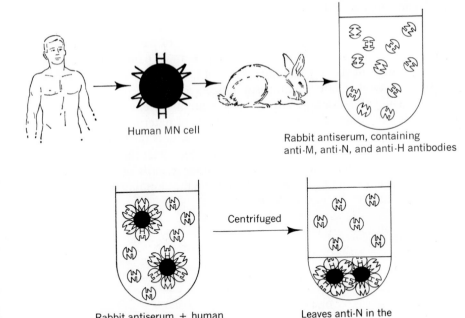

Human MN cell

Rabbit antiserum, containing
anti-M, anti-N, and anti-H antibodies

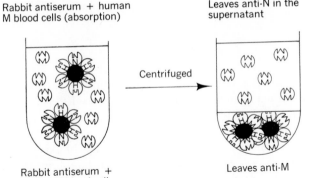

Rabbit antiserum + human
M blood cells (absorption)

Centrifuged

Leaves anti-N in the
supernatant

Rabbit antiserum +
human N blood cells

Centrifuged

Leaves anti-M

FIGURE 20.2
Preparation of anti-N and anti-M serum by injecting blood from an MN individual into a rabbit.
Note that the ubiquitous H antigen is also present, and its anti-H antibody is precipitated.
(Modified from A. M. Srb, R. D. Owen, and R. S. Edgar, *General Genetics* 2nd Ed. W. H.
Freeman and Company. Copyright © 1965.)

The MN, or MNS, system has a somewhat complex form of inheritance.
Contrary to the mechanism for ABO, the antibodies are not naturally found in
human serum, but can be induced to form in rabbits or other animals, as shown
in Figure 20.2. This group does not appear to present medical problems, but
still presents some genetic mysteries: some human populations have an excess
of heterozygotes at this locus compared with expectations based on gene frequency. Antigen S is a sort of honorary blood group, and is associated with the
MN system. S causes the ABO antigens to become water soluble and, thus, detectable in the saliva.

Besides anthropological and medical significance, blood groups also have legal significance, particularly in cases of disputed parentage. Using as many systems as possible makes it feasible to rule on some claims of paternity and to identify the parentage of a child if there has been a mix-up of babies in a hospital.

Clearly, in arguments over paternity, the real father cannot be positively identified. But if a putative father happens to possess a genotype not congruous with that of the child (for example, an AB man cannot, except in the rare case of mutation, be the father of an O child), the claim of paternity can be dismissed. The more systems that are used in the test, the greater is the possibility of exclusion. Thus, R. Race and R. Sanger presented a table under the title "The chances of an Englishman being exonerated by the blood groups of a false charge of paternity brought by an Englishwoman," showing the cumulative effect of adding tests for the different systems as follows:

ABO alone	0.176
adding	
MNS	0.373
Rh	0.531
Kell-Cellano	0.549
Lutheran	0.564
Duffy	0.584
Kidd	0.596

For the U.S. white population, the total probability of exclusion has been computed at 0.716. In many states, laws have been passed that permit the courts to make the results of voluntary tests admissible evidence or to order blood tests in cases for which they would be relevant. It may be noted in passing that where artificial insemination is used in cattle reproduction, blood typing is commonly done to verify paternity. There is one very useful blood-group system in cattle that has more than 250 identifiable antigen combinations.

Blood-group typing has also been used to determine what percentage of children born to married women were conceived in extramarital intercourse. (Such investigations are not just nosy snooping, as this is an important source of error in many human genetic studies, and thus its magnitude must be known.) In one study in a Detroit hospital, in testing for eight systems, it was found that 1.4 percent of white children and 8.9 percent of black children did not have blood groups consistent with those of the legal husbands.

20.2 THE RH SYSTEM

The significant antigen in the Rh (Rhesus) system was first discovered by using Rhesus monkeys. Blood from the monkeys was injected into rabbits and guinea pigs, the antibodies of which then agglutinated some kinds of human blood cells.

(Even more than the ABO system, this is complicated by several groups of similar antigens, and we present here a simplification of the Rh system. There still is some question whether a multiple allelic series or a group of three tightly linked loci is involved in the inheritance of this system.) For practical purposes, we shall be concerned only with three genotypes: Rh-positive homozygote (*DD*), Rh-negative homozygote (*dd*), and the heterozygote (*Dd*), which is Rh-positive. There are considerable differences in the frequencies of these two alleles in different populations. Among white Europeans, the frequency of *d* is approximately 0.4, although in some subgroups, for example, the Basques, it is closer to 0.5. African populations have lower frequencies, centering near 0.25. Asian, American indian, and native Australian populations have *D* frequencies near 1.0–i.e., they are nearly 100 percent Rh positive.

As with ABO, there are mother-fetus incompatibilities in the Rh system: a mating between an Rh-positive man and an Rh-negative woman may result in such incompatibility. The first heterozygous child borne by an Rh-negative woman does not normally suffer any ill effects. However, during pregnancy, there are frequently small ruptures that allow some fetal blood to pass into the mother's bloodstream, and sometimes at birth, larger amounts of fetal blood enter the mother's circulation. She does not have a constitutive antibody against the D antigen, as she does for the antigens of the ABO system. Instead, she slowly becomes sensitized, but usually by the time massive amounts of anti-D are being produced, this first child has been born. (In some cases, an Rh-negative mother has been sensitized prior to her first pregnancy, by, for instance, receiving a transfusion of blood containing Rh-positive cells. Then even the first heterozygous child is in danger.)

When second or later heterozygous fetuses are being carried by such a sensitized Rh-negative mother, a slight leakage of fetal blood into the mother's circulation will set off a massive production of anti-D. These anti-D antibodies diffuse across the placenta and attack the embryo's blood, producing a form of erythroblastosis or *hemolytic jaundice* that is lethal if treatment is not given. Figure 20.3 diagrams events in the origin of erythroblastosis. The incidence of risk in a European white population can be computed from the following table, which gives

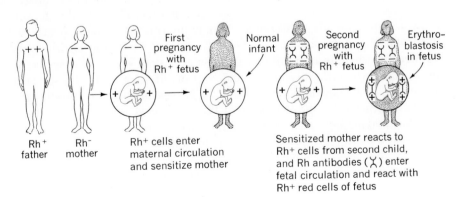

Rh⁺
father

Rh⁻
mother

First
pregnancy
with
Rh⁺ fetus

Rh+ cells enter
maternal circulation
and sensitize mother

Normal
infant

Second
pregnancy
with
Rh⁺ fetus

Erythro-
blastosis
in fetus

Sensitized mother reacts to
Rh+ cells from second child,
and Rh antibodies (Ⅹ) enter
fetal circulation and react with
Rh+ red cells of fetus

FIGURE 20.3
The sequence of events in Rh incompatibility. (Modified from A. M. Srb, R. D. Owen, and R. S. Edgar, *General Genetics,* 2nd Ed. W. H. Freeman and Company. Copyright © 1965.)

the frequencies of the different genotype combinations expected, assuming random mating.

		Frequency of the paternal genotypes		
		DD .36	Dd .48	dd .16
DD	.36	.1296	.1728	.0576
Dd	.48	.1728	.2304	.0768
dd	.16	.0576	.0768	.0256

(Left margin label: Frequency of the maternal genotypes)

The combinations enclosed in boxes are the matings that may lead to Rh incompatibility. Not all of these roughly 13 percent of the matings will produce a diseased child. If the mating is between a *dd* ♀ and a *Dd* ♂, the following sequence, for example, is possible:

first child, *dd,*	no effect;
second child, *Dd,*	possible sensitization of mother;
subsequent children, *dd,*	no effect;
subsequent children, *Dd,*	possible hemolytic jaundice.

Sometimes, for various reasons, an Rh-negative mother may bear more than one heterozygous child before the disease appears in subsequent *Dd* sibs.

The Rh situation provides an example of nongenetic *telegony,* a phenotypic effect of a previous mating on the characteristics of offspring of subsequent matings. The existence of such an effect was widely believed in pre-Mendelian days but then was generally discredited, except for venereal infection, until the discovery of Rh disease.

From the population standpoint, Rh inheritance provides an example of natural selection against heterozygotes. This is not an equilibrium situation, since every time a heterozygous individual dies, the rarer allele suffers proportionately greater loss than the more common allele. This would mean that the *d* alleles should be gradually disappearing from European populations. Whether this is happening, and, for that matter, what the evolutionary history of the alleles at the Rh locus was, is a completely speculative matter. Theoretically, however, Rh disease may provide an example of some interracial crosses that may have detrimental effects. Should migration of Europeans to China be followed by interbreeding, the effect on the new Chinese gene pool would be dysgenic; previously unknown erythroblastosis would start mainfesting itself. But should the reverse migration take place, the effect on the European population would apparently be beneficial, for the frequency of the *D* allele would increase in its gene pool, thus making *dd* × *Dd* or *DD* matings less frequent.

There is an incompletely understood relationship between ABO and Rh incompatibilities, but the following example seems substantially correct. Suppose an Rh-negative woman with type-O blood conceives an Rh-positive baby with type-A blood. Blood of type O contains constitutive anti-A antibodies (Section 20.1). When some fetal red blood cells enter the mother's blood stream, they are immediately attacked and inactivated by the anti-A antibodies before they can reach receptor cells to sensitize the mother against the Rh-positive, or D, antigen they also have. Thus, the next Rh-positive child will not be harmed by anti-D (but the present one could suffer a milder—or even severe—form of damage from the A-O incompatibility).

There are presently three treatments for prevention of Rh-hemolytic disease. The oldest in use is an exchange transfusion at birth, by which the blood of the newborn infant is replaced with blood lacking the antibodies. Some 7,000 such transfusions are performed annually in the United States. Second, techniques for intrauterine transfusion before birth have also been developed. More recently, the injection of certain anti-D gamma globulins into the mother shortly after (or even prior to) birth of the first and all subsequent heterozygous children has been practiced. This may seem strange, as these are the very antibodies that do the damage. But the principle is the same as that suspected for the ABO-Rh inter-action, and was suggested by it, but appears to be more certain and perhaps safer. The antibodies inactivate the fetal Rh+ red blood cells which invaded the mother's blood stream before or during birth, and apparently sensitization is a slow enough process that they can prevent it from going to completion.

20.3 IMMUNITY

It appears that antibodies, like other cell constituents, are under some form of genetic control. Just how it works is still in considerable doubt. Antibodies (also called gamma globulins, or immunoglobulins) are molecules that each include two light chains and two heavy chains (about 214 and 440 amino acids long, respectively) joined by bonds between atoms of sulfur, as shown in Figure 20.4. Mammals can produce an enormous number of different kinds of antibodies, apparently one for each kind of antigen that might possibly invade their bodies. There are three main theories of how this great diversity might be coded for. The *germ line theory* suggests that the genome contains untold numbers of anti-body genes, each specifying a unique antibody (and many coding for antigens

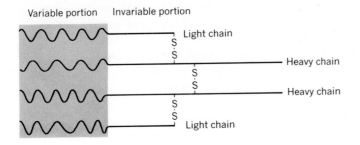

FIGURE 20.4
The four linked chains of the gamma-globulin protein. Different sequences of amino acids in the variable portion endow each molecule with specificity to react against a certain antigen.

never encountered in the evolutionary history of the organism). The *somatic mutation theory* suggests that only a relatively small number of different genes are transmitted through the gametes, but that these are highly mutable during somatic (mitotic) divisions, and thus create thousands of related but different somatic clones. The *somatic recombination theory* is a variant of the second theory, postulating that relatively few genes code for components of the molecule, and that recombinations of these genes code for a very large number of assembly sequences. This latter would probably employ some variation of minute somatic translocations (Section 16.1).

The immune response has two major features: How the body organizes its defense army; and what the army does when it encounters the invaders. As an example of the first, consider an invasion of smallpox viruses in a human body. Sooner or later, one or more of the viruses will encounter a plasma cell with antibodies on its surface that interact with one of the antigens of the virus. This stimulates the plasma cells to proliferate actively, producing a large clone of anti-smallpox cells (Figure 20.5). These cells secrete a soluble form of the antibody, and smallpox viruses throughout the body are encountered and attacked. After the infection is defeated, this large clone of anti-smallpox plasma cells is available to quickly put down any new invasion of smallpox virus, and thus survivors of smallpox attack become immune to further attacks of smallpox. One of the early breakthroughs in the battle against smallpox was the discovery that infection with the relatively harmless cowpox would confer subsequent immunity to smallpox, probably because some one or more of its surface antigens are similar or identical to smallpox antigens.

It now appears that when the antibody army engages the invaders, the antibodies act more as scouts than as infantry. The antibodies have three important functions. They recognize the foreign antigen. They activate a complex group of enzymes, termed *complement*, which can attack the invading molecules. And they attach to the antigenic sites of the invading molecules or cells, and thus provide sites at which the complement can mass for attack on the invader.

The complement is not always perfectly controlled by the antibodies, however, and there may be resulting damage to the body's own cells, which is known as an allergic or hypersensitive reaction. This may be frequent in viral infections, and it may be that many of the symptoms of a virus infection are side effects of the immune response, rather than direct damage caused by the virus. If sufficient damage is done, massive amounts of histamine are released by the body, which may lead to *anaphylactic shock,* and even death. Such anaphylactic reactions of persons sensitive to bee stings or penicillin, or such allergens as ragweed pollen or certain foods, are well known.

Autoimmune diseases, in which the body begins to make antibodies against its own cells, are a particularly serious example of the immune response system out of control. One theory of aging suggests that the body's own cells slowly and progressively begin to combat each other.

Investigation of globulin synthesis and function is of significance to the problems of grafting and the transplantation of tissues and organs. Whether single cells, a suspension of cells, or whole organs are transplanted, an immune response

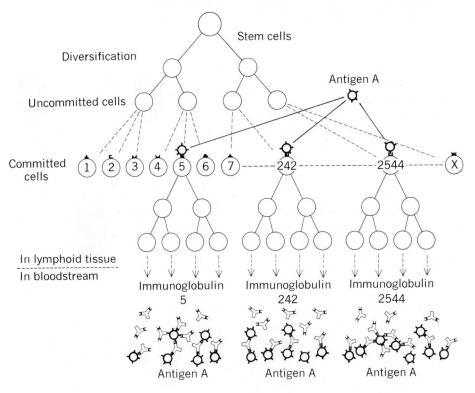

FIGURE 20.5
Stem cells (precursors of antibody-producing cells) perhaps contain information for making all possible antibodies; at some point in embryonic development each is committed to producing a unique immunoglobulin (*numbers*). These committed cells, which are capable of producing antibodies, can interact with various antigens. A single antigen may be recognized by more than one form of antibody-producing cell. Interaction of an antigen with a committed cell stimulates proliferation of the cell and the synthesis of antibody. (From G. M. Edelman, "The Structure and Function of Antibodies." Copyright © 1970 by Scientific American, Inc. All rights reserved.)

against the foreign antigens usually occurs. Some tissues, however, are transplantable. For instance, corneal grafts are accepted by hosts with genotypes different from the donors. The use of immunosuppressive drugs and X-rays before grafting may allow a transplant to be accepted temporarily. In cases of temporary kidney failure, for which a flushing-out of accumulated toxic products is essential, human lives have been saved by transplants of chimpanzee or baboon kidneys. Such transplants may function long enough to do the necessary job. Resort in such cases may also be made to artificial organs, but they are very costly, expensive to maintain, and in short supply.

A very significant area in this field of research deals with the development of immunological tolerance to foreign proteins. Nobel prizes for such investigations have been awarded to the British biologist **Sir Peter Medawar** and the Australian **Sir Macfarlane Burnet,** and we seem to be on the threshold of further important

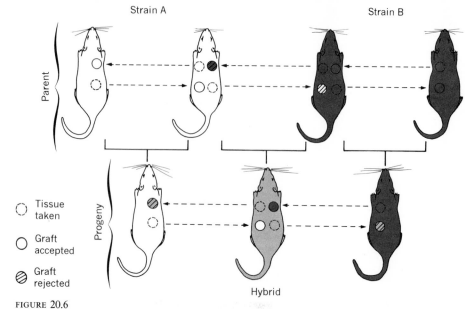

FIGURE 20.6

Grafts exchanged within inbred lines of mice, such as between two mice of strain A or two of strain B, are accepted. Grafts between mice of different strains, however, are vigorously rejected, being destroyed within two weeks. Hybrid offspring will accept grafts from either of the pure parental lines, but grafts in the reverse direction are rejected. (From R. A. Reisfeld and B. D. Kahan, "Markers of Biological Individuality." Copyright © 1972 by Scientific American, Inc. All rights reserved.)

discoveries. It appears that an organism learns to recognize its own antigens by early exposure to them (see also references to autoimmune diseases in Section 3.5 and to cattle twins in Section 11.3). It is as if, sometime before birth, the immune system makes an inventory of all antigens present in the body. These are catalogued as "self," and are not to be molested. But all other antigens are to be vigorously attacked. A mistake in the inventory memory may lead to the autoimmune diseases. If foreign antigens are injected into the embryo at about the time this inventory is being taken, they will also be recognized as "self," and can later be introduced without causing an immune reaction. It has been possible to inject protein from a potential donor animal into an embryo, and later have the individual that developed from the embryo accept tissue from that donor. The potential medical significance of this finding is great.

Since most antigens are probably coded for by genes, two individuals having genes in common will have antigens in common. Thus, it has been demonstrated that organs and tissues can be exchanged without provoking an immune-response rejection between identical twins and within pure lines (Figure 20.6). Thus, one of the early uses of human clones by rich and powerful individuals may be to clone themselves not as an act of ultimate ego, but to establish a dependable source of spare parts, so that their own aging tissues and organs can be replaced as needed.

21

HUMAN
MATING SYSTEMS

This is the final major topic in our overview of formal human genetics. Mating systems, in humans or any other organisms, affect the distribution of genotypes within a gene pool, and of the phenotypes in a population. Since selection acts on phenotypes, the changed probabilities of different phenotypes occurring will thus make selection effective to a greater or lesser degree as a consequence of changes in mating systems.

21.1 KINSHIP AND MATING SYSTEMS

Among different societies there are many varieties of mating systems, which may include mating taboos based on kinship, birthplace, or socioeconomic status. They may also be *prescriptive*, specifically designating who must marry whom. A man may be required to marry his father's mother's brother's son's daughter, or a woman from a particular village or tribe. Anthropologists, as a rule, have been interested in the cultural aspects and consequences of particular marriage systems. To geneticists, the primary interest lies in their biological and evolutionary consequences. Many of the systems are clearly based on property or other economic considerations that have trivial or undetermined genetic consequences.

For instance, the Hindu custom of *niyoga*, in which a sonless man may have a person appointed to beget a child by his wife, has definite economic effects in terms of succession to property. From the standpoint of population genetics, unless selective considerations enter the picture, this custom is no different from the artificial insemination by an anonymous donor's sperm resorted to by childless couples in our own society. By contrast, the systems of *levirate* among ancient Hebrews, in which a childless widow must be married to her late husband's brother, or of *sororate*, in which a man marries his wife's sister, do not seem to have clear-cut, fully worked-out genetic effects, but must be similar to the effects of modern medical practice, which extend the life of the original spouse. In these latter cases, the genes of the spouse are more likely to be passed on to additional children, either from the spouse's relatives, or because the spouse lives longer. Of similar uncertain biological significance was the prohibition of marriage between a man and his deceased wife's sister in Britain, where it was repealed only in our century after an exceedingly prolonged parliamentary wrangle.

On the other hand, the Attic law specifying that the nearest male relative of an heiress whose father had made no will was entitled to divorce his wife in order to marry the heiress, had both economic and biological effects, since the law permitted uncle-niece and even half-sib matings under these circumstances.

In general, prescriptive and prohibitive marriage systems will have genetic consequences when they affect population size or the degree of inbreeding. Perhaps the one marriage taboo that is an ethnological universal is the ban on unions between mother and son. Avoidance of mother-son sexual relations has also been observed in some species of monkeys. Most human cultures have other incest restrictions, but they vary greatly from group to group.

In 1790, Noah Webster wrote the following on consanguinity in America:

> It iz no crime for brothers and sisters to intermarry, except the fatal consequences to society; for were it generally practised, men would become a race of pigmies. It iz no crime for brothers' and sisters' children to intermarry, and this iz often practised; but such near blood connections often produce imperfect children. The common people hav hence drawn an argument to proov such connections criminal; considering weakness, sickness and deformity in the offspring az judgements upon the parents. Superstition iz often awake when reezon iz asleep.

It wasn't until the late nineteenth and early twentieth century that legal prohibitions of consanguineous marriages were enacted by the various individual states, motivated by a mixture of religious and practical objectives that sometimes even included reliable data and scientific reasoning. Among the latter was an 1883 report by Alexander Graham Bell, who was concerned with the inheritance of deafness.

Three types of legal consanguinity are variously recognized and dealt with: lineal consanguinity (marriage of relatives who are in the same direct line of descent, such as father and daughter, or grandmother and grandson; such can have serious genetic consequences, as well as social and economic ones); collateral consanguinity (marriage of relatives who have a common ancestor, but who are not directly descended one from the other, such as brother and sister, or niece and uncle; such can also have serious genetic consequences, and significant social and economic ones); and affinal consanguinity (marriage of individuals who are related by marriage only, such as stepson and stepmother, or a woman and her sister's husband; such have no special deleterious genetic consequences, but may have significant social or economic consequences).

American law in general prohibits marriage between close relatives. As of 1969, 30 of the 49 states plus the District of Columbia and two territories prohibited marriage between first cousins, and North Carolina permitted marriage between first cousins but prohibited it between double-first cousins (produced from the marriages of two full-sibs of one family to two full-sibs of a second family). Some interesting exceptions include Maryland, which rather liberally allows a man to marry either his great-great grandmother, or his great-great-granddaughter. In Wisconsin, first cousins are permitted to marry only if the woman is older than 55. Based on some rather interesting deference to Leviticus 18, other Old Testament tradition, and the Talmud, in Rhode Island uncle-niece marriages are legal only for Jews. In Georgia it is legal for a man to marry his daughter, grandmother, or niece (but illegal for a woman to marry her son or nephew).

In the eleventh century, the Catholic Church prohibited cousin marriage. William the Conqueror thus erred when he married Matilda, his cousin. In order to expiate their sin, they built the Abbaye aux Hommes and the Abbaye aux Dames at Caen. In more recent times, it has been possible to have the prohibition lifted for a particular marriage between cousins by a special dispensation from the Vatican. This makes Italy and other predominantly Roman Catholic countries a fertile source for studies of consanguineous marriages, because of the availability of accurately documented evidence that spans several centuries.

The proportions of marriages that are consanguineous differ a good deal from one society to another. Among the Parsees of India some 13 percent of all marriages are between cousins; the corresponding statistic for Germany at the beginning of the century is less than half of one percent. In the province of Andra Pradesh, on the east coast of India, as many as 25 percent of all marriages are between uncle and niece. The percentage varies with caste from high among shepherds and fishermen, to low among Brahmans as well as among Moslems. In Japanese feudal families, as many as 20 percent of the marriages were consanguineous.

In some societies, special circumstances dictate departures from incest bans. Among the Thai tribe of Black-bellied Yaos, unlike-sexed twins are separated at birth and reared apart, but eventually must marry each other. In some African tribes, royal marriages included a very high proportion of consanguineous mating (Box 21.A).

The XVIII dynasty of Egypt (1580–1350 B.C.) as well as the Ptolemaic dynasty (323–30 B.C.), which ended with Cleopatra, is said to have required that in the royal families reproduction be through sib mating. In fact, however, the pedigrees

BOX 21.A CONSANGUINEOUS MARRIAGES IN BECHUANALAND

The British anthropologist I. Schapera made an exhaustive study of the mating system in Bechuanaland. He found that first-cousin and uncle-niece marriages were quite usual among commoners, and, that the proportion of such marriages was exceedingly high in the families of tribal chiefs. With the spread of Christianity and other western influences, the rate of consanguineous marriage is declining.

Examination of the genealogies of 485 persons who were chiefs, and sons and agnate (tracing through the male) grandsons of chiefs from 1841 to 1940, who between them contracted 777 marriages with one to eight wives each, revealed that over 70 percent married close relatives. The most common form of consanguineous mating was to the father's brother's daughters. Marriages between first cousins, uncles and nieces, nephews and aunts, and three between agnate half-sibs are also on record.

Schapera's interpretation of this unusually high proportion of consanguineous marriages is that such marriages serve in this society to reinforce social ties between different and potentially hostile branches of the royal line. He quotes a native proverb that says, "Child of my father's younger brother, marry me, so that the cattle may return to our kraal."

Other societies have different reasons for consanguineous marriages. A study of such reasons in Japan by the American geneticist W. J. Schull indicates that 40 percent of such marriages are explained by saying that the couple was acquainted when they were very young. The next most common explanation cites economic reasons.

are not at all clear. The interpretation made by Egyptologists of the degree of relationship of successive royal marriages seems to have been rather arbitrary. Furthermore, outbreeding has been found to occur in different generations. Every time this happened, the cumulative increase in homozygosity expected under continued inbreeding would be interrupted and heterozygosity restored, at least in the immediate offspring. Whether the similar sib-mating rule of the later Inca Emperors of Peru was honored mostly in the breach is not known. The Kings of Hawaii had the best of both worlds through a requirement that they each contract two marriages, one with a sister and one with an unrelated woman. Many royal families of Europe also practiced inbreeding. Carlos II (1661–1700), the last Hapsburg king of Spain, was the issue of an uncle-niece marriage. His grandmother, who appears on both the paternal and maternal sides of the pedigree, was also a product of an uncle-niece union, giving Carlos an inbreeding coefficient of 0.196 (Section 21.3).

In general, the origin of incest taboos is clouded in mystery. A number of different theories have been proposed, some of them rather unlikely. The religious view takes it that there is a divine law against incest. This, at best, is a description and not an explanation. Freud and his followers attributed incest taboos to totemistic considerations for which the Oedipus myth provided the basis: murder and incest were the two fundamental crimes in the primitive society, corresponding to the two components in the Oedipus complex. After observing various primitive cultures, many anthropologists have questioned the general validity of this notion.

It has also been suggested that the taboos came into being after observation of biological ill effects. This seems (to some) highly unlikely, since a much greater level of sophistication than is found in many primitive societies that have such taboos is required to detect the biological consequences of inbreeding. Critics of the idea point out that some Australian tribes (which have inbreeding) apparently do not even understand the cause and effect relation of copulation and procreation, although the critics acknowledge there are semantic problems that might obscure the precise meaning of the explanations given by the aborigines to anthropologists on this subject. Supporters of the idea point out that many of the deleterious consequences of human inbreeding are sufficiently dramatic and early in appearance (albinism, mental deficiency, major physical deformations) so that it is likely that a connection between incest and abnormality was noted in most cultures. Roughly one-third of myths involving incest include deformed offspring or infertility as a consequence of the union. Indeed, with a natural tendency to seek mating partners who are nearby, who are similar in appearance, attitudes, values, and background, and who are familiar, it was recognized as necessary to have strong taboos that prohibit the selection of those partners who best fit these criteria. In a recent presidential address to the American Psychological Association, it was strongly suggested that the negative effects of inbreeding were generally understood by peoples of many cultures, but that they were not widely understood, or accepted, by behavioral scientists of late.

Natural selection between populations, some of which have adopted restrictions fortuitously, has also been suggested as a source of taboos. That is to say, small primitive populations in which *endogamy* (inbreeding or marriage within a tribe) was practiced were at a disadvantage in competing with *exogamous* (marrying outside the tribe) populations, both politically and because of the burden of recessive homozygous defectives they had to carry. Whether this hypothesis is true for a given group of populations depends to a great degree on the size of the various isolates or partial isolates concerned, and on other factors not fully known.

Economic natural selection has been suggested, too, as a cause of mating restrictions. Exogamous tribes have marriage connections with others, which may naturally lead to alliances, to agreements about hunting territories and war, and to increased social resources in general. There are also numerous other theories. According to a socialization theory, children's erotic impulses must be directed outside of their families if they are to be motivated to accept their roles in society. A demographic theory postulates that the number of persons surviving to breeding age in primitive families was so small that outbreeding had to be resorted to. More complex hypotheses that include economic and demographic considerations peculiar to the development of family organization in early hunting societies have also been proposed. Finally, it may be noted that mating systems are not permanent in all populations. They have evolved culturally and hence are modifiable. But changes of mating systems can also happen in a species for purely biological reasons. For instance, the South American ancestor of the domestic tomato was cross-fertilized, or outbred. When it was domesticated in the absence of the insect normally responsible for pollination, the tomato adapted itself to reproduce under the highest degree of inbreeding possible, self-fertilization.

21.2 ASSORTATIVE MATING

While there is much romantic writing about opposites attracting, and some social theory couched in such terms as "need-complementarity," there is a great amount of data accumulating on the human animal that suggests, for many physical and behavioral attributes, like prefers and copulates with like. Box 21.B gives some figures on the extent of this phenomenon in Swedish, British, and American populations. The general effect of high positive assortment for monogenically determined traits is to increase the proportions of homozygotes at the expense of heterozygotes, with an eventual subdivision of the population into isolates being (theoretically) possible. But unless assortment is complete, such subdivision is not only an exceedingly slow process, but also one that never reaches the state of complete isolation between groups. For polygenically determined traits, assortative mating acts somewhat like selection, extending the range of character expression. It has been argued, for instance, that the establishment of populations like the Oak Ridge scientific community should lead to increased assortment for a high IQ, as it concentrates people with exceptionally high IQ's in a fairly isolated community.

The causes of assortative-mating preferences of humans generally are not known, although in special cases (for example, assortative mating among the deaf) they may be guessed at. If the statement that there is positive assortment is paraphrased to say that men tend to marry women resembling their mothers, and women tend to marry men who are like their fathers, Freudian overtones are introduced into the search for an explanation of the phenomenon. Several instances of positive assortment in certain birds, for example, the pigeon, the Arctic skua, and the blue-lesser snow goose, have been described. In the blue-lesser snow goose there is polymorphism for color, blue (dominant) and white (recessive), and it appears that males tend to select mates with plumage color like their own and that of their mothers.

In animals, the tendency toward assortative mating is in part explained by the occurrence of **imprinting.** This is a phenomenon that imposes a certain behavior pattern on individuals by very early exposure to a given stimulus. Thus, newly hatched ducks or geese tend to follow the moving object that stimulated them visually during hatching, whether it be a bird or a human being. In Thomas More's *Utopia,* written in 1516, he refers to artificially incubated baby chicks following those who feed them just as the naturally incubated ones follow the hen. Auditory forms of imprinting have also been found: chicks exposed before hatching to a given pattern of noise recognize it after hatching and respond to it but not to others.

BOX 21.B ASSORTATIVE MATING IN HUMANS

The degree of assortment can be measured in metric characters by the correlation between mates, and in all-or-none traits by comparing the proportion of marriages observed in which the partners share the given trait with that expected on the basis of chance. The following comparison, from a Swedish population, in which the data are numbers of couples, shows a significant degree of assortment.

| | | Husbands | |
		Dark eyes	Blue eyes
	Observed	92	117
Dark eyes			
	Expected	73	136
Wives			
	Observed	77	197
Blue eyes			
	Expected	96	178

A similar study among Swedish lapps showed no assortment for eye color, but for hair color it did.

Studies in Britain and the United States by various investigators indicate a high degree of assortment for many environmentally or partially genetically determined traits. Among the statistically significant correlations between mates were the following.

Trait	Correlation
Age	0.76
Memory	0.57
Intelligence	0.47
Neurotic tendency	0.30
Stature	0.28
Eye color	0.26
Weight	0.21

No correlation was found in skull proportions, fingerprint ridge counts, or head length, indicating the unlikelihood of prospective mates going around measuring each other's heads. However, certain facial features and conformation characters showed significant correlation.

Trait	Correlation
Ear-lobe length	0.40
Ear length	0.40
Waist circumference	0.38
Hip circumference	0.22
Neck circumference	0.20

The effect of computer dating on the future of assortative mating has not yet been ascertained.

Imprinting has been observed in insects, in fishes, in birds, and in many mammals that are capable of locomotion almost immediately after birth: for instance, sheep, goats, deer, and buffalo. In cattle, identical twins become each other's objects of imprinting; even if they are separated when very young, they will recognize one another when they are brought together again and will renew their association.

Imprinting develops mating preferences in the subject for the kind of individual resembling the object; yet in some observations on birds, sibs from the same hatch are negatively imprinted: they will not mate together despite early exposure to each other. The sexual preferences are exercised when the individual is offered a choice of mates. It has been suggested that failure of some species of animals to reproduce in zoos is connected with their imprinting on keepers or various objects around them. Although processes much like imprinting have been observed in infant monkeys, it is not known what role, if any, imprinting plays in human biology.

21.3 INBREEDING

A main feature of inbreeding is its power to fix alleles. Under persistent consanguineous mating, heterozygosity declines within a population; if several groups each undergo consanguineous mating, the differences between isolates increase. Plant and animal breeders produce inbred lines through self-fertilization or haploid doubling of plants, and full- or half-sib matings of animals, and use them for crossing to recover hybrid vigor, plus consistent and uniform performance.

The rate of inbreeding or the degree to which an individual is inbred is measurable by the **coefficient of inbreeding F,** devised by Sewall Wright. Its exact formula need not concern us here, but essentially, F may be considered to express the probability that the two alleles at a locus are derived from a common ancestral allele. Alternatively, F may be viewed as measuring the proportionate decline of the average expected number of heterozygous loci in an individual, compared to the average number in its ancestors, or the average decline in population heterozygosity at all loci. In terms of genetic effects, inbreeding is conceptually similar to genetic drift (Section 4.3), and has similar evolutionary significance. It differs from drift in that it is a consequence of nonrandom mating of relatives (or genetic assortative mating), rather than of small population size, and in that its expected effects in increasing homozygosity can be of predictive value for individuals as well as for populations. The value of F may range from zero in a large random-mating population to 1.0 when complete homozygosity at all loci is attained. A negative value of F is also possible, when long-separated populations again exchange genes.

The increase in F under several different systems of inbreeding possible in animal populations is shown in Figure 21.1, which is derived from S. Wright. A single generation of inbreeding between parent and offspring or between full sibs produces an F of $\frac{1}{4}$; between uncle and niece, $\frac{1}{8}$; first cousins, $\frac{1}{16}$; second cousins, $\frac{1}{64}$; and third cousins, $\frac{1}{256}$. Risks of expression of deleterious recessives in matings between relatives can be computed from F values. These risks are proportional to the inbreeding coefficient, and account for the observed **inbreeding depression** in many traits of experimental plants and animals, and especially in those connected with fitness. The causes of inbreeding depression lie partially in the increased possibility of uncovering deleterious recessive alleles, and partially in rendering homozygous certain genes that are advantageous in the heterozygous state.

Consider the example of a first-cousin human mating, such as represented in Figure 21.2. Let us make the assumption, which is probably an underestimate, that every human being carries on the average eight detrimental recessive alleles

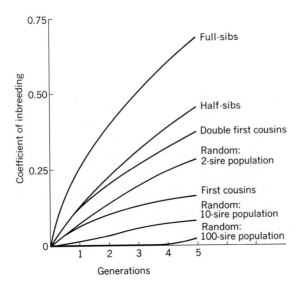

FIGURE 21.1
Increase in homozygosity under inbreeding.

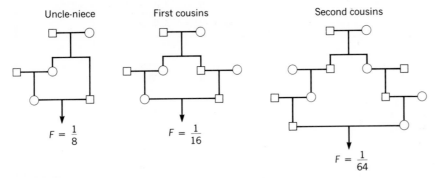

FIGURE 21.2
Examples of consanguineous matings with coefficients of inbreeding of offspring (F). The pedigrees are drawn to emphasize lines of common descent from ancestors who contributed genes to both the mother and the father of the child (arrow).

in a heterozygous state. (Muller called this the population's "genetic load.") Note that children of a first-cousin marriage have only six, rather than the normal complement of eight, great-grandparents. If the great-grandfather common to both sides of the pedigree in Figure 21.2 (the father of the brother and sister whose children married) had his share of such deleterious recessives, which we may number from 1 through 8, and his wife was a carrier of other such alleles, numbered from 9 through 16, the probability that two grandchildren who are first cousins would both be carriers of any of these 16 alleles is $F = \frac{1}{16}$. This means that the probability of their not having it in common is $\frac{15}{16}$. For sixteen loci this becomes $(15/16)^{16}$, or 0.356, meaning that the chance one or more of the 16 being present in both cousins is $1 - 0.356 = 0.644$. Because the risk of expression in

BOX 21.C INBREEDING EFFECTS IN HUMANS

Genetic literature abounds in statistics on the effects of consanguineous marriage on the offspring. There is no doubt that the risks that deleterious recessive genes will be expressed are increased in matings between relatives. From the standpoint of the population as a whole, inbreeding provides an opportunity for natural selection to remove detrimental alleles. At the same time, their manifesta-

tion is a social and personal burden. A sampling of reports from a variety of sources follows. The figures must not be considered as reflecting genetic effects only. There are differences in rates of marriages between relatives in different countries and socioeconomic classes. Note the great variability between the various reports.

HIROSHIMA STUDY (from J. V. Neel)

Trait		Noninbred control	Offspring of first-cousin mating
One or more major	♂	7.4	9.0
defects, percent	♀	9.0	10.3
Height at 10	♂	130	129
years, in cm	♀	130	129
Verbal ability	♂	59	55
test score	♀	57	53
Average grade	♂	3.2	3.0
in mathematics	♀	3.2	3.0

JAPAN STUDY (from W. J. Schull)

	Frequency per 10,000 births		
Abnormality	First-cousin mating	Second-cousin mating	Unrelated
All anomalies	84.3	77.7	46.7

MINNESOTA STUDY ON MENTAL RETARDATION
(from E. W. Reed and S. C. Reed)

	Degree of inbreeding (F)			
Abnormality	$\frac{1}{4}$	$\frac{1}{8}$	$\frac{1}{16}$ or less	General population
Percent mentally retarded among surviving children	60.0	33.3	8.6	10.2

J. V. NEEL COMPILATION

	Percent deaths as of various prereproductive ages	
Population	Offspring of first-cousin marriages	Offspring of unrelated controls
Negroes		
Brazil	46.0	31.2
Tanganyika	32.1	34.3
Asians		
Japan	13.0	9.1
Europeans		
U.S.	16.8	11.6
France	12.2	5.4
Sweden	25.6	31.4
Brazil	31.1	31.1
Germany	32.2	29.5

MICHIGAN MATCHED INBRED AND NONINBRED
CHILDREN AVAILABLE FOR ADOPTION
(from M. S. Adams, R. T. Davidson and P. Cornell)

Measure	Degree of inbreeding (F)	
	$\frac{1}{4}$	0
early survival	83%	100%
early IQ above 100	40%	73%
early IQ below 90	47%	11%
serious physical defect	39%	22%
institutionalization recommended	13%	0%
adoption recommended	53%	83%

any child for each deleterious recessive allele carried by both parents is $\frac{1}{4}$, there is at least a 0.16 chance that one or more of these defects will occur in any child of a first-cousin marriage from inbreeding causes alone. Box 21.C gives examples of observations on consanguineous matings.

We have already discussed a number of examples of the degree to which consanguineous mating is practiced in different human populations, and some further examples follow. Even within a single mating system, however, there may be considerable variation in the proportion of consanguineous mating because of social, economic, and other factors. Thus, within the diocese of Parma in Italy, the proportion of marriages between 1640 and 1965 that required papal dispensation because of consanguinity varied geographically: in cities it was 0.68 percent, in the countryside plains 0.63 percent, in foothill villages 1.64 percent, and in high mountain villages 7.13 percent. Part of the reason for this variation lies in the differences in population size, and hence, in the numbers of eligible mates. Inbreeding in endogamous groups derived from small founder populations may arise merely from the fact that the whole population tends to become highly interrelated.

Several examples of inbred populations in the United States are provided by religious communities established by immigrants from Europe. The founder effects in the Old Order Amish were noted in Section 19.3. This sect originated in Switzerland in 1693 and was established in eastern Pennsylvania, as founder members immigrated between 1720 and 1770. The Amish, numbering today about 45,000 in Pennsylvania, Ohio, and Indiana, are descended from no more than 200 original settlers. The extent of inbreeding that they have undergone can be gathered from the fact that only eight different surnames account for 81 percent of the group in Lancaster County, Pennsylvania, which settled before the American Revolution. Eight other names account for 77 percent of the Amish in Holmes County, Ohio. The history of the Amish is clearly that of sub-isolate formation.

Another similar group is the Hutterites of German origin now living in the Dakotas, Montana, and on the Canadian prairies. The nearly 10,000 Hutterites now alive descended from 101 settling couples. Although inbreeding cannot be assigned the full responsibility (since the Hutterites are a cultural as well as a biological isolate), the Hutterites rank high in the proportion of psychotics among populations compared, even though their rural, nonmaterialistic life styles might be expected to protect its members from psychosis-inducing pressures.

Still another religious community of interest to geneticists is the German Baptist Brethren, or Dunkers. Founded by an original group of fifty families between 1719 and 1729, they numbered 58,000 by 1882. Apparently the exact degree of inbreeding in their population has not been studied; nevertheless, they provide an example of the operation of drift. Not only was their A blood-group frequency (60 percent) different from that of populations either in West Germany (45 percent) or in the United States (40 percent), but so were the M and N frequencies. In both West Germany and the United States, the frequencies are 30 percent for M, 20 percent for N, and 50 percent for the heterozygous type MN; the corresponding figures for the Dunkers are 44.5, 13.5, and 42.0. And not only has there been a startling shift in frequencies of the M and N phenotypes, but the proportion of heterozygotes has declined, as would be expected in a largely endogamous group. These figures show a combination of drift and inbreeding effects. Other traits for which drift can be suspected that have been studied in the Dunkers include hitchhiker's thumb (the ability to bend the thumb backwards at an angle of 50 degrees or less), and presence of hair on the middle segment of the fingers, which is controlled by a single dominant gene.

21.4 BREAKDOWN OF ISOLATES

One of the important phenomena that we are witnessing at present is the gradual breakdown of isolates due to increased human mobility, both geographical and social. Individuals are beginning to have an increasingly wider choice of mates, and the incidence of consanguineous marriages is dropping as a result. This is evident not only in the melting-pot gene pool of the United States, but in many European local populations. Box 21.D shows three examples of this phenomenon. Another reason for decrease in consanguineous marriages lies in birth control and limitation of family size, which reduces the proportion of eligible relatives among prospective mates. Incidentally, the effect of reducing family size is accompanied by the lowering of the average of the age of mothers at the time of each pregnancy, and this in turn leads to a reduction in birth abnormalities correlated with maternal age. Thus, the introduction of legal abortion in Japan after 1947 has produced the following statistics:

Variable	1947	1953	1960
Number of births	2,600,000	1,800,000	1,600,000
Percentage of births that were no later than the third in the family	64	75	90
Deaths from congenital malformation per 1,000	23.7	21.1	19.0

In addition, there was a 40 percent reduction in trisomy 21 and a 50 percent reduction in the incidence of erythroblastosis.

The breakdown of isolates has an effect opposite to that of inbreeding. Among the Hutterites, stature has been shown to decline under inbreeding. But wherever isolates are breaking down, stature is on the increase, as shown in Sweden, Poland, Italy, and the mountain villages of Switzerland. This effect may, in part, be due to improved nutrition, but to some degree it must also be attributed to increased

The illustration at right (after Dahlberg) shows the drop in consanguineous marriages in Prussia.

The next diagram (after Sutter and Tabah) shows an even more dramatic decrease from an initially high frequency of consanguineous marriages in two departments in France. The results shown in this diagram included marriages up to those between second cousins.

The last figure below is taken from the monumental study of Professor Antonio Moroni of Parma, who has analyzed the Vatican records of the province of Reggio Emilia for the last four centuries. The rise in incidence of consanguineous marriages after the Industrial Revolution and its fall after World War I is clearly in evidence. The upper line includes marriages resulting in F values from $\frac{1}{256}$ to $\frac{1}{8}$. Only data up to 1917 have been analyzed. The lower line is confined to F values from $\frac{1}{64}$ to $\frac{1}{8}$.

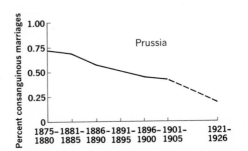

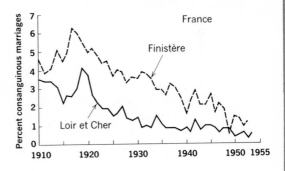

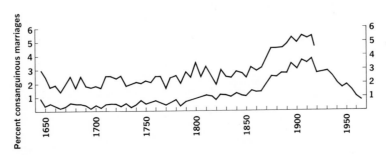

exogamy. In a recent study of Italian Swiss people from the Ticino Canton, men whose two parents had come from different villages were compared with men whose two parents were from the same village. Many people from the Canton migrated to the same region of California. Here, some continued to find mates whose families had come from the same villages as their own, while others married Ticinos from other villages. The following table presents the average heights of men born of exogamous (different village) or endogamous (same village) marriages in Switzerland and in California:

Subjects	Number	Mean height (cm)
Swiss born		
Exogamous	249	168.8 ± 0.4
Endogamous	310	167.0 ± 0.3
California born		
Exogamous	85	172.3 ± 0.7
Endogamous	64	170.4 ± 0.8

Similarly, we might expect a considerable decline in the frequency of expressed genetic abnormalities that are inherited as recessives to accompany the breakdown of isolates. In fact, the halving of the incidence of juvenile amaurotic idiocy observed in the last 30 years in Sweden has been attributed to the breakdown of geographical and social isolation.

The one possible advantage of endogamous marriages may lie in a lower incidence of blood-group incompatibilities. For all other factors, we should benefit by decreasing genetic burdens from generation to generation because of the trend towards a single world gene pool replacing numerous small isolates. It is true that from the standpoint of evolutionary strategy, subdivision into partially isolated demes is a more efficient process. But it is difficult to see why such consideration should apply to human beings. First, convincing arguments that we should accelerate our evolution are lacking. Second, not only the biological, but also the social, and the moral, costs must be considered when dealing with the management of human genetic resources.

22

MANAGEMENT OF THE HUMAN GENE POOL

Having completed the survey of organic evolution and genetics, we turn in the remaining chapters to a number of issues concerning the present and future that this knowledge places before us.

22.1 EUGENIC PROPOSALS

The social climate at the time Mendel's laws were rediscovered was such that eugenic movements could prosper. This was true in England, where eugenics was initiated by Galton. It was true in Germany, where the discipline took the name of "race hygiene" and eventually culminated in sterilization laws and genocide. (Some geneticists and anthropologists who are still active acclaimed Hitler for making a public policy of racism in the guise of science.) It was true in the United States, where intoxication with the elegance of Mendel's laws led some human geneticists to allege they had discovered such improbable examples of single-gene-determined behavior as tendency to nomadism or wild temperament. By the 1920's, American eugenics had degenerated into a mixture of pseudoscience, Bible-belt religion, extreme reactionary politics, and racism, so that the very term became repulsive to geneticists. The same movement still continues to flourish in a mad sort of way among bigots on the fringes of society, who are supported by a few conservative rich. Its principal tenets revolve around selective restriction of immigration to the United States, and compulsory prevention of reproduction of phenotypes that are described as undesirable.

BOX 22.A H. J. MULLER (1890–1967)

H. J. Muller's career spans much of the history of genetics, and he personally contributed or anticipated many of its central ideas. He began his graduate student career at the most active center of genetics research of the time—Columbia University, which housed T. H. Morgan's fly lab. During his first two years, he was not officially admitted to the lab nor financially supported, although he did participate in the lab's discussion sessions. Morgan thought this brash student a bit of a zealot, and Muller thought Morgan and his students were capitalizing on his ideas—a problem he was to have in many an abrasive relationship with colleagues throughout his career.

Muller joined the work that provided early proofs of crossing-over, and then shifted to studies of mutation. In one of his important early papers he recognized error copying to be a fundamental property of the gene, and he later expanded this idea (and therefore the concept of the gene) as being a property of life itself (1929). He thus replaced the study of "heredity *and* variation" with that of the "heredity *of* variation."

He moved from Columbia to positions at Rice and then Texas, where he began work with X-rays and on the development of recombined drosophila chromosomes that would allow accurate detection of new lethal mutations. It was part of his philosophical view to consider his work in evolutionary terms, and he soon began adding warnings in his scientific papers about the dangers of X-rays.

By 1933, Muller had become a socialist, and left Texas for Berlin under a cloud of suspicion for his role in an underground newspaper on the Texas campus. The Berlin stay was short-lived as Hitler and the Nazis rose to power, and he went next to Leningrad and then Moscow at the invitation of Russia's leading geneticist, N. I. Vavilov.

He continued to pursue the nature of the gene, and in 1936, suggested to a Moscow audience of physicists and chemists that the time was ripe for them to work on it. He also became involved in pedigree analysis and twin studies at the Maxim Gorky Institute, perhaps the only place in the

The example of PKU in Section 13.3 has already shown how ineffective a *negative* eugenics program would be. Meanwhile, *positive* eugenic proposals have been put forward on sounder biological grounds. A revival of interest and a considerable amount of public exposure has been given the new eugenics—which by no means should be confused with the movement of the 1920's—by such eminent figures as H. J. Muller (Box 22.A) and the British biologist Sir Julian Huxley.

One eugenic measure, sometimes called *eutelegenesis* to repudiate any connection with the discredited ideas, was first suggested in the Soviet Union by A. S.

world where human heredity and the nature-nurture problem were being investigated with state support.

He wrote *Out of the Night,* strongly advocating positive eugenic proposals, and in his enthusiasm sent a copy and long letter to Stalin. Stalin was not pleased, having already decided to support T. D. Lysenko and to repress genetics in Russia. The eugenics program was rejected, the Institute's program was shattered with charges of racism and class elitism, and for the next four years, Muller devoted himself to polemic attacks on Lysenkoism (Chapter 23) and defense of genetics. He joined the blood transfusion unit of the International Brigade, which was supported by the Soviet government, and gracefully slipped out of the USSR to the Spanish Civil War.

Following the siege of Madrid, Julian Huxley brought him to Edinburgh. During the 7th Genetics Congress at Edinburgh, Muller helped to write and circulate the *Geneticist's Manifesto,* advocating economic security, birth control, removal of racial prejudice, and equal opportunity as essential conditions for voluntary eugenics. While at Edinburgh, he supported Charlotte Auerbach's interests in chemical mutagenesis, which she pioneered following his departure to Amherst and then Indiana.

He received the Nobel Prize in 1946, and soon found himself in heavy demand for consultation and public appearances. He became increasingly concerned that nongeneticists were unable to appreciate the hazards of radiation-caused mutation, due to its cryptic way of spreading the damage over many generations.

He then devoted much energy to writing, producing 9 articles on Lysenkoism, 27 on radiation protection, 14 on evolution, 8 on the human load of mutations, 30 on the relation of science to society, and 10 on eugenics through germinal choice. In opposing the use of anonymous donors from sperm banks, he took the position that the mother, the adopting father, and perhaps even the sperm donor, not the physician, should have the greater say in the quality of life of their child. As an example of a sperm donor's rights, a donor with a probing or inquisitive mind could (on the suspicion that his children might possess irrepressible intellectual curiosity) request that his sperm not be used in families in which conformity or unquestioning belief was rigidly enforced.

Muller's eugenic views brought neither professional nor popular support. Revulsion from race prejudice at home and the Nazi genocide programs overseas created a fear and suspicion of any eugenic program. But rather than adopting the then-popular attitude that values were scientifically meaningless, and thus not in the domain of scientific inquiry, Muller claimed that evolutionary considerations demanded the opposite. He felt we can guide our evolution to create a better distribution of human qualities than we now have, or are likely to get considering the negative forces of mutation we have unleashed on ourselves.

Serebrovsky, and later independently in America. It proposes that sperm of men adjudged to be superior be used for artificial insemination, thereby increasing the frequency of desirable genes in the human gene pool. Many ramifications have been put forth for programs based on this idea. Sperm may now be preserved by freezing (a common practice in cattle breeding), making possible the establishment of human sperm banks, or semen cryobanks. There is a rapid increase in such frozen-sperm facilities for humans, from fewer than 10 in 1969 to 16 (plus 7 branch offices) in 1973. The popularity of vasectomies accounts for some of this: some of the men who have the operation want to store some of their semen in case they later decide to have children. The storage of semen, by the way, is in line with one of Muller's suggestions for positive eugenics. He proposed that men should store semen shortly after puberty. It would then be protected from radiation, caffeine, high temperature, and the many other physical and chemical mutagenic insults the typical young adult testis receives. When children are desired, this semen could be withdrawn and used—and it would have been protected from perhaps as much as half-a-lifetime's share of germinal mutations. A second reason for the increasing demand being placed on semen cryobanks is that among the couples in the United States who want to have children but are unable to conceive, 40-50 percent of the cases are due to the husbands' infertility. Some of these couples have children by making use of artificial insemination with the semen from cryobanks. As of 1973, more than 500 documented normal births had resulted from the use of frozen sperm from cryobanks.

Eventually, egg banks may also be established. We are developing the technology that will allow us to fertilize an egg in a laboratory, check its early development, and then implant it (not necessarily in the biological mother) after determining that it is the desired sex and is free of some specified set of genetic and developmental defects. In one formal scheme for eutelegenesis, a birth certificate of the future has been devised that would list five parents for each child: the legal father and mother, the two genetic parents, and the incubator proxy mother, into whose uterus the artificially fertilized egg had been implanted. Sperm of all men could be deposited in cryobanks, and kept there to be used after an individual's superiority had been established by his achievements or, even better, by progeny tests. Perhaps sperm drives will be held in the future just as blood-bank drives are held today. (It is interesting to note that in 1975, after an eighteen-month study, the British Academy of Sciences has recommended to Parliament that rock stars be prohibited from selling semen to commercial sperm banks. It isn't that they disliked rock stars, but they recognized that the popularity of some rock stars could result in thousands of offspring being produced from a single star's commercially available semen. Inbreeding problems could then occur when these offspring intermarried, possibly not even realizing they were related.)

A further, and as yet somewhat visionary, proposal is to clone superior individuals by growing cells from them in tissue cultures (Section 11.3). This procedure would make possible production of any number of replicates of the future's counterparts of Shakespeare and Leonardo, should they constitute the popular ideals (in a democratic society) or the personal ideals of state managers (in a dictatorship). Such suggestions should be understood to come, not from cranks, but from some exceedingly brilliant, informed, and well-meaning scientists, imbued with ideas of social justice. In their zeal for the preservation and improvement of humankind, however, they do not agree with the views of other well-meaning biologists that there is more than one difficulty and danger in putting their proposals into operation. For example, should the tastes run to the likes of Hitler or Al Capone, they too could be cloned.

First of all, then, there is the problem of values. Perhaps today everybody can agree in some general terms on some standards of value (see Section 17.4). Intelligence, adaptability, mental stability, courage, energy, and perseverance are generally admired, even though we have no way of testing individuals for the genetic components of these virtues. We have no clue as to how relevant these properties will be to the human situation a thousand years hence, which would be short-range planning in a human eugenic program. Indeed, we are actually very much confused about some values rather important to us even today. The Nobel laureate physiologist A. Szent-Györgyi has said that we might reflect on a list of common crimes: murder, rape, robbery, destruction, lying. They are commonly held crimes when committed by individuals within groups. But, ironically, it seems that they become virtues when one's own group commits them against others. Statesmen and diplomats are rewarded for the most outrageous lies; pilots and generals are decorated for wholesale destruction and homicide; annexation of territories after wars adds to national glory; and the rape of the Sabine women is still held to be a heroic chapter of Roman history.

Second, there are no perfect genotypes: each gamete of a great person, who is held great by whatever standard is chosen, carries in it numerous genes for traits undesirable by any standard. With respect to genes controlling physical defects, this is unquestionable. With respect to less tangible traits, there is no warrant to believe that such generally admired traits as musical ability, goodness of character, mental stability, scientific aptitude, and moral probity are genetically positively correlated.

Third, the acceptability of eutelegenesis in our present state of social development may be questioned. Propronents of positive eugenics often point to the success of animal breeders in improving their stocks, but this is a poor analogy, because animal breeders became successful only when they had a Platonic ideal, such as a fat steer, a lilliputian dog, or a fine-wooled sheep, in mind. They selected for uniformity; whereas humankind, perhaps, wants diversity. They used inbreeding, to a degree not generally accepted for humans. They selected by ruthless culling and complete prevention of reproduction of the majority of animals under control. They used progeny testing, which in human terms would mean wholesale reproduction for testing purposes with full knowledge that the great majority of persons so produced would be barred from contributing to the next generation's gene pool.

Fourth, the efficacy of the degree of eutelegenesis that might be acceptable has been brought in question. Thus the British geneticist J. Maynard-Smith made some computations on the basis of several reasonable assumptions, taking IQ-test performance as the desideratum. Let us say that one percent of women will agree to have half of their children by AID (artificial insemination by donor). Let us further assume that the real husbands of these women are a random sample of the population. (Under assortative mating, it likely would be the more intelligent couples who would appreciate the alleged benefits of eutelegenesis to the world and themselves.) Let us further assume that the donors have an average IQ 30 points higher than the mean, which would provide a strong selection pressure. Finally, we can postulate a 50 percent heritability for the IQ. It works out that the expected change in the mean IQ of the population under these circumstances would be 0.08 points per generation. It would take 1,500 years to raise the average by 4 points by this means. On the other hand, data from the Philadelphia school system show that children who entered its schools from poorer school systems in grade 1A scored an average IQ of 93.3 in grade 6B, while children from poorer school systems entering in grades 5 and 6 scored only 88.2. Thus, six years of better schooling appears to be more effective than 1,500 years of selection (although it must be emphasized that the methods are not mutually exclusive).

And, indeed, as Joshua Lederberg has pointed out, there is a negative correlation between the level of acceptability and the efficiency of selection progress of programs of positive eugenics. It may then be asked if eugenics should be abandoned completely. The answer is no. There are many different measures that can have positive effects. We must grant that these effects may not be very great on a per-generation-change basis, but on a worldwide scale, they promise to eliminate much personal grief, and could significantly lighten some of our social burdens.

Among such measures is genetic counseling, already considered in Section 17.4. If use of this service is to be widespread, there must first be dissemination of information on genetics among the medical profession and the public at large. If couples who are first cousins, or known heterozygotes for the same serious defect, choose not to reproduce (they could adopt children instead), the proportion of expressed defects will be reduced. Restriction of family size has been demonstrated (Section 21.4) to lower the frequency of congenital malformation and the incidence of trisomy 21 and of erythroblastosis. Since the use of AID is spreading among childless or incompatible couples, an intelligent choice of donors could be of help. Most donor sperm comes from medical students and interns, and yet, to quote H. J. Muller, this source is used "without regard for the fact that U.S. Army IQ tests have indicated this group to have the lowest mental ratings of all professions tested."

It has been asserted that we now have the ability to control our own evolution. This, as the previous discussion shows, is actually true only in a very limited sense. Without infringement of personal liberty, all that can be done is to reduce somewhat the expressed gross defects in each cohort of babies. We still do not know how to stop recurrence of undesirable mutations. Even if the values of the future could be somehow foreseen, we still know little about measuring genetic worth. Above all, the social organization of a democratic society may be incompatible with exercising whatever powers of controlling our gene pool we may have. Nowhere is this dilemma more evident than in connection with the population explosion (Section 15.1). The right of individuals to have children is held by most segments of our population to be an inalienable one. Yet we are headed for disaster if it is exercised indiscriminatively. Proposals have been made to limit population growth by providing a financial incentive—in other words, by placing a tax on children instead of granting exemptions for them—but should this measure be adopted, the social and psychological consequences for those children whose parents are so irresponsible as to continue having babies without worrying about supporting them, may be a stiff price to pay. All in all, humanity has serious problems on its hands, and no easy solutions are in sight, despite all the advances in biological knowledge.

22.2 EUPHENICS AND GENETICAL ENGINEERING

Euphenics deals with the improvement of the phenotype by biological means. The term was proposed by the Russian biologist N. K. Koltsov (he published an article under this title in a 1929 Soviet medical encyclopedia), and formulated independently in the 1960's by J. Lederberg. Essentially, euphenics involves the incorporation into preventive and therapeutic medical practice, of the broad advances that are being made in molecular biology, immunology, neurophysiology, and other rapidly growing biological fields. Lederberg, in particular, has been a strong advocate of euphenics as a corrective measure for our genetic ills.

He has emphasized the ineffectiveness, unacceptability, and exceeding slowness of action of eugenic measures, whereas investment in euphenics can lead to many immediate benefits. Specific examples would include a crash program for development of artificial organs, and of industrial methodology for the synthesis of hormones, enzymes, antigens, and other proteins. These could be used to supply individuals suffering from genetic or acquired metabolic blocks, to solve problems of transplantation of organs, and to induce immunological tolerance during fetal life. (An eugenic program for primates is also suggested as a means of providing material for compatible grafts.)

While some of these techniques are much faster and, in the short run, more directly effective than reducing the frequency of deleterious genotypes by either natural or directed selection, they may be unacceptable due to financial or energetic costs (Section 15.3). For instance, a recent average annual cost to a patient for 18 hours per week on a kidney machine is over $25,000 per year. It costs about $22,000 per year to keep a hemophiliac on globulin therapy (it is estimated that there are 100,000 hemophiliacs in the United States). These are costs that cannot be borne by the average person, and thus society must decide to pick up the tab, an easy thing to do as long as there are only a few diseases we can counter with this sort of sophisticated biological engineering.

Euphenic changes in kinds of brain function can also be visualized. Thus, H. J. Muller suggested that rapid transmission of abstract thought or telepathic communication may be achieved by providing brains with machinery for direct transmission and reception of light waves. Pharmacological manipulation of the developing brain will certainly be possible before too long, bringing with it the obvious dangers of thought control (see Section 22.3). Thus, the genotype that might be desired when specific brain hormones are isolated would be one with a capacity to respond to brain-stimulating injections, a further instance of the unpredictability of genetic value judgments in advance.

Somewhat less distant are the prospects for genetical engineering, the direct manipulation of the genetic message by changing, subtracting, or adding to the instructions received by the cell (Section 6.6). Transduction and transformation, or directed mutation of the message received by somatic cells could be used to change the genetic instructions in clones of cells in some particular organ. For example, after application of engineering correctives, a PKU child's liver would be able to synthesize the missing enzyme, but the genes such an individual would transmit to the next generation would still be defective. If, however, the correction could be applied to germinal cells, positive eugenic effects would ensue: the genes passed on to the next generation would be normal. Of some appeal is the tongue-in-cheek suggestion, of the Rockefeller University biochemist R. D. Hotchkiss, that problems of human food supply could be solved by incorporating a gene for the ability to digest cellulose into the human genotype. He is particularly attracted to the idea, he claims, because he would then be able, truthfully, to tell a correspondent how much he enjoyed his letter.

A further opportunity for genetical engineers lies in the manipulation of regulator genes (Box 6.D), which could be made to turn on and turn off synthesis of proteins by supplying a cell with the appropriate regulatory substances.

One suggestion made is that even teleduplication, creating replicas of human beings across vast distances, by feeding computers their DNA specifications, and having receivers on other planets reconstruct them instantly, will be possible in the twenty-second century. Problems of personal identity of the subjects of such experiments and operations, and in cases where people emerged from successive transplants with a set of completely new organs, including the brain, would become a worry not only to theologians. We all will have our hands full long before then in having to sort out and put in their rightful places the range of beings that the prophets of genetical engineering confidently promise will be manufactured by introducing various human chromosomes into primate genotypes.

All of this is heady stuff. But whereas only a few years ago these possibilities were out of the scope even of science fiction writers, today they are being advanced and seriously discussed by leading biologists and biochemists as definite prospects for our future. How soon the various possibilities will materialize, no one can say. But in just the last few years, we have synthesized a short (77-nucleotide) but nevertheless complete, double-stranded gene in a laboratory preparation that did not include any DNA as a starting ingredient. Human fibroblasts from a galactosemic person have been infected by a virus carrying the G1PUT gene (Section 12.2) from a bacterium, and G1PUT enzyme activity was then initiated in some of the cells of the culture (this experiment has not yet been reliably repeated). A number of other early tricks flowing from our increasing understanding of the gene, its action, and regulation could be listed here, and others will surely soon be forthcoming.

Clearly, whatever biological problems the wonders of euphenics and genetical engineering may solve, they will create many unprecedented social and ethical problems, for the solution of which much collective wisdom will be needed. The requisite wisdom is unlikely to come from the genetical engineers alone, because it involves moral issues on which they are not experts. The traditional ethical guidelines have come from religion, but the new religion of science and technology that is arising, with its hierarchy of scientists instead of priests, with its sacred language of mathematics instead of Latin, with its sacrifices of traffic casualties instead of heretics, and with space exploration for its Crusades, is as yet not capable of providing any (see Section 24.3). And therein lies still one more reason why the gulf between C. P. Snow's two cultures, or more likely a much greater number of subcultures, must be bridged, if we are to extricate ourselves from the pitfalls that will accompany the blessings that modern biology has brought and will continue to create.

22.3 EUTHENICS, EUPSYCHICS, AND SOCIAL ENGINEERING

Improvement in the environment, **euthenics,** and educational and behavioral engineering, which by analogy we may call **eupsychics,** provide further ways of managing human biological resources. And, similarly, social engineering involves methods of restructuring our society and institutions to meet the problems of the age of atomic physics and molecular biology. We shall discuss these only very briefly, since the details are clearly beyond the scope of the book and the competence of the writers.

Producers of food and fiber practice both eugenics and euthenics on their plants and animals. It is possible, for example, to select lines of animals for heat resistance. At the same time, one can provide cooling devices in the environment of nonresistant genotypes. This is clearly applicable to the human situation. Both unplanned and planned environmental effects have relevance here. With respect to the first, as Lederberg has pointed out, human mobility permitted by the jet airplane has already had incalculably greater consequences to the genetic structure of human populations than any conceivably feasible program of eugenics would.

The planned measures could be exercised effectively in education and in the control of human behavior. Psychologists and psychiatrists have never been known to hide their lights under a bushel so far as their alleged powers to manage human beings are concerned. Freud was not the only one to set an example. The modest pronouncement that the father of behavioral psychology, John B. Watson, made over 40 years ago may be recalled: "Give me a dozen healthy infants, well-formed, and my own specified world to bring them up and I'll guarantee to take any one at random and train him to become any type of specialist I might select—doctor, lawyer, artist, merchant-chief and, yes, even beggarman and thief, regardless of his talents, penchants, tendencies, abilities, vocations, and race of his ancestors."

Watson himself admitted that he was stretching the point. Yet it seems that if differences between genotypes are recognized and educational techniques appropriate to each genotype are devised, it is not impossible that the neo-Watsons of the twenty-first century may be able, if allowed, to do just what Watson claimed.

According to the standards of a good part of humankind, not necessarily including believers in the supremacy of white Nordics or orthodox Jews inalterably wedded to the Chosen People concept, human diversity is one of our basic resources or treasures. One of the reasons that we continue and are likely to continue to be the dominant species on Earth lies in the enormous variation in our gene pool, in the large number of polymorphisms it maintains, and in our control of environmental factors that permit us both to exploit and conserve the natural riches available, and to create any number of artificial ones.

There is little doubt that educational engineering could be used to make similar genotypes into widely different phenotypes by custom-tailoring the educational process. But a concern that we feel is that these methods may be used to nullify differences between genotypes by similar tailoring in the opposite direction, to produce uniformity. To quote the American psychologist R. S. Crutchfield:

> . . . the most striking new factor in the increasing threat to independent thought in a conformist world is the development of far-reaching psychological methods for behavior control—direct electrical stimulation of the brain, high-speed computer control of man-machine systems . . . so-called teaching machines. . . . Just as there is a race between peaceful and destructive uses of nuclear energy and a race between medical advances which reduce death rate and those which control birth rate, so there is a race between the destructive and constructive use of these radically new techniques of behavior control.

Behavior control is not a new phenomenon: there is (or at least there commonly used to be) control by parents, by laws, by customs, by one's own subconscious drives. Presumably in the totality of things these controls are benevolent. The peril lies in the possibility that educational and psychopharmacological or electrical means of control can fall into the hands of antisocial types that Wright placed in group 6 (Section 17.4). An example of genocide through manipulation of behavior by drugs is to be found in the deliberate use of alcohol by the white settlers to destroy resistance of North American indian tribes. And the technology of anxiety-inducing, hallucinogenic, and other drugs that affect information storage and retrieval in the brain is only beginning to develop.

There are other problems of eupsychics which are connected with changes that may be expected from social engineering. As has been noted by many, the views of Western society on private property, murder, rape, and adultery have been very nearly static for three millennia. Sexual behavior patterns may be undergoing a substantial revolution with the introduction of cheap and effective contraceptive devices. In such a revolution, the institution of the family would certainly undergo a change and might even disappear. The psychological effects on the individual and the genetic effects on the gene pool of this change in society are unpredictable, but they will have to be faced if such a revolution takes place. The decline of the family might modify social structure in either one or both of two directions: (1) a centrifugal broadening of the individual's loyalties from parish to nation, and from nation to the whole world, and (2) an increased tendency for personal differentiation and loss of group identity or status.

We have stressed the biochemical uniqueness of the individual. Socially, members of a plumber's union are now practically completely interchangeable, as are members of the fisherman caste in India. In the world of tomorrow this may not be so. Indeed, when enough human loci are known to discriminate between each and every human being, our social identities may be determined by the catalogues of our genes. This may be a distasteful prospect, but might have a somewhat greater significance than discrimination between people by zip codes and social security, telephone, and charge-account numbers.

23

GENETICS AND POLITICS

We have already noted the subversion of genetics in Germany for political uses, and the racist aspects of the 1920's eugenics movement. Perhaps the most bizarre chapter in the political history of genetics was written in the USSR during the last forty-odd years. Its climax came in 1948 when a fanatical charlatan, **Trofim D. Lysenko,** was established as a dictator of genetics. On August 26, 1948, a decree of the Praesidium of the Soviet Academy of Sciences appeared in *Pravda*, basically directing what biological doctrines were to be taught and what kind of research was to be permitted. This decree signalled the end of genetics in the Soviet Union for years to come. Lysenko bore a great deal of responsibility for the dismissal, exile, and execution of a number of Russian geneticists. Only after Khrushchev's resignation did Lysenko lose his power; subsequently he was publicly disgraced and exposed as head of a school of wholesale faking of data. This episode in the history of genetics in the USSR has such an important bearing on the relationship between science and politics and on the vitally important problem of governmental control of science, that it is worth examining at some length.

23.1 THE RISE AND FALL OF LYSENKO

Science in Russia has always followed the generally authoritarian European tradition. Even in prerevolutionary days, dissent within a research institute or a university department was not tolerated. This condition, combined with political dictatorship, set the stage for what was to follow. We can take the year 1925 as a starting point of this chapter of Soviet genetics. In that year, Lysenko, a young agronomist of peasant origin lacking rigorous biological education, started his career as a junior plant breeding specialist in an Azerbaijan agricultural experiment station. At that time genetic research in Russia was just beginning to develop into the enormous enterprise it became within a few years. A little earlier, its direction, and particularly its harnessing to the needs of Soviet agriculture, had been, by Lenin's personal instructions, entrusted to **N. I. Vavilov** (1887–1943). A botanist and plant geographer who had studied abroad, and traveled around the world, Vavilov was a scientist who achieved an international reputation. He was one of the few non-Communists to become a member of the USSR Central Executive Committee, and from all evidence, he was an organizer and administrator of exceptional brilliance.

Under Vavilov's leadership, a network of research institutes and experiment stations, employing eventually more than 20,000 people, was built up. In addition to the botanical work under his immediate direction, genetic research developed rapidly along many other fronts. It would be an exaggeration to say that the Russians were responsible for the majority of genetic advances in the 1930's, but no history of genetics for that period would be complete without footnotes referring to many Russian contributions.

Lysenko had no claims to being a geneticist in 1925. In an autobiographical note, he described his primary interest at that time as being in the physiological problem of the length of the vegetative period in plants. By 1928, he was able to formulate a theory of plant development having a bearing on the nature of the process of maturation. His investigations were reported at a Congress of Genetics, Selection, Plant and Animal Breeding, which was held in Leningrad in January, 1929. Lysenko's report apparently excited no comment. This was understandable, since the audience of 1,400 also listened to 268 other contributions, examined 943 exhibits, and in general was subjected to a menu of genetic fare of a breadth and range probably unsurpassed to that time at a specialized scientific meeting. The general theme of the Congress was the conversion of theory into practice; its slogan, voiced by Vavilov, to "raise the level of our agriculture"; its cardinal problem, to ensure that scientific results and information "reach and spread among the many millioned peasantry" of the Soviet Union.

Somewhat before this meeting, Lysenko had moved to the Odessa Institute of Selection and Genetics (later renamed after him) where he continued his investigations, and what is more significant historically, took his notions of agronomic practice into the field. Among these notions a number that were innovations as far as Russia was concerned (although they appear to have been practiced elsewhere) were put into operation on collective farms. Some were partially successful, such as the summer planting of potatoes; others failed, such as certain methods of pretreatment of seed. In general, improvement in agricultural practice ensued, which ensured personal support for Lysenko on the part of thousands of collective farm workers.

Such support from provincial sources peripheral to science enabled Lysenko eventually to raise the flag of insurrection against Vavilov. In January, 1932, the first issue of a Lysenko-edited journal appeared in Odessa. Roughly one-half of its 80 pages came from the pen of Lysenko himself, while another 10 pages were devoted to a series of resolutions of various local bodies, the main gist of which was that "in spite of the great importance of the original scientific work of comrade Lysenko an unobjective and inimical attitude toward his researches was noted among certain representatives of agricultural science."

In Odessa, Lysenko formed a partnership with a philosopher of the dialectical-materialism persuasion named I. I. Prezent. In 1931, Prezent published one of his first contributions having bearing on biology, a hostile philosophical analysis of the excursion of Filipchenko, a Russian geneticist by then deceased, into the realm of eugenics. Filipchenko's stand on matters of human genetics was related to the eugenics movement of the day. Prezent's attack, which incidentally revealed his magnificent gifts for vituperation and argumentation *ad hominem,* was less on Filipchenko's biology than on his class motivation.

Lysenko and Prezent were able to ignore, distort, and ridicule what were then well established biological facts and principles. An important target was the stability of biological systems, which change only slowly, over evolutionary time. Prezent's understanding of dialectics required that they change much more quickly.

Disentangling the mutual influences of Lysenko and Prezent is difficult, but it is with the formation of this partnership that Lysenko first emerged as a contender for genetic honors. By Lysenko's own admission, it was Prezent who first "cranked-up" the attack on genetics and Mendelism. The collaboration bore its first fruit in the production of a book in which the putative genetic implications of Lysenko's theory of plant development were elaborated.

In the meantime, all was not quiet on the genetic front even outside the Lysenko orbit. The Russian tradition of introspection, and the general European heritage of viewing scientific endeavor as part of a general system of natural philosophy, did not die with the political revolution. Such was the case with Lamarckism (Box 3.A), which had been (up to then) interpreted by most biologists in Russia as an idealistic rather than a materialistic concept. Yet among some scientists, and among more nonscientists, the idea of inheritance of acquired characters was held in favor. It had mass appeal, was easy to understand, and, if true, would give society a much easier way of changing nature than was otherwise available.

There were various other philosophical bones of contention among geneticists. Some were accused of deviation from the party line in the form of "menshevizing idealism"; others were subjected to criticism for advocating the extension of genetic principles to human society. The whole field of human genetics was subjected to attack on the twin grounds that its study led to racism, and that extension of laboratory results with animals to humans degraded humans to the level of beasts. By the end of 1936, all studies on human genetics were suspended in Russia, and concurrently, an event of even greater importance took place.

This was a session of the Lenin All-Union Academy of Agricultural Sciences, a body of which Vavilov was at that time a vice-president, held in Moscow in December, 1936. There the first full-dress attack on Vavilov, on genetic theories, and on genetic practice was launched by Lysenko, Prezent, and their supporters. At that time, Lysenko denied that he was a Lamarckian. He said "it is difficult to find a greater enemy of Lamarckism than I. I. Prezent." He stated that "starting from Lamarckian positions, the work of remaking the nature of plants by 'education' cannot lead to positive results. If, however, we are successful in remaking, by means of appropriate education of plants, the nature of their heredity in the desired direction, this already speaks for the fact that we are not Lamarckians and do not set out from the Lamarckian position." But Lysenko insisted on the indivisibility of an organism, denying that it is possible by any means to separate hereditary from environmental influences. His theory assigned what essentially amounts to free will to individual plants, who not only can select their nutrients, but enter "love marriages," wherein a female gamete selects in fertilization a particular pollen grain, or "lad," as Prezent quaintly described it. Heredity was most often defined by Lysenko in a somewhat meaningless way as "the property of the living body to demand definite conditions for its life, its development, and to react definitely to these or other conditions." "The conservatism of the nature of organisms"—in other words, their genetic endowment—can in Lysenko's view be liquidated. Genes, or units of heredity, do not exist. Mendelian segregation and recombination are statistical phenomena without biological significance. The whole structure of genetics is void of any but metaphysical meaning. Above all, Lysenko raised the question whether genetics contributed anything to the development of agriculture.

The 1936 conference was officially declared a draw, but it nevertheless presented a clear-cut victory of the Lysenko forces. The duties of Vavilov were sharply circumscribed, and Lysenko moved to Moscow to take over direction of a considerable amount of agricultural research in the whole Soviet Union. There was apparently no suppression of genetics in fields outside of agriculture and medicine, and a great deal of useful and important work in many phases of genetics was carried out for another ten years.

Even in agriculture, Lysenko's control was as yet incomplete. Although rumors and denials flew thickly in the West about the safety of Vavilov's person, and that of others, little authentic information was at hand. An International Congress of Genetics originally scheduled to be held in Moscow in 1937 was postponed by the Russians and later cancelled. When the Congress was eventually held in Edinburgh, just at the time that Germany marched into Poland, no Russians attended, even though Vavilov had been elected President of the Congress. Shortly thereafter (October, 1939), the second formal round of the battle took place, when the party journal *Under the Banner of Marxism* arranged another conference. This time the meeting was a melancholy and tragic affair resembling a trial more than a scientific conclave. Vavilov and others offered an honest defense of their scientific views, but in their philosophical argumentation were forced to proceed according to Zeno's unfortunate theory that knavery is the best defense against knaves. There is little point in relating the details of this encounter. Tempers were short, any pretense of impersonal discussion was abandoned, and both the attackers and defenders relinquished dignity and logic to establish their respective points. Mendelian laws of segregation were attacked most severely, together with other valid genetic ideas, and many more ideas that no geneticists shared.

Meanwhile, the Stalinist purges of 1938 took their toll of many scientists. Vavilov was arrested as a British spy in 1940, sentenced to death, and died in prison three years later. The demands of war halted the onslaught on genetics, but after the war was over, the battle was joined again. In 1948 came the climactic conference, a surrealistic inquisition of the surviving Mendelians. On its last day, Lysenko produced a bombshell. To quote from *Pravda,* he said: "'Before starting my concluding speech I must answer a note sent to the Praesidium. I am asked: what is the attitude of the Central Committee of the Party towards the report I gave at this session. I reply: the Central Committee [of the Party] has examined my report *The situation in biological science* [an outline of his theories] and approved it.'" The *Pravda* article went on to report: "The communication of the president elicited universal animation among the participants of the session. In a united burst of enthusiasm all those present rose from their seats and a tempestuous, lengthy ovation ensued in honor of the Central Committee of the Party of Lenin-Stalin, in honor of the wise leader and teacher of the Soviet people, the greatest scientist of our era, comrade Stalin."

This statement initiated Lysenko's full control over agriculture and most of biology, and the extermination of genetics as a discipline. By now, Lysenko's Lamarckism was no longer concealed. Said he: ". . . the well-known Lamarckian propositions, which recognize the active role of the conditions of external environment in the formation of the living body and the inheritance of acquired characters, in contrast to the metaphysics of Neo-Darwinism . . . are indeed not faulty, but on the contrary perfectly correct and entirely scientific."

While textbooks were being rewritten, and names and pictures were ordered blacked out in the books still in use, Vavilov disappeared from history precisely as if put down the memory hole in Orwell's *1984*: all mention of his existence was erased. The less brave of the remaining geneticists indulged in public recantation, admission of error, and pledging of support to Lysenko. The more courageous ones refused to knuckle under, were sent to work camps, became ornithologists, entered cancer or antibiotic research, or were forced to be laboratory technicians, librarians, or farmers on the Afghanistan border. How many perished is not yet ascertained.

Lysenko was elevated to semidivine status. His pronouncements and claims, no matter how unbelievable, were unchallengeable. He was decorated as much as any war hero, with three Stalin prizes, six orders of Lenin, and the order of the Red Banner. He was proclaimed a Hero of Socialist Labor, became a deputy and vice-president of the Supreme Soviet and of the Central Committee of the Party. His portraits were hung in scientific institutions. Art stores sold busts and bas-reliefs of Lysenko. In some cities, monuments were erected to him. The State Chorus had in its repertory a hymn, in which they sang "of the eternal glory of Academician Lysenko."

In the USSR and the satellite countries, Lysenko's rule was supreme. Indeed, China still seems to be oriented towards Lysenkoism. As Lysenko took over the genetic reins, Soviet investigators were completely sealed off from the Western world of genetics. The whole corpus of genetical knowledge—painstakingly accumulated in thousands of experiments throughout the world in the course of half a century—was discarded by simply stating that it came from the capitalist world. Of all the living and dead western biologists, only Darwin appears to have been granted honorary Soviet citizenship, largely because he had been vouched for by Marx and Engels. The nationalistic criterion of scientific validity reached absurd heights in the statement of one of Lysenko's supporters, whose work had been criticized abroad: "This attitude of foreign scientists not only does not worry us but, on the contrary, fills us with joy. Apparently, we are on the right track, if we are abused by the other side."

After the post-Stalin thaw, Lysenko supporters began appearing at international meetings, invariably bringing back home reports of the interest in their work and enthusiasm with which it has been received. In fact, however, outside of a handful of iconoclastic geneticists, the interest largely lay in seeing what breed of men these self-styled scientists were. When an invitation was issued for Lysenkoites to demonstrate their experiments in the United States (ironically enough, their expenses were to be borne by the Rockefeller Foundation, which is satirized in the Soviet cartoons that are reproduced as Figure 23.1), it was rejected. Similarly, a suggestion to Lysenko by the Soviet physical chemist N. Semenov (later a Nobel laureate) to permit western geneticists to observe the experiments in Lysenko's laboratory was turned down out of hand. Nevertheless, attempts to establish Lysenkoism as a respectable doctrine abroad continued for a number of years without success. Only after Lysenko's dismissal did Soviet geneticists really rejoin the international family of geneticists.

Lysenko's reign lasted until well after Stalin's death in 1953. The dramatic disclosures by Khruschev of Stalin's misdeeds permitted some relaxation. A few geneticists returned to their former pursuits, but in the guise of doing medical, radiation, or biochemical research. Khruschev, a personal friend of Lysenko, continued his support of the genetic dictator. Several attempts at scientific insurrection were not too successful. By 1961, busts of Lysenko were still available in stores, but at triple discount. In 1964 more voices began to be heard; the failures of Soviet agriculture were becoming too serious a matter to be ignored. Reliance on Lysenko in these matters was at least one of the reasons claimed for the downfall of Khruschev. In any case, soon after this event, Lysenko was dismissed from his administrative positions. Investigating committees established

FIGURE 23.1

Four views of western genetics from the pen of the famed Soviet caricaturist Boris Efimov, which appeared in the 1949 *Ogonyok,* a popular journal, to illustrate an article entitled "Flylovers—Manhaters." The inscription on the flag reads "The banner of pure science."

wholesale fraud, not only in reports claiming increased agricultural yields, but also in the presumably scientifically conducted experiments providing evidence of the validity of Lysenko's theories.

Posthumous rehabilitations followed. Vavilov's reputation was rescued and postal envelopes bearing his portrait appeared. A revision of school curricula to include the teaching of Mendelism followed, and translations of western text-books on genetics were ordered. Teaching of biology in high schools was suspended for a year while textbooks were being revised. Mendel, no longer officially portrayed as a sinister tool of the Catholic Church, became a hero. A spate of articles vilifying Lysenko began appearing in scientific and popular journals of all kinds, whether they dealt with sport, education, science, or literature. Whole-sale conversion to Mendelism by philosophers, teachers, writers, and agronomists ensued. New institutes, research programs, and journals on genetics are now established.

In the decade since the fall of Lysenko, Soviet biologists have attempted to catch up with Western science, especially in the area of molecular genetics. In agriculture, the baneful influence of Lysenkoism still lingers. Many administrators, planners, and party officials still incorporate vestiges of the Lysenko era of Soviet biology in their management of scientific resources and endeavors. Lysenko's fate was happier than Vavilov's: he still has friends in high places, and runs his own research institute (although there is no published record of any experimental results produced since his disgrace). Nevertheless, it is still true that the atmosphere of Soviet science limits the freedom of research endeavor. The authoritarian tradition still dominates, and a struggle for pecking order among the senior geneticists is clearly evident. One can only hope for the best.

23.2 LYSENKOISM AND ITS METHODS

A coherent account of Lysenkoism is difficult because of its very irrationality. Its essence included the rejection of the role of the chromosomes and of DNA in heredity, belief in inheritance of acquired characters, and in the possibility of spontaneous origin of one species from acellular material of another. Some of the experimental claims made by Lysenkoites are remarkable for their effrontery. They include miraculous transformations of wheat into rye, barley, oats, and even cornflowers; beets into cabbage, pine into fir, a tree of the hornbeam group into a forest walnut (using doctored photographs as evidence), and even the hatching of cuckoos from eggs laid by warblers, as well as the origin of mammalian cells from cereals.

Numerous experiments purporting to show the successful and permanent transformation of small white fowl into large black ones by blood transfusions were reported. When a French publication claimed that injection of DNA from one breed of ducks into another produced heritable changes in the descendents of recipients, there was both jubilation and consternation in the Lysenko camp. The first came from the alleged confirmation of the possibility of modifying heredity by such means; the second from the fact that DNA was the transforming agent. Since chicken blood cells have nuclei and hence DNA, the previous Lysenko experiments would implicate nucleic acids in the hereditary machinery. Immediately results of new experiments were reported, in which it was supposedly found that serum lacking blood cells (and hence DNA) was even more efficient in producing transformations.

As it happened, the French experiment was found not to be repeatable. Its results may well have originated in the hybrid origin of the recipients. Many other large-scale attempts outside of the Soviet orbit to repeat the reported results were negative, with two exceptions, neither of which appears to have adequate controls. Since Lysenko's fall, Soviet geneticists themselves failed to confirm his findings. This is not to say that means of inserting genetic messages into higher organisms and incorporating them by transformation do not exist, but injection of blood or of pure DNA into the body does not seem to be one of them.

Another of Lysenko's tenets concerns the impropriety of mathematical analysis to biology. Thus, one of his assistants published a great number of F_2 Mendelian ratios from individual crosses, pointing out that the proportion in them was hardly ever exactly 3:1. When Kolmogorov, one of Russia's most brilliant statisticians, showed that on statistical grounds the data not only did not disprove the existence of the ratio, but provided one of the best confirmations of it, he was answered with a chain of intricate syllogisms. Kolmogorov's notions of probability were founded on those of the German (later American) von Mises; he in turn was a follower of the Austrian philosopher Mach, whose concepts in their turn were allegedly demolished by Lenin because they derived from the idealistic views of Bishop Berkeley. Therefore, it was said, Mendelian ratios do not exist. One experiment pointing out that sex ratios do not approximate 100 but can be as low as 40 was later found to be based on a single litter of rabbits consisting of two males and five females.

Experiments of the Lysenko school usually lacked controls. Their outcome was predetermined by the experimenter's *a priori* notions of the results desired. A direct quote from Lysenko says: "to obtain a certain result, one must wish to obtain such a particular result: if you want a particular result you will obtain it." And talking of selecting his staff, he said: "I need only people who will obtain what I require."

Alogical discourse (that is, discourse which is outside of any logical system) and circular reasoning are other attributes of Lysenko's methodology. Usually it was explained that certain results could be obtained only "under particular conditions," and the conditions were then defined as those under which the particular results were obtained. Simplification of reasoning was involved in most of Lysenko's philosophy. His attitude was: If the Hardy-Weinberg equilibrium is too difficult to understand, let us handle the matter by denying its existence. Examples of similar fuzzy thinking can be found closer to home, as in the attempt by a Kansas legislator to make π an easier number to deal with in Kansas by legally changing it to a rational number.

Especially in the dialectic of debate did Lysenko's school indulge in spectacular methods. Windmills were readily attacked. Some assertions of western geneticists were put to question in 1948 by citing articles from the 1917 edition of the *Encyclopedia Britannica,* as if Einstein could be proved wrong by attacking the cosmology of Ptolemy. In the Soviet intellectual climate, associative arguments have a powerful weight. Given the infallibility and canonical status of Engels, Marx, Lenin, Stalin, and eventually of Lysenko himself, any reference to their views was tantamount to proof.

Ridicule, caricature, and the equation of Mendelism with Fascism (Figure 23.1) were freely resorted to. The *argumentum ad hominem* was constantly employed: Mendel was a priest, hence his laws were invalid (to which charge Vavilov was impelled to reply that the English geneticist Bateson was an agnostic). At the first of the international meetings where Lysenkoites appeared, a startled American geneticist asked what the effect of the discovery of transforming properties of blood would have for human blood transfusions. The reply was immediate and unblushing. The speaker had no information on the subject since in the civilized Soviet Union, experiment with humans is prohibited. And then he expressed surprise that the question was raised by an American: were there not separate blood banks for Negroes and whites during the war? Now, the American's question was not whether bigotry, ignorance, and prejudice exist in the United States. The original question had been addressed to biological issues.

The Lysenko propaganda machine also worked at full speed in agricultural practice. The post-Lysenko-era description of its operation by Semenov may be cited:

> If now we turn to T. D. Lysenko's own recommendations, their usual history was as follows. First a certain promise was made which was widely advertised. Assurances were given that grandiose successes would be attained in an exceedingly short time and at very meager cost. After a while, a report would appear that the promise had been basically fulfilled, and that the methods worked out must be incorporated into practice on the widest scale. As a rule, all this was accompanied by noise about still newer achievements. Gradually, however, the method was being less and less practiced or mentioned in the press. . . . But the flop was masked by a boom around the new promise, the history of which differs from the one described only in details.

The question arises why a society, presumably dedicated to the betterment of the human condition, which has done so much to ameliorate the lot of the peasant and the worker, which brought in education, and which has the capability of advanced space exploration, would wilfully proceed in this way to its own detriment both in theory and in practice.

We must seek explanation not in philosophy or genetics as such, but probably in the political outlook and the internal struggles for power. Up to now the USSR has been a monocultural society, one organized as a unity according to explicitly formulated principles involving abolition of capitalism. Bonapartism and personality cults can flourish in such an atmosphere. Lysenkoism was successful in precisely the way Stalinism was. The slogan of the revolution, "who is not with us, is against us," played as much of a role in one as in the other. There are some exact parallels in the rise of Lysenko and Stalin: oversimplification of the issues with appeal to uninformed masses, promise of rapid improvement in conditions, distrust of capitalist ideas and motives, appeal to chauvinism and national pride,* elevation to a position of undisputed authority, and ruthless extermination of opposition and rivals.

Now the monoculture appears to be developing cracks. Repression of religion and of the literature of dissent, however, still exists. The winds in genetics may shift again, but, at least there are many encouraging signs that the rulers of the Soviet Union have learned the lesson and are aware of the costs of tampering

---

*To the extent of a claim that even though Mendel was wrong, he had really been first appreciated by a Russian; see Section 7.2.

with the freedom of research and of the need to keep inquiry free from ideological and political interference. And it might be hoped that the western societies, including our own, will take heed from this lesson too. As Sir Julian Huxley, in discussing Lysenkoism, said: "The battle of Soviet genetics will not have been fought . . . in vain if over the great majority of the world the scientific movement will . . . have become fully conscious of itself and its social functions, of the vital importance, but at the same time the limitations, of scientific method, of the equal importance of a proper degree of scientific autonomy, and of the rights and duties of science in relation to other higher activities of man, to the State, and to human society as a whole."

24
ENVOI

We are beginning to understand just how complex life, and human behavior, are. But we are far from understanding that complexity. Even if or when we do, *understanding* does not automatically allow *prediction,* nor does prediction automatically lead to *control.* We know what causes rain, but our weather forecasts still lack precision, and control of the weather is still largely beyond our grasp. How much more so is human destiny.

24.1 THE SHAPE OF THINGS TO COME

Predictions about the future and normative descriptions of what human society should be like have been with us since Plato's *Republic* and *Laws* (fourth century B.C.), and through Thomas More's *Utopia* (1516), Edward Bellany's *Looking Backward* (1888), Aldous Huxley's *Brave New World* (1932), H. G. Wells' *The Shape of Things to Come* (1933), George Orwell's *1984* (1949) and many other books. More recently, exploration of the future has become a widespread occupation of natural scientists, technologists, social engineers, and government bodies.

Most new administrations, be they state, federal, or local, create dozens of Task Forces, staffed with high-powered people, to deal with the urgent problems of our societies. (There is a rumor that there is now a Task Force on Task Forces, to find out what happens to them, as most are never heard from again.) A large number of research groups, subsidized by both federal and private funds, are investigating what scientific and technological advances are likely to be made, time tables for their achievement, and the effects they will have on war and peace, on social structure, and on nearly all other aspects of human existence. Planning in our society is no longer a simple matter of looking ahead five years, although too often, the people who actually make decisions are making them on that time scale, or on time scales that are even shorter. The inability to handle time as a dimension is one of the great failings in our modern culture, and it creates great risk for our future.

Predictions and descriptions of some genetic potentialities of the future are scattered throughout the previous chapters of this book. Some more general possibilities that have biological bearing, gathered from a variety of sources, are presented here. Many of them have already been alluded to, albeit in oblique fashion.

The first question about our future is, of course, whether there will be one. The biggest immediate threat to our existence is the breaking out of another world war. Some experts compute the probability of a major war before the end of our century at 20 percent. Thermonuclear weapons and other weapons whose effects are spread over a wide area, such as nerve gas or biological agents, would probably be used in such a war. Opinions about the consequences to Earth's present inhabitants range through a wide spectrum. One suggests that mammals would be completely wiped out, and insects, which are more resistant and have tremendous reproductive potential, would take over the world. A less (or more?) pessimistic outlook is that a band of resistant human survivors would need to re-build civilization from barbarism, while the human gene pool would have to be purged of numerous deleterious mutations. Development of a highly authoritarian state following a nuclear holocaust has also been predicted. More optimistic prognosticators hope that if such weapons are actually ever used, the powers-that-be, shocked into a full realization of the consequences, would then form an effective world organization and effect general disarmament. And these optimists even hope that the movement towards such an organization will win the race against chauvinism and nationalism before war breaks out. They believe that international politics and the tensions they cause will fade away in the face of a supergovernment, and the wonders that can ensue from the harnessing of atomic and biomolecular powers will be enjoyed by all humans, forever at peace with each other.

But even in a peaceful world, problems beyond those already mentioned must be solved. They include air and water pollution, the use of pesticides, and mutation from various causes.

As scientists, we are sensitive to the special satisfaction associated with discovery. There may be (and probably are) a finite number of nontrivial unknowns.

During the past several decades, we have been converting unknowns to knowns at an unprecedented and increasing rate. DNA, once discovered and understood, cannot again become unknown while our civilization lasts. Future generations may look back on the scientists of this century as we now look back on the wastrel lumber barons of our country's past. Significant unknowns are thus a nonrenewable resource, and the conservation of some for future minds to probe and to convert to knowns at a more measured pace (so that similar minds will have time to appreciate, even savor, the conversion) has not yet become a part of the general conservation ethic.

A greater participation of biologists, politicians, and citizens in many other walks of life in the development of procedures, regulations, or laws governing the search for and application of genetic techniques and principles seems needed and is likely to become a reality. Senator John V. Tunney, in an address printed in the 23 May 1972 *Congressional Record*, said: "The political impact of these possibilities, however, might be just as powerful as the scientific impact. . . . The ethical questions raised by the possibilities implicit in genetic engineering are no less fundamental than the issues of free choice, the quality of life, the community of man, and the future of man himself." Senator Tunney spoke of six aspects of human genetic engineering that need attention now—or will in the future: amniocentesis and therapeutic abortion; mass genetic screening, which may be accompanied by the invasion of personal rights; monogenic gene therapy, which will probably have restrictive application and be enormously costly; fertilization outside of the womb and artificial insemination, which will require new legal definitions of the status of the offspring; cloning, the very idea of which provokes a highly variable public response; and polygenic gene therapy, which seems so far away as not yet to warrant concern.

Among the legal problems is the clarification of legal responsibility at various levels. For instance, if amniocentesis becomes common and accepted medical practice and a woman who is not offered amniocentesis then gives birth to a Down's syndrome child, is her obstetrician liable for malpractice? (He probably would be.) If the woman is offered amniocentesis and declines, is the genetically handicapped child correct in suing the parents for negligence? (To date there have been no test cases whose outcomes give a clear indication of what the judgment in such a suit might be.)

A. G. Motulsky suggests that the media frequently present lurid stories likely to titillate their clientele when discussing the effects of medical practice and genetic engineering. We hope we have not overdone it in this book. But there are significant concerns that will need informed attention from a broad spectrum of people. E. A. Carlson suggests that if we fail to relate science to values, and thereby allow science to be abused by society, we keep our objectivity but may lose much more than that.

We geneticists scared ourselves recently (1974), when it was reported that potentially dangerous genes had been experimentally inserted into the bacterium *Escherichia coli*. *E. coli* commonly reside in the human intestine, and they are capable of exchanging genetic information with other bacteria, some of which are pathogenic to humans. The scientists conducting the research voluntarily deferred further experiments until the effects of such systems were better understood. Furthermore, committees have been formed to evaluate similar risks from types of biological research that might release an unprecedented epidemic—that is, for example, if some new molecular construct escaped from a test tube and was unwittingly transported by a researcher out into the world.

Of the many marvels of the near future foreseen by experts, we may list development of fully acceptable synthetic foods, individual worldwide telecommunication, transformation of ocean areas into marine-life farms, complete climate control, efficient desalinization of sea water, development of robots to do factory work and housework, establishment of permanent manned artificial satellites and lunar and planetary installations, with interplanetary travel, even if confined to the Solar System, becoming commonplace.

To mention but a few of the predicted changes in the biological and psychological life of humans: Successful methods of inducing hibernation for short periods of time, or for years, are promised. So are euphenic techniques for changing one's sex in mid-life (which, in fact, are to some extent already available) and measures for overcoming forms of deafness and blindness that cannot be corrected now. The possibility of completely wiping out communicable disease is a debatable point, but further and substantial increases in life expectancy, postponement of aging, advances in the technology of controlling human suffering, and limited rejuvenation are all probable achievements of the near future.

One of the biggest problems to be faced is how people will use the leisure time that prospects for the future promise. The expected removal of limitations on the expression of intelligence and creative ability could help to solve this. But it is also possible that our future lies in a pleasure-oriented society, full of what has been called "wholesome degeneracy," in which people would devote most of their time to programmed dreams. As Lord Snow, and others, have suggested, if something becomes technically possible, it will happen. We have a bewildering number of technically possible futures. It is wise, but very difficult, to prepare for them.

24.2 DIRECTIONS OF FUTURE RESEARCH

Much has already been said in the various chapters of this book on the topic of this section. This material will not be recapitulated here. Neither is it feasible or necessary to list all that we do not know about genetics and evolution. Hence, only a partial listing of problems toward which genetic research is, or should be, directed is given. It must be emphasized that the listing does not contain all of the most important problems. Many of these have already been noted. Rather it is a supplement, presenting items that are somewhat neglected in the earlier chapters. It is based to some extent on committee reports to United States government agencies, but these committees are in no way responsible for the selection of items included here.

The pace of research expansion has been slowing, and research output may soon actually begin to decrease. In addition to the fact that nothing can continue to grow forever, there are several shorter-term reasons for this slowing. The volume of scientific information that already exists, the increasing specialization, and the rate at which scientific and technological practitioners become obsolete, are all feedbacks serving to slow research progress, largely by clogging communication channels both within and between fields. The incidence of unexpected side-effects is shifting research emphasis toward interdisciplinary work, and it takes longer for such research to progress and to train people to participate effectively. Many of the easier problems have been solved, and many of those remaining require more knowledge or more expensive and complicated equipment—or both—than were required to solve the easier problems. Finally, science is no longer the sacred cow it once was, and adulation is being replaced by a growing hostility, felt by many people including some of the scientists.

Many problems of genetic chemistry pertaining to structure and function of nucleic acids, proteins, and whole chromosomes remain to be solved. Even though the genetic code has been cracked, much is still unknown about replication, damage, and repair of DNA and RNA, and the details of transcription and translation. Actual codon sequences of important genes should be worked out. Genetical engineering will need to have such information at its command. Determination of the structure of many proteins and of their amino acid sequences, so useful in the reconstruction of the evolutionary past and of taxonomic relations, is also an important need. Instrumentation and automation for the mass production of euphenic substitutes for natural substances are yet to be developed.

Moving from molecular to cellular studies, investigation of regulator mechanisms in higher organisms should have a high priority. All too little is known of the fine physiological details of mitosis and meiosis. Only a formal description of nondisjunction is now available: its causes and precise mechanics are still to be determined. Automation of procedures for screening and diagnosing of cytological abnormalities is still in its infancy.

Almost every aspect of developmental biology, and, in particular, the mechanisms of differentiation in individual development deserves intensive study. Investigation of metabolic pathways between immediate gene products and their phenotypic expression can be carried out for some parts of the process in tissue cultures. Autoimmune diseases should share the research stage with immunological tolerance. Studies on various gene-environment interactions and, especially, that between the genotype of the fetus and uterine environment, should be extended.

The causes of spontaneous mutation are very imperfectly understood. A search for antimutagens should be undertaken. Methods for more accurate estimation of human mutation rates are badly needed. In formal genetics, little is known about human linkage groups. Biochemical identification of heterozygotes is possible only for a handful of human genes.

The extent of heterozygous advantage in human polymorphisms is still at issue, and only guesses can be hazarded in the majority of instances of the nature of selection between polymorphs. We are also ignorant of the equilibrium status at most loci. Accumulation of familial data, development of genetic registers for human diseases, adaptation of census material for genetic use, extension of studies on human phylogeny, and genetic investigations of isolates provide goals for human population genetics research. The investigation of isolates has some urgency as civilization sweeps forward, and human isolates are separated from their traditional environments and incorporated in the general human gene pool.

Only a few methods of increasing genetic variability in plant and animal stocks have been exhaustively tested. Comparative biochemical studies combined with hybridization experiments may be able eventually to increase food supplies by producing new forms of plants and animals. Somatic-cell hybridization, followed by recovery of an entire organism from the cell culture, may allow gene recombinations not possible by normal sexual means. It has even been suggested that extinct forms of life, such as Siberian mammoths found frozen in glaciers, could be thawed and propagated from cultured cells.

The nervous system is a major frontier of biological inquiry. The gaps in our knowledge of behavioral genetics of humans and other animals, and of the inheritance of psychological traits, must be only too apparent. More studies of monozygotic twins reared apart are needed. An increased reciprocal understanding and cooperation between geneticists on the one hand, and psychologists and psychiatrists on the other, is a prime desideratum. And beyond this mutual understanding, a real interpenetration of thought between geneticists and social scientists, virtually absent today, is needed.

24.3 GENETICS AND HUMAN CONCERNS

By now the inescapable involvement of genetics with medicine and public health, law and politics, agriculture and industry, nuclear warfare and space exploration, and with religion and social relations should be clear. As a coda, this section reemphasizes the serious impact of the explosion of scientific and technological advances on our continuing search for a body of beliefs and values.

Long the province of theologians and philosophers, ethical systems and the bases of individual and group behavior more recently have become within bounds to social and behavioral scientists, and now to natural scientists. Here we will not discuss in detail either the history or the status of the ethical foundations of human mores, but rather express personal viewpoints bearing on the purposes of this book.

In the past, some systems of ethics were based on the assumption that there are either supernatural sanctions or natural laws to which human beings must conform. Others were constructed from purely hedonistic justifications, or based on criteria of personal happiness and well-being or of the maximum common good. Still others, alleged to have an objective basis, rested on such criteria of value as energy production, or minimal psychological conflict within a society, or life expectancy, or even the second law of thermodynamics, that is, that increase in order is good. The dubiousness of the system based on the second law of thermo-dynamics was demonstrated by *reductio ad absurdum* by one critic who pointed out that since life has a degree of order so much higher than that of the inorganic world, it should be encouraged. Hence, we should breed drosophila by the million. Further, since humans are (supposed to be) the highest form of life, we should promote human reproduction to the limit of its capacity. The first conclusion is ridiculous; the second is monstrous.

Among biologists, the debate on the possibility of deriving objective ethics based on the evolutionary mode of thought is an unresolved one. In particular, Sir Julian Huxley and C. H. Waddington have proposed such systems; G. G. Simpson and Th. Dobzhansky do not share their belief that human values and wisdom are derivable from the principles of organic evolution. A summary of the various views on the matter may be found in Dobzhansky's book *Mankind Evolving*. Much of what follows is drawn from Dobzhansky and from the writings of Simpson.

It should be clarified that argumentation here is within the naturalistic frame-work of reference. This framework can be defined by quoting the late American anthropologist, Clyde Kluckhohn: "Philosophers tell us that there have been four main approaches to the problem of value: the Platonic view that values are 'eternal objects'; the position of subjectivism or of radical ethical relativity; the assumption held in common by certain Marxists, logical positivists, and 'linguistic' philosophers that judgments of value are merely 'emotional' or 'verbal' assertions altogether removed from the categories of truth and falsity; the naturalistic approach which holds that values are accessible to the same methods of enquiry and canons of validity applied to all forms of empirical knowledge."

The naturalistic approach is the valid one from the evolutionary viewpoint. The human species, in Waddington's phrase, is *the ethical animal,* and the process of ethicizing is a biological adaptation necessary for its welfare. The other ap-proaches seem to be nonadaptive. But the question whether an actual system of ethics is objectively derivable from biological evolution rather than from cultural evolution is still at issue.

A variety of evolutionary criteria have been proposed to provide the objective basis of ethics. They include, with and without attempts at precise definition, increased individual and gene-pool integration, complexity or homeostatic control, maximization of metabolism, minimization of effort, survival (of the species, since no individual ever survived), improvement or progress (however defined), and increased richness of experience. Yet since it is a reasonable assumption that protists lack ethical or moral values, we, with Simpson, must reject all of these as being irrelevant to the only ethics we know, human ethics. These human ethics involve not only the moral ideal of goodness, but the scientific ideal of perfect knowledge, the aesthetic ideal of beauty, and even the economic ideal of abundance, only the last of which can possibly be located in prehuman biological history.

This is not to say that ethics is solely a cultural phenomenon. There is a feed-back between biology and culture, as has been already pointed out on several occasions, so the ethical beliefs have evolutionary consequences, just as ongoing human evolution has ethical sequelae. Indeed, ethical systems must undergo evolution themselves if they are to function. The new theological thought realizes it. Among Roman Catholics, Teilhard de Chardin, in a mystical and profoundly unscientific way, has seen the relevance of evolution to human ideals and aspirations. Among Protestant theologians, the Lutheran Philip Hefner has recently discussed the new doctrine of human life that emphasizes the two aspects of evolution that have been stressed in this book: change and diversity. The first refers to the historical dimension of our biological existence, a dimension that has not been abolished, even if cultural evolution is now the more rapidly operating force. The emphasis on diversity abandons the view of our having been created after an archetypal image. Both ideas would have been totally unacceptable to last century's Christian theology.

In a 1973 paper presented to the World Council of Churches' Consultation on Genetics and the Quality of Life, C. Birch presented his view of genetic inequality and moral responsibility, to which we strongly subscribe:

> If I am more or less normal, that is through no merit of my own. That others have genetic deficiencies is no fault of theirs. . . . The fact that others have not, through no fault of theirs, changes for me the character of my having. It drives me to a realization of both the unity of mankind and the cost of creation. All are called to share the cost. The genetic inequality of man makes me realize my responsibility to mankind. Through chance some must suffer more than others. How can I make their suffering less?

Whatever the new ethics of our day are, Hefner calls "attention to the imperative that the life sciences seem to place before man to assume ever more intelligent and responsible control over . . . nature . . . , as well as over his own evolutionary process within it." The fruit of the tree of scientific knowledge that we have now tasted places an obligation on us to exercise our growing powers in distinguishing between good and evil, not as absolutes established by divine sanction, not as characteristics that can be derived from science itself, but as forces operating between people, forces whose meaning we have to define for ourselves.

The decision-making discussed in the introduction to this book is of an individual and more-or-less trivial sort. The collective decision of what we want to do with our species is not. The development of an effective and just method for collective ethical decision-making has a high priority on society's agenda. Only an informed society can accomplish this task. That is why an understanding of genetics, of the new biology, and of the evolutionary outlook, are indispensable ingredients in the cultural baggage of every person.

ADDITIONAL READING

There are many books that deal individually with the various issues which have been discussed. The ones we have selected to list below have been published in paperbound editions. Most of them call for no more biological knowledge than is provided in this book and many have further bibliographical information.

In addition to the books there are several hundred offprints from *Scientific American* that are available at 30 cents each from W. H. Freeman and Company, San Francisco. References to many of these are scattered throughout this book. Some of them have been collected into books of readings under the titles: *Psychobiology, Thirty-nine Steps to Biology, From Cell to Organism, Facets of Genetics, Communication, Human Variation and Origins, Ecology, Evolution, and Population Biology, Biology and Culture in Modern Perspective, The Chemical Basis of Life, Science, Conflict, and Society,* and *Cellular and Organismal Biology.*

Baer, A. S. *Heredity and Society.* Macmillan.
Bajema, G. J. (ed.) *Natural Selection in Human Populations.* Wiley.
Baldwin, R. E. *Genetics.* Wiley.

Beadle, G. and M. *The Language of Life.* Doubleday.

Bonner, D. M., and S. Mills. *Heredity.* Prentice-Hall.

Brewbaker, J. L. *Agricultural Genetics.* Prentice-Hall.

Brink, R. A. (ed.) *Heritage from Mendel.* University of Wisconsin Press.

Cain, A. J. *Animal Species and Their Evolution.* Harper and Row.

Campbell, B. G. (ed.) *Human Evolution.* Aldine.

Connell, J. H., D. B. Mertz, and W. W. Murdock (eds.) *Readings in Ecology and Ecological Genetics.* Harper and Row.

Count, E. W. *Being and Becoming Human.* Van Nostrand, Reinhold.

Daniels, R., and H. H. L. Kitumo. *American Racism.* Prentice-Hall.

Darwin, C. *Origin of Species.* Atheneum; Collier: New American Library; Washington Square Press.

Dawson, P. S., and C. E. King (eds.) *Readings in Population Biology.* Prentice-Hall.

Dobzhansky, Th. *Evolution, Genetics, and Man.* Wiley.

Dobzhansky, Th. *Heredity and the Nature of Man.* New American Library.

Dobzhansky, Th. *Mankind Evolving.* Yale University Press.

Dunn, L. C. *Heredity and Evolution in Human Populations.* Atheneum.

Ehrlich, P. R. *The Population Bomb.* Ballantine.

Francoeur, R. T. *Utopian Motherhood.* Perpetua.

Goldsby, R. A. *Race and Races.* Macmillan.

Haldane, J. B. S. *Causes of Evolution.* Cornell University Press.

Handler, P. *Biology and the Future of Man.* Oxford University Press.

Hardin, G. (ed.) *Population, Evolution and Birth Control,* 2nd ed. W. H. Freeman and Company.

Hamilton, T. H. *Process and Pattern in Evolution.* Macmillan.

Hartman, P. E., and S. R. Suskind. *Gene Action.* Prentice-Hall.

Herskowitz, I. H. *Basic Principles of Molecular Genetics.* Little, Brown.

Jensen, A. R. *How Much Can We Boost IQ and Scholastic Achievement?* Harvard Education Review, No. 39.

Keosian, J. *Origin of Life.* Reinhold.

Levine, R. P. *Genetics.* Holt, Rinehart and Winston.

Lewontin, R. C. *The Genetic Basis of Evolutionary Change.* Columbia U. Press.

Loehlin, J. C., G. Lindzey, and J. N. Spuhler. *Race Differences in Intelligence.* W. H. Freeman and Company.

Loewy, A. G., and P. Siekevitz. *Cell Structure and Function.* Holt, Rinehart and Winston.

Markert, C. *Developmental Genetics.* Prentice-Hall.

McKusick, V. A. *Human Genetics.* 2nd Ed. Prentice-Hall.

Mettler, L. E., and T. G. Gregg. *Population Genetics and Evolution.* Prentice-Hall.

Molnar, S. *Races, Types, and Ethnic Groups.* Prentice-Hall.

Moore, J. A. *Heredity and Development.* Oxford University Press.

N.A.S. *Genetic Vulnerability of Major Crops.* National Academy of Sciences.

Olson, E. C., and J. Robinson. *Concepts of Evolution.* Merrill.

Ovenden, M. W. *Life in the Universe: A Scientific Discussion.* Doubleday.

Pai, A. C. *Foundations of Genetics.* McGraw-Hill.

Pedder, I. J., and E. G. Wynne. *Genetics, A Basic Guide.* Norton.

Peters, J. A. (ed.) *Classic Papers in Genetics.* Prentice-Hall.

Ramsey, P. *Fabricated Man.* Yale U. Press.

Rhodes, F. H. T. *Evolution of Life.* Penguin.

Roller, A. *Discovering the Basis of Life.* McGraw-Hill.

Simpson, G. G. *The Meaning of Evolution.* Yale University Press.

Smith, J. M. *Theory of Evolution.* Penguin.

Spuhler, J. N. *Diversity and Human Behavior.* Aldine Press.

Stebbins, G. L. *Processes of Organic Evolution.* Prentice-Hall.

Stern, C., and E. R. Sherwood (eds.) *The Origin of Genetics: A Mendel Source Book.*
W. H. Freeman and Company.

Swanson, C. P. *Cytogenetics.* Prentice-Hall.

U.S. Pub. Health Ser. *Report on XYY Chromosomal Abnormality.* Pub. 2103.

Volpe, E. P. *Human Heredity and Birth Defects.* Pegasus.

Wallace, B. *Chromosomes, Giant Molecules and Evolution.* Norton.

Wallace, B., and A. M. Srb. *Adaptation,* 2nd Ed. Prentice-Hall.

Watson, J. D. *Molecular Biology of the Gene.* 2nd Ed. Benjamin.

Wilson, G. B. *Cell Division and the Mitotic Cycle.* Van Nostrand, Reinhold.

Winchester, A. M. *Human Genetics.* Merrill.

Woese, C. R. *The Genetic Code.* Harper and Row.

Young, L. B. (ed.) *Evolution of Man.* Oxford University Press.

CREDITS

Many ideas and some data were presented in the various sections of this book without direct citation of sources. The following listing acknowledges the contributors whose work was so used. Any responsibility for misinterpretation or misrepresentation is ours. The Teacher's Manual for this book has bibliographic references for many of the names and to other sources of information.

1.3 G. W. Beadle, G. G. Simpson
2.1 A. Szent-Györgi
2.2 A. Gibor
2.4 H. G. Baker, P. H. Raven, C. H. Waddington
2.5 B. G. Campbell, B. S. Kraus, D. Mainardi, G. G. Simpson, L. B. Slobodkin
2.6 E. S. Barghoorn and J. W. Schopf, N. H. Horowitz, J. Keosian, C. Ponnamperuma, G. G. Simpson, M. J. D. White
2.7 S. -S. Huang, I. S. Shklovski and C. Sagan, H. Spinrad
3.1 Th. Dobzhansky, T. S. Kuhn, P. B. Medawar, G. G. Simpson
3.2 D. Shapere, S. Toulmin
3.3 B. S. Blumberg
3.5 W. M. Fitch, M. P. Kambysellis, M.-C. King. R. Logan, E. Margoliash, L. R. Maxon, V. M. Sarich, F. Sherman, A. C. Wilson

3.6   R. A. Fisher, H. W. Norton, G. Wald

4.1   T-Y. Ho, A. Seilacher, G. G. Simpson, S. Wright

4.2   R. B. Cowles, B. Kurten

4.3   K. Keeler, M. Scriven

5.1   B. G. Campbell, J. Cronin, R. B. Eckhardt, M.-C. King, J. T. Robinson, V. M. Sarich, E. L. Simons, S. L. Washburn, A. C. Wilson

5.2   R. J. Andrew, C. Birch, B. G. Campbell, R. L. Holloway, S. L. Washburn

5.3   W. Catton, Th. Dobzhansky, J. B. Lancaster, D. A. Livingstone, S. L. Washburn, P. Wylie

6.1   A. W. Ravin, J. F. Sambrook, et al., G. S. Stent, A. Tomasz

6.2   R. S. Edgar and W. B. Wood, E. B. Lewis

6.3   F. R. Babich et al., E. H. Davidson et al., J. W. Fristrom, H. N. Guttman, A. L. Hartry et al., J. T. King and R. W. Briggs, M. Luttges et al., D. H. Malin, F. Rosenblatt et al., G. Ungar and L. N. Irwin

6.4   J. D. Cooper, C. Denniston, R. T. Jones et al., D. S. Kleinman, C. C. Mabry et al., M. Murayama, R. S. Stevenson and C. C. Huntley, C. J. Witkop, Jr.

6.5   N. Arnheim, R. E. Dickerson, T. H. Jukes, M. Kimura, J. L. King, R. C. Richmond, C. E. Taylor

6.6   H. J. Curtis, S. M. Gershenson, J. Lederberg, S. Rogers, R. Stanier

7.1   S. W. Brown, R. G. Edwards, L. Hayflick

7.2   V. A. McKusick

7.3   C. Stern

7.4   A. C. Allison, S. L. Culliton, M.-C. King, T. M. Powledge, H. E. Sutton, S. L. Wiesenfeld

8.2   W. E. Kerr, U. Mittwoch

8.3   D. W. Cooper, G. Pincus, D. R. Robertson, C. Stern, E. O. Wilson, F. Wilt

8.4   H. Grüneberg, V. A. McKusick, U. Mittwoch, P. Pearson, L. B. Russell

8.5   M. D. Casey et al., A. H. Child, E. H. Y. Chu, M. M. Cohen et al., A. de la Chapelle et al., C. E. Ford, E. B. Hook, P. A. Jacobs et al., L. Moor, S. A. Shah and D. S. Borgaonkar, H. E. Sutton, T. R. and I. Tegenkamp, H. C. Thuline and D. E. Nordby

8.6   R. M. Dawes, A. W. Edwards, R. G. Edwards, W. D. Hamilton, J. McDonald, K. Oishi, A. S. Parkes, C. Stern, J. A. Weir

8.7   S. Chandra, G. Klein, M. W. Olsen and E. G. Buss

8.8   C. M. Berg, W. Warren

9.2   H. G. Baker, V. A. McKusick, N. E. Morton, M. Nabholz, J. R. Platt, F. H. Ruddle, T. B. Shows, W. K. Silvers and R. E. Billingham, C. P. Swanson, J. D. Watson

9.3   Th. Dobzhansky, K. Kojima and K. M. Yarbrough, E. Mayr

9.4   H. Laven, R. Sager, T. M. Sonneborn

10.1   K. Pearson

10.2   J. A. Reeds

10.3   L. S. Penrose

11.1   J. Hirsch

11.3   R. F. Stettler

11.4   M. G. Bulmer, Th. Dobzhansky, H. P. Donald, L. Eisenberg, L. Erlenmeyer-Kimling and L. F. Jarvik, J. Gurdon, A. F. Guttmacher, M. M. Haller, B. Harvald and M. Hauge, R. C. Johnson, H. Kalmus, N. Morton, H. H. Newman et al., N. Pastore, J. Shields, C. Stern, J. S. Thompson and M. W. Thompson

12.2 P. L. Broadhurst, D. S. Falconer, D. W. Fulker and J. Wilcock, I. I. Gottesman, L. L. Heston, D. Hoefnagel, J. S. Huxley et al., D. D. Jackson, F. J. Kallman, K. K. Kidd and L. Cavalli-Sforza, G. E. McClearn and J. C. DeFries, W. L. Nyhan, E. W. Reed and S. C. Reed

12.3 S. Benzer, P. L. Broadhurst, D. W. Fulker and J. Wilcock, L. Kelley

12.4 P. L. Broadhurst, L. Erlenmeyer-Kimling and W. Paradowski, D. W. Fulker and J. Wilcock, J. L. Fuller and W. R. Thompson, J. Hirsch, G. E. McClearn, J. R. Nichols and S. Hsiao, J. P. Scott and J. L. Fuller, W. C. Rothenbuhler, R. Shuter, C. Stern, S. G. Vandenberg

12.5 C. J. Bajema, C. O. Carter, R. B. Cattell, J. Cravioto, K. Davis, E. R. De-Licardie and H. G. Birch, Th. Dobzhansky, B. K. Eckland, R. Hernstein, J. Hirsch, K. Hutton and C. Carter, N. Irons, A. R. Jensen, D. M. Johnson, D. D. Krech and R. S. Crutchfield, G. E. McClearn and J. C. DeFries, L. S. Penrose, E. W. Reed and S. C. Reed, S. Scarr-Salapatek, J. N. Spuhler and G. Lindzey, J. H. Waller, C. J. Witkop

12.6 M. F. Giluta and D. N. Daniels, S. Washburn

13.1 K. Keeler

13.3 Th. Dobzhansky, R. C. Lewontin, J. L. Lush, R. Milkman, J. Price, H. E. Sutton.

14.1 W. C. Boyd, J. F. Crow, F. Fenner, E. R. Nye, J. N. Spuhler, T. E. Reed, C. Stern, F. Vogel and M. R. Chakravartti, T. Watanabe

14.2 J. F. Crow, R. H. Post, C. J. Witkop

14.3 D. S. Falconer, W. B. Jackson and D. Kaukeinen. T. H. Savory, R. Stonecypher

15.1 S. N. Agarawala, B. Bertram, H. Brown, A. Champagnat, G. C. Darwin, K. Davis, P. Ehrlich, B. A. Hamburg, G. Hardin, G. Leach, J. Lederberg, C. Markert, O. E. Nelson et al, R. Ravelle, W. Schockley

15.2 D. Bjorkman and J. Berry, N. E. Borlaug, A. Champagnat, C. D. Darlington, R. H. Dyson, Jr., G. Hardin, E. Hyams, D. H. Janzen, R. Slack, R. H. Smith and R. C. von Borstel, T. A. Wertire

15.3 E. Bennett, J. A. Browning, O. Frankel, J. R. Harlan, J. Harper, J. V. Neel, K. Sakai, R. Waller, H. G. and S. Wilkes

16.1 M. S. Al-Aish et al., M. Benazzi, R. K. Dhadia, Th. Dobzhansky, L. S. Penrose, A. Robinson and T. T. Puck, M. Shaw, O. Smithies, D. Soudek, A. Stoller, H. E. Sutton, S. W. Wright

16.2 J. F. Crow, M. Kimura, E. G. Leigh, Jr., E. T. Mørch, J. V. Neel, G. Schlager and M. M. Dickie, J. N. Spuhler

16.3 H. Harris and J. F. Watkins, C. B. Kerr, J. D. Watson, M. C. Weiss and H. Green

16.4 Y. Ahuja, C. Auerbach, L. Ehrenberg et al., D. Grahn, D. Grahn and J. Kratchman, H. Grüneberg, S. Mittler et al., A. Novick, R. B. Webb and M. M. Malina

17.1 J. F. Crow and S. Abrahamson

17.2 J. Cummings, R. E. Balzhiser, J. P. Holdren, E. B. Lewis, A. Maimoni, K. Z. Morgan, J. V. Neel, H. B. Newcombe and J. F. McGregor, W. J. Schull et al., C. H. Waddington, B. Wallace and Th. Dobzhansky

17.3 H. Abplanalp et al., S. I. Alikhanian, B. L. Astaurov, R. D. Brock, Å. Gustafson, E. F. Knipling, J. Monro, R. E. Scossiroli, Y. Tazima

17.4 B. Ames, W. Loughman

17.5 F. J. Ayala, J. W. Crenshaw, J. F. Crow and S. Abrahamson, Th. Dobzhansky, L. S. Penrose

18.1 J. R. Baker, W. C. Boyd, G. de Beer, Th. Dobzhansky, L. C. Dunn, S. M. Gartler, S. B. Holt, J. V. Neel and W. J. Schull, W. B. Provine

18.2 W. C. Boyd, R. C. Connelly and A. Abdalla, H. Gershowitz, R. G. Harrison, M. Smith, S. Yada

18.3 H. F. Blum, Th. Dobzhansky and M. F. H. A. Montagu, K. Keeler, N. Kretchner, W. F. Loomis, M. T. Newman, R. H. Tuttle

18.4 B. Glass and C. C. Li, B. E. Ginsburg and W. S. Laughlin, D. M. Heer, J. Lederberg, T. E. Reed, R. Severo, H. Slatis, C. Stern

18.5 R. C. Lewontin

19.1 V. A. McKusick, A. G. Motulsky

19.2 A. C. Allison and K. G. McWhirter, J. W. Fristrom, G. Hardin, H. Kalmus, V. A. McKusick, J. Price, C. R. Shaw, G. Skude

19.3 R. J. Berry, G. A. Chase, C. A. Clark, G. Dean, V. A. McKusick, I. MacAlpine and R. Hunter, N. C. Myrianthopoulos

19.4 M. Alter, C. Birch, L. L. Heston and T. T. Gottesman, R. C. Juberg, P. Kelley, A. G. Motulsky and F. Hecht, S. C. Reed

20.1 W. Bodmer, W. C. Boyd, C. S. Chang et al., H. Leven and R. E. Rosenfield, J. Moor-Jankowski and A. S. Wiener, L. E. Schacht and H. Gershowitz

20.2 P. Parsons, O. Smithies

20.3 R. E. Billingham, A. Comfert, G. M. Edelman, M. M. Mayer, G. J. V. Nossal and G. Ada, A. L. Notkins, R. R. Porter, J. D. Watson

21.1 D. F. Aberle, et al., C. D. Darlington, M. G. Farrow, M. Fortes, H. D. Kitto, J. Nada, S. C. Reed, L. D. Sanghvi, S. Sugiyama and W. J. Schull, S. L. Washburn and C. S. Lancaster

21.2 L. Beckmann, F. Cooke and F. G. Cooch, H. P. Donald, H. Heidger, E. H. Hess, H. Kalmus, H. Kalmus and S. M. Smith, D. Mainardi, P. O'Donald, M. B. Seiger, W. Sluckin, J. N. Spuhler

21.3 J. W. Eaton and R. J. Weil, B. Glass, V. A. McKusick et al, A. Moroni

21.4 J. B. S. Haldane and S. D. Jayakar, F. S. Hulse, A. P. Mange, E. Matsunaga, S. Rayner

22.1 R. Bellman, C. O. Carter, F. H. C. Crick, R. W. Day, L. C. Dunn, M. S. Frankel, J. B. S. Haldane, C. W. Kline

22.2 B. L. Astaurov, R. Bellman, R. G. Edwards, M. R. Geier, J. Lyman, C. R. Merril, J. C. Petricciari, M. R. Rosenzweig et al., J. M. Tanner

22.3 H. Hoagland, R. S. Morison

23.1 Zh. Medvedev

23.2 C. H. Waddington

24.1 B. Commoner, B. Glass, J. B. S. Haldane, A. Hoppe, H. Kahn and A. J. Weiner, J. Lederberg, G. Stent, *Time*

24.2 B. Glass, R. D. Owen et al., O. Smithies et al.

24.3 R. L. Ackoff, A. Stander

INDEX

The appearance of a page number in **boldface** type indicates that the term or name is defined or introduced in context on that page. Furthermore, it is likely to be a term the student will wish to understand, or a person or place of considerable importance to the development of the ideas or knowledge presented in this book.